Student Solutions Manual and Study Guide

for

Serway and Jewett's

Physics for Scientists and Engineers

Sixth Edition

Volume One

John R. Gordon
James Madison University

Ralph V. McGrew
Broome Community College

Raymond A. Serway

THOMSON
BROOKS/COLE

Australia • Canada • Mexico • Singapore • Spain • United Kingdom • United States

For more information about our products,
contact us at:
Thomson Learning Academic Resource Center
1-800-423-0563

For permission to use material from this text,
contact us by:
Phone: 1-800-730-2214
Fax: 1-800-730-2215
Web: http://www.thomsonrights.com

Brooks/Cole—Thomson Learning
10 Davis Drive
Belmont, CA 94002-3098
USA

Asia
Thomson Learning
5 Shenton Way #01-01
UIC Building
Singapore 068808

Australia/New Zealand
Thomson Learning
102 Dodds Street
Southbank, Victoria 3006
Australia

Canada
Nelson
1120 Birchmount Road
Toronto, Ontario M1K 5G4
Canada

Europe/Middle East/South Africa
Thomson Learning
High Holborn House
50/51 Bedford Row
London WC1R 4LR
United Kingdom

Latin America
Thomson Learning
Seneca, 53
Colonia Polanco
11560 Mexico D.F.
Mexico

Spain/Portugal
Paraninfo
Calle/Magallanes, 25
28015 Madrid, Spain

Preface

This <u>Student Solution Manual and Study Guide</u> has been written to accompany the textbook **Physics for Scientists and Engineers**, Sixth Edition, by Raymond A. Serway and John W. Jewett, Jr. The purpose of this Student Solution Manual and Study Guide is to provide the students with a convenient review of the basic concepts and applications presented in the textbook, together with solutions to selected end-of-chapter problems from the textbook. This is not an attempt to rewrite the textbook in a condensed fashion. Rather, emphasis is placed upon clarifying typical troublesome points, and providing further drill for methods of problem solving.

Each chapter is divided into several parts, and every textbook chapter has a matching chapter in this book. Very often, reference is made to specific equations or figures in the textbook. Every feature of this Study Guide has been included to ensure that it serves as a useful supplement to the textbook. Most chapters contain the following components:

- **Equations and Concepts:** This represents a review of the chapter, with emphasis on highlighting important concepts, and describing important equations and formalisms.

- **Suggestions, Skills, and Strategies:** This offers hints and strategies for solving typical problems that the student will often encounter in the course. In some sections, suggestions are made concerning mathematical skills that are necessary in the analysis of problems.

- **Review Checklist:** This is a list of topics and techniques the student should master after reading the chapter and working the assigned problems.

- **Answers to Selected Conceptual Questions:** Suggested answers are provided for approximately fifteen percent of the conceptual questions.

- **Solutions to Selected End-of-Chapter Problems:** Solutions are shown for approximately twenty percent of the problems from the text, chosen to have odd numbers and to illustrate the important concepts of the chapter.

We sincerely hope that this Student Solution Manual and Study Guide will be useful to you in reviewing the material presented in the text, and in improving your ability to solve problems and score well on exams. We welcome any comments or suggestions which could help improve the content of this study guide in future editions; and we wish you success in your study.

John R. Gordon
Harrisonburg, Virginia

Ralph McGrew
Binghamton, New York

Raymond A. Serway
Leesburg, Virginia

Acknowledgments

It is a pleasure to acknowledge the excellent work of Laura and Michael Rudmin, Richardas Preksat, and Aleksandras Urbonas of DSC Publishing, whose attention to detail in the preparation of the camera-ready copy did much to enhance the quality of this Sixth Edition of the Student Solutions Manual and Study Guide to accompany Physics for Scientists and Engineers. Their graphics skills and technical expertise combined to produce illustrations for earlier editions which continue to add much to the appearance and usefulness of this volume.

Special thanks go to Susan Dust Pashos, Senior Developmental Editor, and Rebecca Heider, Associate Developmental Editor at Brooks/Cole Thomson Learning, for managing all phases of this project. James McLean of the State University College at Geneseo, New York, served as accuracy reviewer for this volume and made many helpful suggestions. Finally, we express our appreciation to our families for their inspiration, patience, and encouragement.

Suggestions for Study

Very often we are asked "How should I study this subject, and prepare for examinations?" There is no simple answer to this question, however, we would like to offer some suggestions which may be useful to you.

1. It is essential that you understand the basic concepts and principles before attempting to solve assigned problems. This is best accomplished through a careful reading of the textbook before attending your lecture on that material, jotting down certain points which are not clear to you, taking careful notes in class, and asking questions. You should reduce memorization of material to a minimum. Memorizing sections of a text, equations, and derivations does not necessarily mean you understand the material. Perhaps the best test of your understanding of the material will be your ability to solve the problems in the text, or those given on exams.

2. Try to solve as many problems at the end of the chapter as possible. You will be able to check the accuracy of your calculations to the odd-numbered problems, since the answers to these are given at the back of the text. Furthermore, detailed solutions to approximately half of the odd-numbered problems are provided in this study guide. Many of the worked examples in the text will serve as a basis for your study.

3. The method of solving problems should be carefully planned. First, read the problem several times until you are confident you understand what is being asked. Look for key words which will help simplify the problem, and perhaps allow you to make certain assumptions. You should also pay special attention to the information provided in the problem. In many cases a simple diagram is a good starting point; and it is always a good idea to write down the given information before proceeding with a solution. After you have decided on the method you feel is appropriate for the problem, proceed with your solution. If you are having difficulty in working problems, we suggest that you again read the text and your lecture notes. It may take several readings before you are ready to solve certain problems, though the solved problems in this Study Guide should be of value to you in this regard. However, your solution to a problem does not have to look just like the one presented here. A problem can sometimes be solved in different ways, starting from different principles. If you wonder about the validity of an alternative approach, ask your instructor.

4. After reading a chapter, you should be able to define any new quantities that were introduced, and discuss the first principles that were used to derive fundamental formulas. A review is provided in each chapter of the Study Guide for this purpose, and the marginal notes in the textbook (or the index) will help you locate these topics. You should be able to correctly associate with each physical quantity the symbol used to represent that quantity (including vector notation if appropriate) and the SI unit in which the quantity is specified. Furthermore, you should be able to express each important formula or equation in a concise and accurate prose statement.

5. We suggest that you use this Study Guide to review the material covered in the text, and as a guide in preparing for exams. You should also use the **Equations and Concepts** to focus in on any points which require further study. Remember that the main purpose of this Study Guide is to improve upon the efficiency and effectiveness of your study hours and your overall understanding of physical concepts. However, it should not be regarded as a substitute for your textbook or individual study and practice in problem solving.

6. **A note concerning significant figures**. When the statement of a problem gives data to three significant figures, we state the answer to three significant figures. The last digit is uncertain; it can for example depend on the precision of the values assumed for physical constants and properties. When a calculation involves several steps, we carry out intermediate steps to many digits, but we write down only three. We 'round off' only at the end of any chain of calculations, never anywhere in the middle.

Problem Solving

Besides what you might expect to learn about physics concepts, a very valuable skill you should hope to take away from your physics course is the ability to solve complicated problems. The way physicists approach complex situations and break them down into manageable pieces is extremely useful. We have developed a general problem-solving strategy that will help guide you through the steps. To help you remember the steps of the strategy, they are called *Conceptualize, Categorize, Analyze,* and *Finalize*.

General Problem-Solving Strategy

Conceptualize

- The first thing to do when approaching a problem is to *think about* and *understand* the situation. Study carefully any diagrams, graphs, tables, or photographs that accompany the problem. Imagine a movie, running in your mind, of what happens in the problem.

- If a diagram is not provided, you should almost always make a quick drawing of the situation. Indicate any known values, perhaps in a table or directly on your sketch.

- Now focus on what algebraic or numerical information is given in the problem. Carefully read the problem statement, looking for key phrases such as "starts from rest" ($v_i = 0$), "stops" ($v_f = 0$), or "freely falls" ($a_y = -g = -9.80 \text{ m/s}^2$).

- Now focus on the expected result of solving the problem. Exactly what is the question asking? Will the final result be numerical or algebraic? If it is numerical, what units will it have? If it is algebraic, what symbols will appear in it?

- Don't forget to incorporate information from your own experiences and common sense. What should a reasonable answer look like? What should its order of magnitude be? You wouldn't expect to calculate the speed of an automobile to be 5×10^6 m/s.

Categorize

- Once you have a really good idea of what the problem is about, you need to *simplify* the problem. Remove the details that are not important to the solution. For example, model a moving object as a particle. If appropriate, ignore air resistance or friction between a sliding object and a surface.

- Once the problem is simplified, it is important to *categorize* the problem. How does it fit into a framework of ideas that you construct to understand the world? Is it a simple *plug-in problem*, such that numbers can be simply substituted into a definition? If so, the problem is likely to be finished when this substitution is done. If not, you face what we can call an *analysis problem* — the situation must be analyzed more deeply to reach a solution.

- If it is an analysis problem, it needs to be categorized further. Have you seen this type of problem before? Does it fall into the growing list of types of problems that you have solved previously? Being able to classify a problem can make it much easier to lay out a plan to solve it. For example, if your simplification shows that the problem can be treated as a particle moving under constant acceleration and you have already solved such a problem (such as the examples in Section 2.5), the solution to the present problem follows a similar pattern.

Analyze

- Now, we need to analyze the problem and strive for a mathematical solution. Because you have already categorized the problem, it should not be too difficult to select relevant equations that apply to the type of situation in the problem. For example, if your categorization shows that the problem involves a particle moving under constant acceleration, Equations 2.9 to 2.13 are relevant.

- Use algebra (and calculus, if necessary) to solve symbolically for the unknown variable in terms of what is given. Substitute in the appropriate numbers, calculate the result, and round it to the proper number of significant figures.

Finalize

- This is the most important part. Examine your numerical answer. Does it have the correct units? Does it meet your expectations from your conceptualization of the problem? What about the algebraic form of the result—before you substituted numerical values? Does it make sense? Try looking at the variables in it to see whether the answer would change in a physically meaningful way if they were drastically increased or decreased or even became zero. Looking at limiting cases to see whether they yield expected values is a very useful way to make sure that you are obtaining reasonable results.

- Think about how this problem compares with others you have done. How was it similar? In what critical ways did it differ? Why was this problem assigned? You should have learned something by doing it. Can you figure out what? Can you use your solution to expand, strengthen, or otherwise improve your framework of ideas? If it is a new category of problem, be sure you understand it so that you can use it as a model for solving future problems in the same category.

When solving complex problems, you may need to identify a series of subproblems and apply the problem-solving strategy to each. For very simple problems, you probably don't need this strategy at all. But when you are looking at a problem and you don't know what to do next, remember steps in the strategy and use them as a guide.

Table of Contents

Chapter 1
PHYSICS AND MEASUREMENT

EQUATIONS AND CONCEPTS

The **density** of any substance is defined as the ratio of mass to volume. The SI units of density are kg/m^3. *Density is an example of a derived quantity.*

$$\rho \equiv \frac{m}{V} \tag{1.1}$$

The mass per atom for a given element is found by using Avogadro's number, N_A.

$$m_{\text{atom}} = \frac{\text{molar mass of element}}{N_A}$$

$$N_A = 6.02 \times 10^{23} \; \frac{\text{atoms}}{\text{mole}}$$

SUGGESTIONS, SKILLS, AND STRATEGIES

A general strategy for problem solving will be described in Chapter 2.

The following are some important skills and mathematical techniques:

- Using powers of ten in expressing large and small numerical values (e.g. $0.00058 = 5.8 \times 10^{-4}$). Appendix B.1 gives a brief review of such notation, and the algebraic operations of numbers using powers of ten.

- Basic algebraic operations such as factoring, handling fractions, solving quadratic equations, and solving linear equations. Some of these techniques are reviewed in Appendix B.2.

- The fundamentals of plane and solid geometry — including the ability to graph functions, calculate the areas and volumes of standard geometric figures or solids. Also, recognize the equations and graphs of a straight line, a circle, an ellipse, a parabola and a hyperbola.

- Using basic trigonometry — definitions and properties of the sine, cosine, and tangent functions; the Pythagorean Theorem, the law of cosines, the law of sines, and some of the basic trigonometric identities. Reviews of geometry and trigonometry are given in Appendix D of the text.

1

REVIEW CHECKLIST

You should be able to:

▷ Describe the basis for the fundamental units of length, mass and time and the standards for these quantities in SI units. (Section 1.1)

▷ Convert units from one system to another and perform a dimensional analysis of an equation containing physical quantities whose individual units are known. (Sections 1.4 and 1.5)

▷ Carry out order-of-magnitude calculations or guesstimates. (Section 1.6)

▷ Identify and properly use prefixes and mathematical notations such as the following: ∝ (is proportional to), < (is less than), ≈ (is approximately equal to), Δ (change in value), etc.

▷ Express calculated values with the correct number of significant figures. (Section 1.7)

ANSWERS TO SELECTED QUESTIONS

2. Suppose that the three fundamental standards of the metric system were length, **density**, and time rather than length, **mass**, and time. The standard of density in this system is to be defined as that of water. What considerations about water would you need to address to make sure that the standard of density is as accurate as possible?

Answer A number of considerations would be necessary. There are the environmental details related to the water — a standard temperature would have to be defined, as well as a standard pressure. Another consideration is the quality of the water, in terms of defining an upper limit of impurities. Another problem with this scheme is that density cannot be measured directly with a single measurement, as can length, mass and time. As a combination of two measurements (mass, and volume, which itself involves **three** measurements!), it has higher inherent uncertainty than a single measurement.

□   □   □   □

5. Suppose that two quantities A and B have different dimensions. Determine which of the following arithmetic operations **could** be physically meaningful: (a) $A+B$ (b) A/B (c) $B-A$ (d) AB.

Answer   One cannot add or subtract different dimensions, because the resultant dimension would be unknown. However, different dimensions can be meaningfully multiplied and divided. Therefore, (b) and (d) **could** be physically meaningful.

☐   ☐   ☐   ☐

SOLUTIONS TO SELECTED END-OF-CHAPTER PROBLEMS

7.   Calculate the mass of an atom of (a) helium, (b) iron, and (c) lead. Give your answers in grams. The atomic masses of these atoms are 4.00 u, 55.9 u, and 207 u, respectively.

Solution

Conceptualize: The mass of an atom of any element is essentially the mass of the protons and neutrons that make up its nucleus since the mass of the electrons is negligible (less than a 0.05% contribution). Since most atoms have about the same number of neutrons as protons, the atomic mass is approximately double the atomic number (the number of protons). We should also expect that the mass of a single atom is a very small fraction of a gram ($\sim 10^{-23}$ g) since one mole (6.02×10^{23}) of atoms has a mass on the order of several grams.

Categorize: An atomic mass unit is defined as 1/12 of the mass of a carbon-12 atom. The mass in grams can be found by multiplying the atomic mass by the mass of one atomic mass unit (u):

$$1\,u = 1.66 \times 10^{-24}\,g$$

Analyze:

For He, $m = 4.00\,u = (4.00\,u)(1.66 \times 10^{-24}\,g\,/\,u) = 6.64 \times 10^{-24}\,g$ ◊

For Fe, $m = 55.9\,u = (55.9\,u)(1.66 \times 10^{-24}\,g\,/\,u) = 9.28 \times 10^{-23}\,g$ ◊

For Pb, $m = 207\,u = (207\,u)(1.66 \times 10^{-24}\,g\,/\,u) = 3.44 \times 10^{-22}\,g$ ◊

Finalize: As expected, the mass of the atoms is larger for higher atomic numbers. If we did not know the conversion factor for atomic mass units, we could use the mass of a proton as a close approximation:

$$1\,u \approx 1.67 \times 10^{-24}\,g$$

13. The position of a particle moving under uniform acceleration is some function of time and the acceleration. Suppose we write this position as $s = ka^m t^n$, where k is a dimensionless constant. Show by dimensional analysis that this expression is satisfied if $m = 1$ and $n = 2$. Can this analysis give the value of k?

Solution

The solution to this problem appears in the chapter text, in Quick Quiz 1.2 and the paragraph preceding it, with just the symbols m and n reversed. You would do better to study the chapter text as preparation for doing problems. We give the solution here as well:

For the equation to be valid, we must choose values of m and n to make it dimensionally consistent. Since s is a position, its dimensions are those of length (L). The acceleration, a, is a length divided by the square of a time (L/T^2). The variable t has dimensions of time (T), and the constant k has no dimensions. Substituting these dimensions into the equation yields:

$$(L) = \left(\frac{L}{T^2}\right)^m (T)^n = (L)^m (T)^{-2m} (T)^n \qquad \text{or} \qquad (L)^1 (T)^0 = (L)^m (T)^{n-2m}$$

Note that the factor (T)^0 introduced on the left side of the second equation is equal to 1. This equation can be true only if the powers of length (L) are the same on the two sides of the equation and, simultaneously, the powers of time (T) are the same on both sides. Indeed, if another basic unit such as mass (M) were present, we would also require that its powers be identical on the two sides. Thus, we obtain a set of two simultaneous equations:

$$1 = m \qquad \text{and} \qquad 0 = n - 2m$$

The solutions are therefore seen to be:

$$m = 1 \qquad \text{and} \qquad n = 2 \qquad \qquad \lozenge$$

This gives no information about possible values of the dimensionless constant k. $\qquad \lozenge$

===========

15. Which of the following equations are dimensionally correct?

(a) $v_f = v_i + ax$

(b) $y = (2 \text{ m}) \cos(kx)$ where $k = 2 \text{ m}^{-1}$

Solution

(a) Write out dimensions for each quantity in the equation $v_f = v_i + ax$

The variables v_f and v_i are expressed in units of m/s, so $[v_f] = [v_i] = LT^{-1}$

The variable a is expressed in units of m/s^2 and $[a] = LT^{-2}$

while the variable x is expressed in meters. Therefore $[ax] = L^2T^{-2}$

Consider the right-hand member (RHM) of equation (a): $[RHM] = LT^{-1} + L^2T^{-2}$

Quantities in a sum must be dimensionally the same.
Therefore, equation (a) **is not** dimensionally correct.   ◊

(b) Write out dimensions for each quantity in the equation $y = (2 \text{ m}) \cos(kx)$

For y, $[y] = L$

for 2 m, $[2 \text{ m}] = L$

and for (kx), $[kx] = [(2 \text{ m}^{-1})x] = L^{-1}L$

For the left-hand member (LHM) and the right-hand member (RHM) of the equation,

$$[LHM] = [y] = L \qquad\qquad [RHM] = [2 \text{ m}][\cos(kx)] = L$$

These are the same, so equation (b) **is** dimensionally correct.   ◊

---

21. A rectangular building lot is 100 ft by 150 ft. Determine the area of this lot in m^2.

Solution

Conceptualize: We must calculate the area and convert units. Since a meter is about 3 feet, we should expect the area to be about $A \approx (30 \text{ m})(50 \text{ m}) = 1\,500 \text{ m}^2$.

Categorize: Area = Length $\times$ Width . Use the conversion $1 \text{ m} = 3.281 \text{ ft}$.

Analyze: $A = L \times w = (100 \text{ ft})\left(\dfrac{1 \text{ m}}{3.281 \text{ ft}}\right)(150 \text{ ft})\left(\dfrac{1 \text{ m}}{3.281 \text{ ft}}\right) = 1390 \text{ m}^2$ ◊

Finalize: Our calculated result agrees reasonably well with our initial estimate and has the proper units of m^2. Unit conversion is a common technique that is applied to many problems.

---

25. A solid piece of lead has a mass of 23.94 g and a volume of 2.10 cm³. From these data, calculate the density of lead in SI units (kg/m³).

Solution

Conceptualize: From Table 1.5, the density of lead is 1.13×10^4 kg / m³, so we should expect our calculated value to be close to this number. This density value tells us that lead is about 11 times denser than water, which agrees with our experience that lead sinks.

Categorize: Density is defined as $\rho = m/V$. We must convert to SI units in the calculation.

Analyze: $\rho = \left(\dfrac{23.94 \text{ g}}{2.10 \text{ cm}^3} \right) \left(\dfrac{1 \text{ kg}}{1000 \text{ g}} \right) \left(\dfrac{100 \text{ cm}}{1 \text{ m}} \right)^3 = 1.14 \times 10^4 \text{ kg / m}^3$　　　　◊

Finalize: At one step in the calculation, we note that **one million** cubic centimeters make one cubic meter. Our result is indeed close to the expected value. Since the last reported significant digit is not certain, the difference in the two values is probably due to measurement uncertainty and should not be a concern. One important common sense check on density values is that objects which sink in water must have a density greater than 1 g/cm³, and objects that float must be less dense than water.

---

29. At the time of this book's printing, the US national debt is about $6 trillion. (a) If payments were made at the rate of $1 000/s, how many years would it take to pay off the debt, assuming no interest were charged? (b) A dollar bill is about 15.5 cm long. If six trillion dollar bills were laid end to end around the Earth's equator, how many times would they encircle the planet? Take the radius of the Earth at the equator to be 6 378 km. (**Note:** Before doing any of these calculations, try to guess at the answers. You may be very surprised.)

Solution

(a) **Conceptualize:** $6 trillion is certainly a large amount of money, so even at a rate of $1 000/second, we might guess that it will take a lifetime (~ 100 years) to pay off the debt.

Categorize: The time to repay the debt will be calculated by dividing the total debt by the rate at which it is repaid.

Analyze: $T = \dfrac{\$6 \text{ trillion}}{\$1\,000 / \text{s}} = \dfrac{\$6 \times 10^{12}}{(\$1\,000 / \text{s})\left(3.16 \times 10^7 \text{ s / yr}\right)} = 190 \text{ yr}$　　　◊

Finalize: Our guess was a bit low. $6 trillion really is a lot of money!

(b) **Conceptualize:** We might guess that 6 trillion bills would encircle the Earth at least a few hundred times, maybe more since our first estimate was low.

Categorize: The number of bills can be found from the total length of the bills placed end to end divided by the circumference of the Earth.

Analyze: $N = \dfrac{L}{C} = \dfrac{L}{2\pi r} = \dfrac{\left(\$6 \times 10^{12}\right)(15.5 \text{ cm} / \$)(1 \text{ m} / 100 \text{ cm})}{2\pi(6.378 \times 10^6 \text{ m})} = 2.3 \times 10^4$ times ◊

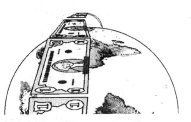

Finalize: OK, so again our guess was low. Knowing that the bills could encircle the earth more than 20 000 times, it might be reasonable to think that 6 trillion bills could cover the entire surface of the earth, but the calculated result shows that a surprisingly small fraction of the earth's surface area (a band around the equator a mile wide) is covered!

---

31. One gallon of paint (volume $= 3.78 \times 10^{-3}$ m^3) covers an area of 25.0 m^2. What is the thickness of the paint on the wall?

Solution We assume the paint keeps the same volume in the can and on the wall. We model the film on the wall as a rectangular solid, with its volume given by its surface area multiplied by its uniform thickness t: $V = At$

Therefore, $t = \dfrac{V}{A} = \dfrac{3.78 \times 10^{-3} \text{ m}^3}{25.0 \text{ m}^2} = 1.51 \times 10^{-4}$ m ◊

---

37. The diameter of our disk-shaped galaxy, the Milky Way, is about 1.0×10^5 lightyears (ly). The distance to Messier 31, which is Andromeda, the spiral galaxy nearest to the Milky Way, is about 2.0 million ly. If a scale model represents the Milky Way and Andromeda galaxies as dinner plates 25 cm in diameter, determine the distance between the two plates.

Solution The scale used in the "dinner plate" model is

$$S = \frac{1.0 \times 10^5 \text{ lightyears}}{25 \text{ cm}} = 4.00 \times 10^3 \ \frac{\text{lightyears}}{\text{cm}}$$

The distance to Andromeda in the dinner plate model will be

$$D = \frac{2.00 \times 10^6 \text{ lightyears}}{4.00 \times 10^3 \text{ lightyears} / \text{cm}} = 5.00 \times 10^2 \text{ cm} = 5.00 \text{ m}$$ ◊

---

39. One cubic meter (1.00 m^3) of aluminum has a mass of $2.70 \times 10^3 \text{ kg}$, and 1.00 m^3 of iron has a mass of $7.86 \times 10^3 \text{ kg}$. Find the radius of a solid aluminum sphere that will balance a solid iron sphere of radius 2.00 cm on an equal-arm balance.

Solution   We require equal masses: $m_{Al} = m_{Fe}$ or $\rho_{Al} V_{Al} = \rho_{Fe} V_{Fe}$

Therefore, $\rho_{Al}\left(\frac{4}{3}\pi r_{Al}^3\right) = \rho_{Fe}\left(\frac{4}{3}\pi(2.00 \text{ cm})^3\right)$

$$r_{Al}^3 = \left(\frac{\rho_{Fe}}{\rho_{Al}}\right)(2.00 \text{ cm})^3 = \left(\frac{7.86 \text{ kg} / \text{m}^3}{2.70 \text{ kg} / \text{m}^3}\right)(2.00 \text{ cm})^3 = 23.3 \text{ cm}^3$$

$$r_{Al} = 2.86 \text{ cm}$$

$\Diamond$

---

41. Find the order of magnitude of the number of Ping-Pong balls that would fit into a typical-size room (without being crushed). In your solution state the quantities you measure or estimate and the values you take for them.

Solution   Since the volume of a typical room is much larger than a Ping-Pong ball, we should expect that a very large number of balls (maybe a million) could fit in a room. Since we are only asked to find an estimate, we do not need to be too concerned about how the balls are arranged. Therefore, to find the number of balls we can simply divide the volume of an average-size living room (perhaps $15 \text{ ft} \times 20 \text{ ft} \times 8 \text{ ft}$) by the volume of an individual Ping-Pong ball. Using the approximate conversion $1 \text{ ft} = 30 \text{ cm}$, we find

$$V_{Room} = (15 \text{ ft})(20 \text{ ft})(8 \text{ ft})(30 \text{ cm} / \text{ft})^3 \approx 7 \times 10^7 \text{ cm}^3$$

A Ping-Pong ball has a diameter of about 3 cm, so we can estimate its volume as a cube:

$$V_{ball} = (3 \text{ cm})(3 \text{ cm})(3 \text{ cm}) \approx 30 \text{ cm}^3$$

The number of Ping-Pong balls that can fill the room is

$$N \approx \frac{V_{Room}}{V_{ball}} \approx 2 \times 10^6 \text{ balls} \sim 10^6 \text{ balls}$$

$\Diamond$

So a typical room can hold about a million Ping-Pong balls. This problem gives us a sense of how large a quantity "a million" really is.

---

47. To an order of magnitude, how many piano tuners are there in New York City? The physicist Enrico Fermi was famous for asking questions like this on oral Ph.D. qualifying examinations. His own facility in making order-of-magnitude calculations is exemplified in Problem 45.48.

Solution

Assume a total population of 10^7 people. Also, let us estimate that one person in one hundred owns a piano. In addition, assume that in one year a single piano tuner can service about 1 000 pianos (about 4 per day for 250 weekdays) assuming each piano is tuned once per year.

Therefore,

$$\text{The number of tuners} = \left(\frac{1 \text{ tuner}}{1000 \text{ pianos}}\right)\left(\frac{1 \text{ piano}}{100 \text{ people}}\right)(10^7 \text{ people}) \sim 100 \text{ tuners} \qquad \Diamond$$

59. The consumption of natural gas by a company satisfies the empirical equation $V = 1.50t + 0.008\,00t^2$, where V is the volume in millions of cubic feet and t the time in months. Express this equation in units of cubic feet and seconds. Assign proper units to the coefficients. Assume that a month is equal to 30.0 days.

Solution

Write "millions of cubic feet" as 10^6 ft^3, and use the given units of time and volume to assign units to the equation:

$$V = \left(1.50 \times 10^6 \text{ ft}^3/\text{mo}\right)t + \left(0.008\,00 \times 10^6 \text{ ft}^3/\text{mo}^2\right)t^2$$

To convert the units to seconds, use $1 \text{ month} = 2.59 \times 10^6 \text{ s}$, to obtain:

$$V = \left(1.50 \times 10^6 \frac{\text{ft}^3}{\text{mo}}\right)\left(\frac{1 \text{ mo}}{2.59 \times 10^6 \text{ s}}\right)t + \left(0.008\,00 \times 10^6 \frac{\text{ft}^3}{\text{mo}^2}\right)\left(\frac{1 \text{ mo}}{2.59 \times 10^6 \text{ s}}\right)^2 t^2$$

$$V = \left(0.579 \text{ ft}^3/s\right)t + \left(1.19 \times 10^{-9} \text{ ft}^3/s^2\right)t^2 \qquad \Diamond$$

61. A high fountain of water is located at the center of a circular pool as in Figure P1.61. Not wishing to get his feet wet, a student walks around the pool and measures its circumference to be 15.0 m. Next, the student stands at the edge of the pool and uses a protractor to gauge the angle of elevation of the top of the fountain to be 55.0°. How high is the fountain?

Solution Define a right triangle whose legs represent the height and radius of the fountain.

From the dimensions of the triangle and fountain, the circumference is

$$C = 2\pi r \quad \text{and} \quad \tan\theta = h/r$$

Therefore, $h = r\tan\theta = \left(\dfrac{C}{2\pi}\right)\tan\theta$

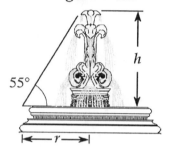

Figure P1.61

or $\qquad h = \dfrac{15.0 \text{ m}}{2\pi}\tan 55.0° = 3.41 \text{ m}$ ◊

---

63. There are nearly $\pi \times 10^7$ s in one year. Find the percentage error in this approximation, where "percentage error" is defined as

$$\text{Percentage error} = \frac{\left|\text{assumed value} - \text{true value}\right|}{\text{true value}} \times 100\%$$

Solution First evaluate the "true value". Remember that every fourth year is a leap year; therefore there are 365.25 days in an average year.

$$1 \text{ yr} = (1 \text{ yr})\left(\frac{365.25 \text{ d}}{1 \text{ yr}}\right)\left(\frac{24 \text{ h}}{1 \text{ d}}\right)\left(\frac{3600 \text{ s}}{1 \text{ h}}\right) = 3.1558 \times 10^7 \text{ s}$$

$$\text{Percentage error} = \frac{\left|\text{Assumed value} - \text{true value}\right|}{\text{True value}} \times 100\%$$

$$\text{Percentage error} = \frac{(3.155\,8 - 3.141\,6) \times 10^7}{3.155\,8 \times 10^7} \times 100\% = 0.449\% \qquad ◊$$

---

67. Assume that there are 100 million passenger cars in the United States and that the average fuel consumption rate is 20 mi/gal of gasoline. If the average distance traveled by each car is 10 000 mi/yr, how much gasoline would be saved per year if average fuel consumption could be increased to 25 mi/gal?

Solution   We define an average fuel consumption rate based upon the total miles driven, and suppose that this rate is currently at 20 mi/gallon.

$$\text{fuel consumed} = \frac{\text{total miles driven}}{\text{average fuel consumption rate}}$$

or $$f = \frac{s}{c}$$

Since the total number of miles driven is the same in each case,

At 20 mi/gal, $$f = \frac{\left(100 \times 10^6 \text{ cars}\right)\left(10^4 \text{ (mi / yr) / car}\right)}{20 \text{ mi / gal}} = 5 \times 10^{10} \text{ gal / yr}$$

At 25 mi/gal, $$f = \frac{\left(100 \times 10^6 \text{ cars}\right)\left(10^4 \text{ (mi / yr) / car}\right)}{25 \text{ mi / gal}} = 4 \times 10^{10} \text{ gal / yr}$$

Thus, we estimate $$\Delta f = 1 \times 10^{10} \text{ gal / yr} \qquad \Diamond$$

Chapter 2
MOTION IN ONE DIMENSION

EQUATIONS AND CONCEPTS

The **displacement** Δx of a particle moving from position x_i to position x_f equals the final coordinate minus the initial coordinate. Displacement is an example of a vector quantity.

$$\Delta x \equiv x_f - x_i \tag{2.1}$$

Distance traveled is the length of path followed by a particle and should not be confused with displacement. When $x_f = x_i$, the displacement is zero; however, if the particle leaves x_i, travels along a path, and returns to x_i, the distance traveled will not be zero.

The **average velocity** of an object during a time interval Δt is the ratio of the total displacement Δx to the time interval during which the displacement occurs.

$$\bar{v}_x \equiv \frac{\Delta x}{\Delta t} \tag{2.2}$$

The **average speed** of a particle is a scalar quantity defined as the ratio of the total distance traveled to the time required to travel that distance. Average speed has no direction and carries no algebraic sign. *The magnitude of the average velocity is not the average speed; although in certain cases they may be numerically equal.*

$$\text{Average speed} = \frac{\text{total distance}}{\text{total time}} \tag{2.3}$$

The **instantaneous velocity** v is defined as the limit of the ratio $\Delta x/\Delta t$ as Δt approaches zero. This limit is called the derivative of x with respect to t. The instantaneous velocity at any time is the slope of the position-time graph at that time. As illustrated in the figure, the slope can be positive, negative, or zero.

$$v_x \equiv \lim_{\Delta t \to 0} \frac{\Delta x}{\Delta t} = \frac{dx}{dt} \tag{2.5}$$

12

The **instantaneous speed** is the magnitude of the instantaneous velocity.

The **average acceleration** of an object is defined as the ratio of the change in velocity to the time interval during which the change in velocity occurs. Equation 2.6 gives the average acceleration of a particle in one-dimensional motion along the x-axis.

$$\bar{a}_x \equiv \frac{\Delta v_x}{\Delta t} = \frac{v_{xf} - v_{xi}}{t_f - t_i} \qquad (2.6)$$

The **instantaneous acceleration** a is defined as the limit of the ratio $\Delta v_x / \Delta t$ as Δt approaches zero. This limit is the derivative of the velocity along the x direction with respect to time. *A negative acceleration does not necessarily imply a decreasing speed.*

$$a_x \equiv \lim_{\Delta t \to 0} \frac{\Delta v_x}{\Delta t} = \frac{d v_x}{dt} \qquad (2.7)$$

The acceleration can also be expressed as the second derivative of the position with respect to time. This is shown in Equation 2.8 for the case of a particle in one-dimensional motion along the x-axis.

$$a_x = \frac{d^2 x}{dt^2} \qquad (2.8)$$

The **kinematic equations, 2.9 - 2.13**, can be used to describe one-dimensional motion along the x axis with constant acceleration. Note that each equation shows a different relationship among physical quantities: initial velocity, final velocity, acceleration, time, and position. *Remember, the relationships stated in Equations 2.9 - 2.13 are true only for cases in which the acceleration is constant.*

$$v_{xf} = v_{xi} + a_x t \qquad (2.9)$$

$$\bar{v}_x = \frac{1}{2}\left(v_{xi} + v_{xf}\right) \qquad (2.10)$$

$$x_f = x_i + \frac{1}{2}\left(v_{xi} + v_{xf}\right)t \qquad (2.11)$$

$$x_f = x_i + v_{xi} t + \frac{1}{2} a_x t^2 \qquad (2.12)$$

$$v_{xf}^2 = v_{xi}^2 + 2a_x\left(x_f - x_i\right) \qquad (2.13)$$

A **freely falling object** is any object moving under the influence of the gravitational force alone. Equations 2.9 - 2.13 can be modified to describe the motion of freely falling objects by denoting the motion to be along the y axis (defining "up" as positive) and setting $a_y = -g$. *A freely falling object experiences an acceleration that is directed downward regardless of its actual motion.*

$$v_{yf} = v_{yi} - gt$$

$$y_f = y_i + \frac{1}{2}\left(v_{yi} + v_{yf}\right)t$$

$$\bar{v}_y = \frac{1}{2}\left(v_{yi} + v_{yf}\right)$$

$$y_f = y_i + v_{yi}t - \frac{1}{2}gt^2$$

$$v_{yf}^2 = v_{yi}^2 - 2g\left(y_f - y_i\right)$$

SUGGESTIONS, SKILLS, AND STRATEGIES

Organize your problem-solving by considering each step of the **Conceptualize, Categorize, Analyze,** and **Finalize** protocol described in your textbook.

REVIEW CHECKLIST

▷ For each of the following pairs of terms, define each quantity and state how each is related to the other member of the pair: distance and displacement; instantaneous and average velocity; speed and instantaneous velocity; instantaneous and average acceleration. (Sections 2.1, 2.2 and 2.3)

▷ Construct a graph of position versus time (given a function such as $x = 5 + 3t - 2t^2$) for a particle in motion along a straight line. From this graph, you should be able to determine the value of the average velocity between two points, t_1 and t_2 and the instantaneous velocity at a given point. Average velocity is the slope of the cord between the two points and the instantaneous velocity at a given time is the slope of the tangent to the graph at that time. (Section 2.2)

▷ Describe what is meant by a freely falling body (one moving under the influence of gravity — where air resistance is neglected). Recognize that the equations of kinematics apply directly to a freely falling object and that the acceleration is then given by by $a = -g$ (where $g = 9.80$ m / s^2). (Section 2.6)

▷ Apply the equations of kinematics to any situation where the motion occurs under constant acceleration. (Section 2.5)

▷ Be able to interpret graphs of one-dimensional motion showing position vs. time, velocity vs. time, and acceleration vs. time.

ANSWERS TO SELECTED QUESTIONS

3. If the average velocity of an object is zero in some time interval, what can you say about the displacement of the object for that interval?

Answer The displacement is **zero**, since the displacement is proportional to average velocity.

☐　　☐　　☐　　☐

9. Two cars are moving in the same direction in parallel lanes along a highway. At some instant, the velocity of car A exceeds the velocity of car B. Does this mean that the acceleration of A is greater than that of B? Explain.

Answer No. If Car A has been traveling with cruise control, its velocity will be high (60 mph), but its acceleration will be close to zero. If Car B is pulling onto the highway, its velocity is likely to be low (30 mph), but its acceleration will be high.

☐　　☐　　☐　　☐

11. Consider the following combinations of signs and values for velocity and acceleration of a particle with respect to a one-dimensional x axis:

	(a)	(b)	(c)	(d)	(e)	(f)	(g)	(h)
Velocity	+	+	+	–	–	–	0	0
Acceleration	+	–	0	+	–	0	+	–

Describe what a particle is doing in each case, and give a real life example for an automobile on an east–west one-dimensional axis, with east considered the positive direction.

Answer

(a) The particle is moving to the right (which we take as the $+x$ direction) since the velocity is positive, and its speed is increasing, because the acceleration is in the same direction as the velocity. This would be the case in an automobile that is moving toward the east, starting up after waiting for a red light.

15

(b)   The particle is moving to the right, since the velocity is positive, and its speed is decreasing, because the acceleration is in the opposite direction to the velocity. An automobile is moving toward the east, slowing down in preparation for a red light.

(c)   The particle is moving to the right, since the velocity is positive, and its speed is constant, because the acceleration is zero. An automobile is moving toward the east, moving at constant speed on a freeway.

(d)   The particle is moving to the left (in the $-x$ direction), since the velocity is negative, and its speed is decreasing, because the acceleration is in the opposite direction to the velocity. An automobile is moving toward the west, slowing down in preparation for a red light.

(e)   The particle is moving to the left (in the $-x$ direction), since the velocity is negative, and its speed is increasing, because the acceleration is in the same direction as the velocity. An automobile is moving toward the west, starting up after waiting for a red light.

(f)   The particle is moving to the left (in the $-x$ direction), since the velocity is negative, and its speed is constant, because the acceleration is zero. An automobile is moving toward the west, moving at constant speed on a freeway.

(g)   The particle is momentarily at rest, and, in the next instant, will be moving to the right. This situation can only exist for an instant, since as soon as the particle begins moving, the velocity will no longer be zero. This is the situation for an automobile that has been coasting uphill toward the west, just when it comes to a stop before beginning to roll downhill to the east.

(h)   The particle is momentarily at rest, and, in the next instant, will be moving to the left. This situation can only exist for an instant, since as soon as the particle begins moving, the velocity will no longer be zero. This is the situation for an automobile that has been coasting uphill toward the east, just when it comes to a stop before beginning to roll downhill to the west.

□   □   □   □

15.   A student at the top of a building of height h throws one ball upward with a speed of v_i and then throws a second ball downward with the same initial speed, v_i. How do the final velocities of the balls compare when they reach the ground?

Answer   They are the same. After the first ball reaches its apex and falls back downward past the student, it will have a downward velocity with a magnitude equal to v_i. This velocity is the same as the velocity of the second ball, so after they fall through equal heights their impact speeds will also be the same.

□   □   □   □

16

SOLUTIONS TO SELECTED END-OF-CHAPTER PROBLEMS

3. The position versus time for a certain particle moving along the x axis is shown in Figure P2.3. Find the average velocity in the time intervals (a) 0 to 2 s, (b) 0 to 4 s, (c) 2 s to 4 s, (d) 4 s to 7 s, (e) 0 to 8 s.

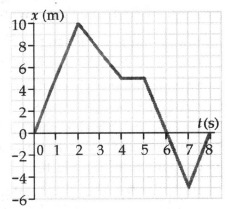

Figure P2.3

Solution

On this graph, we can tell positions to two significant figures:

(a) $\qquad x = 0 \qquad$ at $\qquad t = 0$

$\quad$ and $\quad x = 10$ m $\quad$ at $\quad t = 2$ s: $\quad \overline{v}_x = \dfrac{\Delta x}{\Delta t} = \dfrac{10 \text{ m} - 0}{2 \text{ s} - 0} = 5.0 \text{ m / s}$ $\qquad \Diamond$

(b) $\qquad x = 5.0$ m $\quad$ at $\qquad t = 4$ s: $\quad \overline{v}_x = \dfrac{\Delta x}{\Delta t} = \dfrac{5.0 \text{ m} - 0}{4 \text{ s} - 0} = 1.2 \text{ m / s}$ $\qquad \Diamond$

(c) $\qquad\qquad\qquad\qquad\qquad\qquad \overline{v}_x = \dfrac{\Delta x}{\Delta t} = \dfrac{5.0 \text{ m} - 10 \text{ m}}{4 \text{ s} - 2 \text{ s}} = -2.5 \text{ m / s}$ $\qquad \Diamond$

(d) $\qquad\qquad\qquad\qquad\qquad\qquad \overline{v}_x = \dfrac{\Delta x}{\Delta t} = \dfrac{-5.0 \text{ m} - 5.0 \text{ m}}{7 \text{ s} - 4 \text{ s}} = -3.3 \text{ m / s}$ $\qquad \Diamond$

(e) $\qquad\qquad\qquad\qquad\qquad\qquad \overline{v}_x = \dfrac{\Delta x}{\Delta t} = \dfrac{0.0 \text{ m} - 0.0 \text{ m}}{8 \text{ s} - 0 \text{ s}} = 0 \text{ m / s}$ $\qquad \Diamond$

---

5. A person walks first at a constant speed of 5.00 m/s along a straight line from point A to point B and then back along the line from B to A at a constant speed of 3.00 m/s. What is (a) her average speed over the entire trip? (b) her average velocity over the entire trip?

Solution

(a) The average speed during any time interval is equal to the total distance of travel divided by the total time:

$$\text{average speed} = \frac{\text{total distance}}{\text{total time}} = \frac{d_{AB} + d_{BA}}{t_{AB} + t_{BA}}$$

$\quad$ But $\quad d_{AB} = d_{BA} = d, \quad t_{AB} = d / v_{AB}, \quad$ and $\quad t_{BA} = d / v_{BA}$

so   average speed $= \dfrac{d + d}{(d/v_{AB}) + (d/v_{AB})} = \dfrac{2(v_{AB})(v_{BA})}{v_{AB} + v_{BA}}$

and   average speed $= 2 \left[\dfrac{(5.00 \text{ m / s})(3.00 \text{ m / s})}{5.00 \text{ m / s} + 3.00 \text{ m / s}} \right] = 3.75 \text{ m / s}$ ◊

(b)   The average velocity during any time interval equals total displacement divided by elapsed time.

$$\overline{\mathbf{v}} = \dfrac{\Delta \mathbf{x}}{\Delta t}$$

Since the walker returns to the starting point, $\Delta \mathbf{x} = 0$ and $\overline{\mathbf{v}} = 0$.   ◊

---

7.   A position-time graph for a particle moving along the x axis is shown in Figure P2.7.   (a) Find the average velocity in the time interval $t = 1.50$ s to $t = 4.00$ s. (b) Determine the instantaneous velocity at $t = 2.00$ s by measuring the slope of the tangent line shown in the graph. (c) At what value of t is the velocity zero?

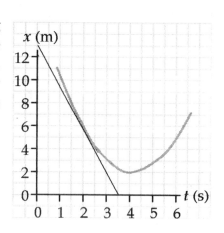

Figure P2.7

Solution

(a)   From the graph:

   At $t_1 = 1.5$ s, $x = x_1 = 8.0$ m

   At $t_2 = 4.0$ s, $x = x_2 = 2.0$ m

   Therefore, $\overline{v}_{1 \to 2} = \dfrac{\Delta x}{\Delta t} = \dfrac{2.0 \text{ m} - 8.0 \text{ m}}{4.0 \text{ s} - 1.5 \text{ s}} = -2.4 \text{ m / s}$ ◊

(b)   Choose two points along a line which is tangent to the curve at $t = 2.0$ s. We will use the two points

$$(t_i = 0.0 \text{ s} , \; x_i = 13.0 \text{ m}) \quad \text{and} \quad (t_f = 3.5 \text{ s} , \; x_f = 0.0 \text{ m})$$

Instantaneous velocity equals the slope of the tangent line,

so $v = \dfrac{x_f - x_i}{t_f - t_i} = \dfrac{0.0 \text{ m} - 13.0 \text{ m}}{3.5 \text{ s} - 0.0 \text{ s}} = -3.7 \text{ m / s}$

The negative sign shows that the **direction** of **v** is along the negative x direction. ◊

(c)   The velocity will be zero when the slope of the tangent line is zero. This occurs for the point on the graph where x has its minimum value.

   Therefore, $v = 0$ at $t = 4.0$ s   ◊

---

15. A particle moves along the x axis according to the equation $x = 2.00 + 3.00t - 1.00\,t^2$, where x is in meters and t is in seconds. At $t = 3.00$ s, find (a) the position of the particle, (b) its velocity, and (c) its acceleration.

Solution With the position given by $x = 2.00 + 3.00t - t^2$, we can use the rules for differentiation to write expressions for the velocity and acceleration as functions of time:

$$v = \frac{dx}{dt} = 3.00 - 2t \qquad \text{and} \qquad a = \frac{dv}{dt} = -2$$

Now we can evaluate x, v, and a at $t = 3.00$ s.

(a) $x = 2.00 + 3.00(3.00) - (3.00)^2 = 2.00$ m ◊

(b) $v = 3.00 - 2(3.00) = -3.00$ m / s ◊

(c) $a = -2.00$ m / s^2 ◊

---

21. An object moving with uniform acceleration has a velocity of 12.0 cm/s in the positive x direction when its x coordinate is 3.00 cm. If its x coordinate 2.00 s later is –5.00 cm, what is the magnitude of its acceleration?

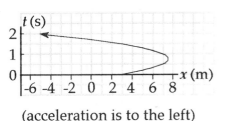

(acceleration is to the left)

Solution

Take $t = 0$ to be the time when $x_i = 3.00$ cm

and $v_{xi} = 12.00$ cm / s

Also, at $t = 2.00$ s $x_f = -5.00$ cm

Use the kinematic equation $x_f = x_i + v_{xi}t + \frac{1}{2}at^2$

and solve for a.:

$$a = \frac{2[x_f - x_i - v_{xi}t]}{t^2}$$

$$a = \frac{2[-5.00 \text{ cm} - 3.00 \text{ cm} - (12.0 \text{ cm} / \text{s})(2.00 \text{ s})]}{(2.00 \text{ s})^2}$$

$$a = -16.0 \text{ cm} / \text{s}^2 \qquad \diamond$$

---

27. A jet plane lands with a speed of 100 m/s and can accelerate at a maximum rate of –5.00 m/s^2 as it comes to rest. (a) From the instant the plane touches the runway, what is the minimum time needed before it can come to rest? (b) Can this plane land at a small tropical island airport where the runway is 0.800 km long?

Solution The negative acceleration of the plane as it lands is normally called **deceleration**; however, physicists tend to use the single term **acceleration** for both cases.

(a) Assume that the acceleration of the plane is **constant** at the maximum rate, so that the plane can be modeled as a particle under constant acceleration:

$$a_x = -5.00 \text{ m} / \text{s}^2$$

Given $v_{xi} = 100 \text{ m} / \text{s}$ and $v_{xf} = 0$, use the equation $v_{xf} = v_{xi} + a_x t$ and solve for t:

$$t = \frac{v_{xf} - v_{xi}}{a_x} = \frac{0 - 100 \text{ m} / \text{s}}{-5.00 \text{ m} / \text{s}^2} = 20.0 \text{ s} \qquad \lozenge$$

(b) Find the required stopping distance and compare this to the length of the runway. Taking x_i to be zero, we get

$$v_{xf}^2 = v_{xi}^2 + 2a_x(x_f - x_i)$$

or $\qquad \Delta x = x_f - x_i = \dfrac{v_{xf}^2 - v_{xi}^2}{2a_x} = \dfrac{0 - (100 \text{ m} / \text{s})^2}{2(-5.00 \text{ m} / \text{s}^2)} = 1\,000 \text{ m}$

The stopping distance is greater than the length of the runway; the plane **cannot land.** $\qquad \lozenge$

---

31. For many years Colonel John P. Stapp, USAF, held the world's land speed record. On March 19, 1954, he rode a rocket-propelled sled that moved down a track at 632 mi/h. He and the sled were safely brought to rest in 1.40 s. Determine (a) the negative acceleration he experienced and (b) the distance he traveled during this negative acceleration.

Solution We assume the acceleration remains constant during the 1.40 s period of negative acceleration.

$$v_{xi} = 632 \; \frac{\text{mi}}{\text{h}} = 632 \; \frac{\text{mi}}{\text{h}} \left(\frac{1609 \text{ m}}{1 \text{ mi}} \right)\left(\frac{1 \text{ h}}{3600 \text{ s}} \right) = 282 \text{ m} / \text{s}$$

(a)   Taking

$$v_{xf} = v_{xi} + a_x t \text{ with } v_{xf} = 0,$$

$$a_x = \frac{v_{xf} - v_{xi}}{t} = \frac{0 - 282 \text{ m / s}}{1.40 \text{ s}} = -202 \text{ m / s}^2 \qquad \lozenge$$

This is approximately 20*g* !

(b)

$$x_f - x_i = \tfrac{1}{2}(v_{xi} + v_{xf})t = \tfrac{1}{2}(282 \text{ m / s} + 0)(1.40 \text{ s}) = 198 \text{ m} \qquad \lozenge$$

---

43.   A student throws a set of keys vertically upward to her sorority sister, who is in a window 4.00 m above. The keys are caught 1.50 s later by the sister's outstretched hand. (a) With what initial velocity were the keys thrown? (b) What was the velocity of the keys just before they were caught?

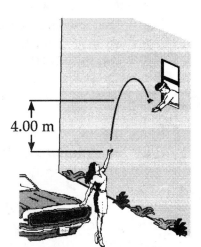

4.00 m

Solution

We model the keys as a particle under the constant free-fall acceleration.

Taking the student's position to be $y_i = 0$

and given that at $t = 1.50$ s $y = 4.00$ m

we find   (with $a = -9.80 \text{ m / s}^2$)

(a) $y_f = y_i + v_{yi}t + \tfrac{1}{2}a_y t^2$ or $v_{yi} = \dfrac{y_f - y_i - \tfrac{1}{2}a_y t^2}{t}$

$$v_{yi} = \frac{4.00 \text{ m} - \tfrac{1}{2}(-9.80 \text{ m / s}^2)(1.50 \text{ s})^2}{1.50 \text{ s}} = 10.0 \text{ m / s} \qquad \lozenge$$

(b)   The velocity at any time $t > 0$

is given by $v_{yf} = v_{yi} + a_y t$

Therefore, at $t = 1.50$ s, $v_{yf} = 10.0 \text{ m / s} - \left(9.80 \text{ m / s}^2\right)(1.50 \text{ s}) = -4.68 \text{ m / s} \quad \lozenge$

The negative sign means that the keys are moving **downward** just before they are caught.

---

47. A baseball is hit so that it travels straight upward after being struck by the bat. A fan observes that it requires 3.00 s for the ball to reach its maximum height. Find (a) its initial velocity and (b) the height it reaches.

Solution

Conceptualize:

The initial speed of the ball is probably somewhat greater than the speed of the pitch, which might be about 60 mph (~30 m/s), so an initial upward velocity off the bat of somewhere between 20 and 100 m/s would be reasonable. We also know that the length of a ball field is about 300 ft. (~100 m), and a pop-fly usually does not go higher than this distance, so a maximum height of 10 to 100 m would be reasonable for the situation described in this problem.

Categorize:

Since the ball's motion is entirely vertical, we can use the equations for free fall to find the initial velocity and maximum height from the elapsed time.

Analyze:

After leaving the bat, the ball is in free fall for $t = 3.00$ s and has constant acceleration, $a_y = -g = -9.80 \text{ m} / \text{s}^2$.

Solve the equation $v_{yf} = v_{yi} + a_y t$ with $a_y = -g$ to obtain v_{yi} when $v_{yf} = 0$, and the ball reaches its maximum height.

(a) $v_{yi} = v_{yf} + gt = 0 + (9.80 \text{ m} / \text{s}^2)(3.00 \text{ s}) = 29.4 \text{ m} / \text{s}$ (upward) ◊

(b) The maximum height in the vertical direction is $y_f = v_{yi} t - \frac{1}{2} g t^2$:

$y_f = (29.4 \text{ m} / \text{s})(3.00 \text{ s}) - \frac{1}{2}(9.80 \text{ m} / \text{s}^2)(3.00 \text{ s})^2 = 44.1 \text{ m}$ ◊

Finalize:

The calculated answers seem reasonable since they lie within our expected ranges, and they have the correct units and direction. We must remember that it is possible to solve a problem like this correctly, yet the answers may not seem reasonable simply because the conditions stated in the problem may not be physically possible (e.g., a time of 10 seconds for a pop fly would not be realistic).

49. A daring ranch hand sitting on a tree limb wishes to drop vertically onto a horse galloping under the tree. The constant speed of the horse is 10.0 m/s, and the distance from the limb to the level of the saddle is 3.00 m. (a) What must be the horizontal distance between the saddle and limb when the ranch hand makes his move? (b) How long is he in the air?

Solution We do part (b) first.

(b) Consider the vertical motion of the man after leaving the limb (with $v_i = 0$ at $y_i = 3.00$ m) until reaching the saddle (at $y_f = 0$).

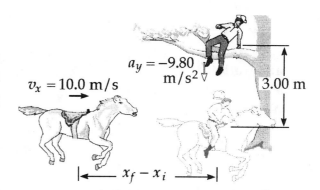

Modeling the man as a particle under constant acceleration, we find his time of fall from

$$y_f = y_i + v_{yi}t + \frac{1}{2}a_y t^2$$

When $v_i = 0$, $\quad t = \sqrt{\dfrac{2(y_f - y_i)}{a_y}} = \sqrt{\dfrac{2(0 - 3.00 \text{ m})}{-9.80 \text{ m}/\text{s}^2}} = 0.782 \text{ s}$ ◊

(a) During this time interval, the horse is modeled as a particle under constant velocity in the horizontal direction.

$$v_{xi} = v_{xf} = 10.0 \text{ m}/\text{s}$$

so $\quad x_f - x_i = v_{xi}t = (10.0 \text{ m}/\text{s})(0.782 \text{ s}) = 7.82 \text{ m}$

and the ranch hand must let go when the horse is 7.82 m from the tree.   ◊

53. Automotive engineers refer to the time rate of change of acceleration as the "jerk." If an object moves in one dimension such that its jerk J is constant, (a) determine expressions for its acceleration $a_x(t)$, velocity $v_x(t)$, and position $x(t)$, given that its initial acceleration, velocity, and position are a_{xi}, v_{xi}, and x_i, respectively. (b) Show that $a_x^2 = a_{xi}^2 + 2J(v_x - v_{xi})$.

Solution   We are given $J = da_x/dt = \text{constant}$, so we know that $da_x = J\,dt$.

(a)   Integrating,
$$a_x = J\int dt = Jt + c_1$$

When $t = 0$, $a_x = a_{xi}$:
$$c_1 = a_{xi}$$

Therefore,
$$a_x = Jt + a_{xi} \qquad \Diamond$$

From $a_x = dv_x/dt$,
$$dv_x = a_x dt$$

Integration yields
$$v_x = \int a_x dt = \int (Jt + a_{xi})dt = \tfrac{1}{2}Jt^2 + a_{xi}t + c_2$$

When $t = 0$, $v_x = v_{xi}$:
$$c_2 = v_{xi}$$

and
$$v_x = \tfrac{1}{2}Jt^2 + a_{xi}t + v_{xi} \qquad \Diamond$$

From $v_x = dx/dt$, $dx = v_x dt$:
$$x = \int v_x dt = \int \left(\tfrac{1}{2}Jt^2 + a_{xi}t + v_{xi}\right)dt$$

and
$$x = \tfrac{1}{6}Jt^3 + \tfrac{1}{2}a_{xi}t^2 + v_{xi}t + c_3$$

Since $x = x_i$ when $t = 0$,
$$c_3 = x_i$$

Therefore,
$$x = \tfrac{1}{6}Jt^3 + \tfrac{1}{2}a_{xi}t^2 + v_{xi}t + x_i \qquad \Diamond$$

(b)   Squaring the acceleration,
$$a_x^2 = (Jt + a_{xi})^2 = J^2t^2 + a_{xi}^2 + 2Ja_{xi}t$$

Solving,
$$a_x^2 = a_{xi}^2 + 2J(\tfrac{1}{2}Jt^2 + a_{xi}t)$$

The expression for v was
$$v_x = \tfrac{1}{2}Jt^2 + a_{xi}t + v_{xi}$$

So
$$(v_x - v_{xi}) = \tfrac{1}{2}Jt^2 + a_{xi}t$$

Therefore,
$$a_x^2 = a_{xi}^2 + 2J(v_x - v_{xi}) \qquad \Diamond$$

---

65.   Setting a new world record in a 100-m race, Maggie and Judy cross the finish line in a dead heat, both taking 10.2 s. Accelerating uniformly, Maggie took 2.00 s and Judy 3.00 s to attain maximum speed, which they maintained for the rest of the race. (a) What was the acceleration of each sprinter?   (b) What were their respective maximum speeds? (c) Which sprinter was ahead at the 6.00-s mark, and by how much?

Solution

(a) Maggie moves with constant positive acceleration a_M for 2.00 s, then with constant speed (zero acceleration) for 8.20 s, covering a distance of $x_{M1} + x_{M2} = 100$ m.

The two component distances are $\quad x_{M1} = \frac{1}{2} a_M (2.00 \text{ s})^2$

and $\qquad\qquad\qquad\qquad\qquad\qquad x_{M2} = v_M (8.20 \text{ s})$

where v_M is her maximum speed, $\quad v_M = 0 + a_M (2.00 \text{ s})$,

by substitution $\qquad\qquad \frac{1}{2} a_M (2.00 \text{ s})^2 + a_M (2.00 \text{ s})(8.20 \text{ s}) = 100$ m

$$a_M = 5.43 \text{ m} / \text{s}^2 \qquad\qquad \diamond$$

Similarly, for Judy, $\qquad x_{J1} + x_{J2} = 100$ m

with $\qquad\qquad x_{J1} = \frac{1}{2} a_J (3.00 \text{ s})^2; \quad x_{J2} = v_J (7.20 \text{ s}); \quad v_J = a_J (3.00 \text{ s})$

$$\frac{1}{2} a_J (3.00 \text{ s})^2 + a_J (3.00 \text{ s})(7.20 \text{ s}) = 100 \text{ m}$$

$$a_J = 3.83 \text{ m} / \text{s}^2 \qquad\qquad \diamond$$

(b) Their speeds after accelerating are

$$v_M = a_M (2.00 \text{ s}) = \left(5.43 \text{ m} / \text{s}^2\right)(2.00 \text{ s}) = 10.9 \text{ m} / \text{s} \qquad \diamond$$

and $\qquad\qquad v_J = a_J (3.00 \text{ s}) = \left(3.83 \text{ m} / \text{s}^2\right)(3.00 \text{ s}) = 11.5 \text{ m} / \text{s} \qquad \diamond$

(c) In the first 6.00 s, Maggie covers a distance

$$\frac{1}{2} a_M (2.00 \text{ s})^2 + v_M (4.00 \text{ s}) = \frac{1}{2}\left(5.43 \text{ m} / \text{s}^2\right)(2.00 \text{ s})^2 + (10.9 \text{ m} / \text{s})(4.00 \text{ s}) = 54.3 \text{ m}$$

and Judy has run a distance

$$\frac{1}{2} a_J (3.00 \text{ s})^2 + v_J (3.00 \text{ s}) = \frac{1}{2}\left(3.83 \text{ m} / \text{s}^2\right)(3.00 \text{ s})^2 + (11.5 \text{ m} / \text{s})(3.00 \text{ s}) = 51.7 \text{ m}$$

So Maggie is ahead by $\quad 54.3 \text{ m} - 51.7 \text{ m} = 2.62 \text{ m} \qquad\qquad \diamond$

69. An inquisitive physics student and mountain climber climbs a 50.0-m cliff that overhangs a calm pool of water. He throws two stones vertically downward, 1.00 s apart, and observes that they cause a single splash. The first stone has an initial speed of 2.00 m/s. (a) How long after release of the first stone do the two stones hit the water? (b) What initial velocity must the second stone have if they are to hit simultaneously? (c) What is the speed of each stone at the instant the two hit the water?

Solution   Set $y_i = 0$ at the top of the cliff, and find the time required for the first stone to reach the water using the particle under constant acceleration model:

$$y_f = y_i + v_{yi}t + \frac{1}{2}a_y t^2$$

or in quadratic form, $-\frac{1}{2}a_y t^2 - v_{yi}t + y_f - y_i = 0$

(a)   If we take the direction downward to be negative,

$$y_f = -50.0 \text{ m}, \ v_{yi} = -2.00 \text{ m / s, and } a_y = -9.80 \text{ m / s}^2.$$

Substituting these values into the equation, we find

$$\left(4.90 \text{ m / s}^2\right)t^2 + (2.00 \text{ m / s})t - 50.0 \text{ m} = 0$$

Using the quadratic formula and noting that only the positive root describes this physical situation,

$$t = \frac{-2.00 \text{ m / s} \pm \sqrt{(2.00 \text{ m / s})^2 - 4(4.90 \text{ m / s}^2)(-50.0 \text{ m})}}{2(4.90 \text{ m / s}^2)} = 3.00 \text{ s} \qquad ◊$$

(b)   For the second stone, the time of travel is $t = 3.00 \text{ s} - 1.00 \text{ s} = 2.00 \text{ s}$.

Since $y_f = y_i + v_{yi}t + \frac{1}{2}a_y t^2$,

$$v_{yi} = \frac{\left(y_f - y_i\right) - \frac{1}{2}a_y t^2}{t} = \frac{-50.0 \text{ m} - \frac{1}{2}\left(-9.80 \text{ m / s}^2\right)(2.00 \text{ s})^2}{2.00 \text{ s}} = -15.3 \text{ m / s} \qquad ◊$$

The negative value indicates the downward direction of the initial velocity of the second stone.

(c)   For the first stone, $v_{1f} = v_{1i} + a_1 t_1 = -2.00 \text{ m / s} + \left(-9.80 \text{ m / s}^2\right)(3.00 \text{ s})$

$$v_{1f} = -31.4 \text{ m / s} \qquad ◊$$

For the second stone, $v_{2f} = v_{2i} + a_2 t_2 = -15.3 \text{ m / s} + \left(-9.80 \text{ m / s}^2\right)(2.00 \text{ s})$

$$v_{2f} = -34.8 \text{ m / s} \qquad ◊$$

73. Kathy Kool buys a sports car that can accelerate at the rate of 4.90 m/s^2. She decides to test the car by racing with another speedster, Stan Speedy. Both start from rest, but experienced Stan leaves the starting line 1.00 s before Kathy. If Stan moves with a constant acceleration of 3.50 m/s^2 and Kathy maintains an acceleration of 4.90 m/s^2, find (a) the time at which Kathy overtakes Stan, (b) the distance she travels before she catches him, and (c) the speeds of both cars at the instant she overtakes him.

Solution

(a) Let the times of travel for Kathy and Stan be t_K and t_S where $t_S = t_K + 1.00$ s.

Both start from rest ($v_{xi,K} = v_{xi,S} = 0$), so the expressions for the distances traveled are:

$$x_K = \tfrac{1}{2}a_{x,K}t_K^2 = \tfrac{1}{2}(4.90 \text{ m / s}^2)t_K^2$$

$$x_S = \tfrac{1}{2}a_{x,S}t_S^2 = \tfrac{1}{2}(3.50 \text{ m / s}^2)(t_K + 1.00 \text{ s})^2$$

When Kathy overtakes Stan, the two distances will be equal; setting $x_K = x_S$,

$$\tfrac{1}{2}(4.90 \text{ m / s}^2)t_K^2 = \tfrac{1}{2}(3.50 \text{ m / s}^2)(t_K + 1.00 \text{ s})^2$$

This can be simplified and written in the standard form of a quadratic as

$$t_K^2 - 5.00t_K \text{ s} - 2.50 \text{ s}^2 = 0$$

and solved using the quadratic formula to find

$$t_K = 5.46 \text{ s} \qquad \Diamond$$

(b) Use the equation from part (a) for distance of travel,

$$x_K = \tfrac{1}{2}a_{x,K}t_K^2 = \tfrac{1}{2}(4.90 \text{ m / s}^2)(5.46 \text{ s})^2 = 73.0 \text{ m} \qquad \Diamond$$

(c) Remembering that $v_{xi,K} = v_{xi,S} = 0$, the final velocities will be:

$$v_{xf,K} = a_{x,K}t_K = (4.90 \text{ m / s}^2)(5.46 \text{ s}) = 26.7 \text{ m / s} \qquad \Diamond$$

$$v_{xf,S} = a_{x,S}t_S = (3.50 \text{ m / s}^2)(6.46 \text{ s}) = 22.6 \text{ m / s} \qquad \Diamond$$

75. Two objects, A and B, are connected by a rigid rod that has a length L. The objects slide along perpendicular guide rails, as shown in Figure P2.75. If A slides to the left with a constant speed v, find the velocity of B when $\alpha = 60.0°$.

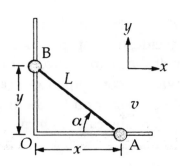

Figure P2.75

Solution We translate from a pictorial representation through a geometric model to a mathematical representation by observing that the distances x and y are always related by $x^2 + y^2 = L^2$. Differentiating this equation with respect to time, we have

$$2x\,\frac{dx}{dt} + 2y\,\frac{dy}{dt} = 0$$

Now the unknown velocity of B is

$$\frac{dy}{dt} = v_B$$

and

$$\frac{dx}{dt} = -v$$

So the differentiated equation becomes

$$\frac{dy}{dt} = -\frac{x}{y}\left(\frac{dx}{dt}\right) = \left(-\frac{x}{y}\right)(-v) = v_B$$

But

$$\frac{y}{x} = \tan\alpha$$

so

$$v_B = \left(\frac{1}{\tan\alpha}\right)v$$

When $\alpha = 60.0°$

$$v_B = \frac{v}{\tan 60.0°} = \frac{v}{\sqrt{3}} = 0.577v \qquad \diamond$$

Chapter 3
VECTORS

EQUATIONS AND CONCEPTS

The **location of a point P in a plane** can be specified by either cartesian coordinates, x and y, or polar coordinates, r and θ. *If one set of coordinates is known, values for the other set can be calculated.*

$$x = r\cos\theta \qquad (3.1)$$

$$y = r\sin\theta \qquad (3.2)$$

$$\tan\theta = \frac{y}{x} \qquad (3.3)$$

$$r = \sqrt{x^2 + y^2} \qquad (3.4)$$

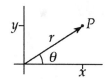

The **commutative law of addition** states that when two or more vectors are added, the sum is independent of the order of addition. To add vector **A** to vector **B** using the graphical method, first construct **A**, and then draw **B** such that the tail of **B** starts at the head of **A**. The sum of **A** + **B** is the vector that completes the triangle by connecting the tail of **A** to the head of **B**.

$$\mathbf{A} + \mathbf{B} = \mathbf{B} + \mathbf{A} \qquad (3.5)$$

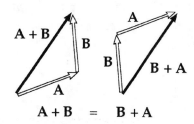

The **associative law of addition** states that when three or more vectors are added the sum is independent of the way in which the individual vectors are grouped.

$$\mathbf{A} + (\mathbf{B} + \mathbf{C}) = (\mathbf{A} + \mathbf{B}) + \mathbf{C} \qquad (3.6)$$

In the **graphical or geometric method** of vector addition, the vectors to be added (or subtracted) are represented by arrows connected head-to-tail in any order. The resultant or sum is the vector which joins the tail of the first vector to the head of the last vector. The length of each arrow must be proportional to the magnitude of the corresponding vector and must be along the direction which makes the proper angle relative to the others. *When two or more vectors are to be added, all of them must represent the same physical quantity — that is, have the same units.*

$$R = A + B + C + D$$

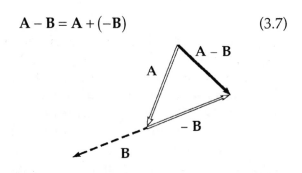

The **operation of vector subtraction** utilizes the definition of the negative of a vector. The vector $(-A)$ has a magnitude equal to the magnitude of A, but acts or points along a direction opposite the direction of A. *The negative of vector A is defined as the vector that when added to A gives zero for the vector sum.*

$$A - B = A + (-B) \qquad (3.7)$$

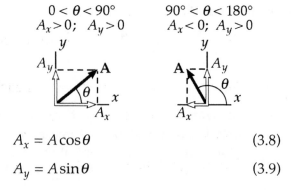

The **rectangular components** of a vector are the projections of the vector onto the respective coordinate axes. As illustrated in the figures, the projection of A onto the x axis is the x component of A; and the projection of A onto the y axis is the y component of A. *The angle θ is measured relative to the positive x axis and the algebraic sign of the components will depend on the value of θ.*

$$0 < \theta < 90° \qquad 90° < \theta < 180°$$
$$A_x > 0; \quad A_y > 0 \qquad A_x < 0; \quad A_y > 0$$

$$A_x = A \cos\theta \qquad (3.8)$$

$$A_y = A \sin\theta \qquad (3.9)$$

The magnitude of A and the angle, θ, which the vector makes with the positive x axis can be determined from the values of the x and y components of **A**.

$$A = \sqrt{A_x^2 + A_y^2} \qquad (3.10)$$

$$\theta = \tan^{-1}\left(\frac{A_y}{A_x}\right) \qquad (3.11)$$

Unit vectors are dimensionless and have a magnitude of exactly 1. A vector **A** lying in the xy plane, having rectangular components A_x and A_y, can be expressed in unit vector notation. *Unit vectors specify the directions of the vector components.*

$$\mathbf{A} = A_x\hat{\mathbf{i}} + A_y\hat{\mathbf{j}} \qquad (3.12)$$

The **resultant, R,** of adding two vectors **A** and **B** can be expressed in terms of the components of the two vectors.

$$\mathbf{R} = \left(A_x + B_x\right)\hat{\mathbf{i}} + \left(A_y + B_y\right)\hat{\mathbf{j}} \qquad (3.14)$$

The **magnitude and direction** of a resultant vector can be determined from the values of the components of the vectors in the sum.

$$R = \sqrt{R_x^2 + R_y^2}$$
$$= \sqrt{(A_x + B_x)^2 + (A_y + B_y)^2} \qquad (3.16)$$

SUGGESTIONS, SKILLS, AND STRATEGIES

When two or more vectors are to be added, the following step-by-step procedure is recommended:

- Select a coordinate system.

- Draw a sketch of the vectors to be added (or subtracted), with a label on each vector.

- Find the x and y components of each vector.

- Find the algebraic sum of the components of the individual vectors in both the x and y directions. These sums are the components of the resultant vector.

- Use the Pythagorean theorem to find the magnitude of the resultant vector.

- Use a suitable trigonometric function to find the angle the resultant vector makes with the x axis.

31

REVIEW CHECKLIST

You should be able to:

▷ Locate a point in space using both rectangular coordinates and polar coordinates.

▷ State and use the basic properties of vectors such as the rules of vector addition and construct graphical solutions for addition of two or more vectors.

▷ Resolve a vector into its rectangular components. Determine the magnitude and direction of a vector from its rectangular components.

▷ Use unit vectors to express any vector in unit vector notation.

▷ Determine the magnitude and direction of a resultant vector in terms of the components of individual vectors which have been added or subtracted. Express the resultant vector in unit vector notation.

ANSWERS TO SELECTED QUESTIONS

5. A vector **A** lies in the xy plane. For what orientations of **A** will both of its components be negative? For what orientations will its components have opposite signs?

Answer The vector **A** will have both rectangular components negative in quadrant III, when the angle of **A** is between π rad (180°) and $3\pi/2$ rad (270°). The vector **A** will have components with opposite signs in two cases: first, in quadrant II, when the angle of **A** is between $\pi/2$ rad (90°) and π rad (180°), and second, in quadrant IV, when the angle of **A** is between $3\pi/2$ rad (270°) and 2π rad (360°).

☐   ☐   ☐   ☐

12. Is it possible to add a vector quantity to a scalar quantity? Explain.

Answer Vectors and scalars are distinctly different and cannot be added to each other. Remember that a vector defines a quantity **in a certain direction**, while a scalar only defines a quantity with no associated direction.

☐   ☐   ☐   ☐

SOLUTIONS TO SELECTED END-OF-CHAPTER PROBLEMS

1. The polar coordinates of a point are $r = 5.50$ m and $\theta = 240°$. What are the Cartesian coordinates of this point?

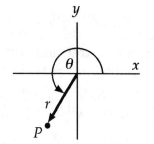

Solution When the polar coordinates (r, θ) of a point P are known, the cartesian coordinates can be found:

$$x = r \cos \theta \qquad\qquad y = r \sin \theta$$

$$x = (5.50 \text{ m}) \cos 240° = -2.75 \text{ m}$$

$$y = (5.50 \text{ m}) \sin 240° = -4.76 \text{ m} \qquad\qquad \Diamond$$

---

3. A fly lands on one wall of a room. The lower left-hand corner of the wall is selected as the origin of a two-dimensional Cartesian coordinate system. If the fly is located at the point having coordinates (2.00, 1.00) m, (a) how far is it from the corner of the room? (b) What is its location in polar coordinates?

Solution

(a) Assume the wall is in the xy plane so that the coordinates are $x = 2.00$ m and $y = 1.00$ m; and the fly is located at point P. The distance between two points in the xy plane is

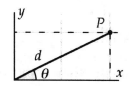

$$d = \sqrt{(x_2 - x_1)^2 + (y_2 - y_1)^2}$$

$$d = \sqrt{(2.00 \text{ m} - 0)^2 + (1.00 \text{ m} - 0)^2} = 2.24 \text{ m} \qquad\qquad \Diamond$$

(b) From the figure, $\theta = \tan^{-1}\left(\dfrac{y}{x}\right) = \tan^{-1}\left(\dfrac{1.00}{2.00}\right) = 26.6°$ and $r = d$

Therefore, the polar coordinates of the point P are (2.24 m, 26.6°) $\qquad\qquad \Diamond$

---

7. A surveyor measures the distance across a straight river by the following method: starting directly across from a tree on the opposite bank, she walks 100 m along the riverbank to establish a baseline. Then she sights across to the tree. The angle from her baseline to the tree is 35.0°. How wide is the river?

Solution

Make a sketch of the area as viewed from above. Assume the river-banks are straight and parallel, show the location of the tree, and the original location of the surveyor. The width w between tree and surveyor must be perpendicular to the riverbanks, as the surveyor chooses to start "directly across" from the tree. Now draw the 100-meter baseline, b, and the line showing the "line of sight" to the tree.

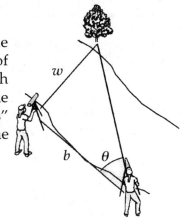

From the figure, $\tan\theta = \dfrac{w}{b}$

and $w = b\tan\theta = (100\text{ m})\tan 35.0° = 70.0\text{ m}$ ◊

---

11. A skater glides along a circular path of radius 5.00 m. If he coasts around one half of the circle, find (a) the magnitude of the displacement vector and (b) how far the person skated. (c) What is the magnitude of the displacement if he skates all the way around the circle?

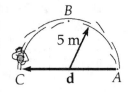

Solution

See the sketch above and to the right.

(a) $|\mathbf{d}| = |-10.0\hat{\imath}| = 10.0\text{ m}$ ◊

since the displacement is a straight line from point A to point C.

(b) The actual distance skated is not equal to the straight-line displacement. The distance follows the curved path of the semicircle (ABC).

$s = \dfrac{1}{2}(2\pi r) = 5.00\pi\text{ m} = 15.7\text{ m}$ ◊

(c) If the circle is complete, $\mathbf{d}$ begins and ends at point A.

Hence, $|\mathbf{d}| = 0$ ◊

---

15. Each of the displacement vectors **A** and **B** shown in Figure P3.15 has a magnitude of 3.00 m. Find graphically (a) **A** + **B**, (b) **A** − **B**, (c) **B** − **A**, (d) **A** − 2**B**. Report all angles counterclockwise from the positive x axis.

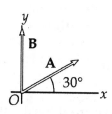

Figure P3.15

Solution To find these vectors graphically, we draw each set of vectors as indicated by the drawings below. Measurements of the results can be taken using a ruler and protractor.

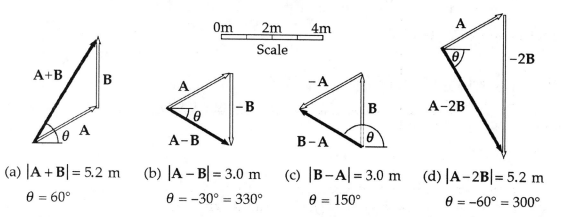

(a) $|\mathbf{A} + \mathbf{B}| = 5.2$ m (b) $|\mathbf{A} − \mathbf{B}| = 3.0$ m (c) $|\mathbf{B} − \mathbf{A}| = 3.0$ m (d) $|\mathbf{A} − 2\mathbf{B}| = 5.2$ m

$\theta = 60°$ $\theta = −30° = 330°$ $\theta = 150°$ $\theta = −60° = 300°$

Remember: when adding vectors graphically, they are connected "head-to-tail," represented by arrows whose lengths correspond to their magnitudes. Also, the relative directions of the vectors must be maintained. When subtracting vectors, remember also that **A** − **B** = **A** + (−**B**). ◊

---

17. A roller coaster car moves 200 ft horizontally and then rises 135 ft at an angle of 30.0° above the horizontal. It then travels 135 ft at an angle of 40.0° downward. What is its displacement from its starting point? Use graphical techniques.

Solution Your sketch when drawn to scale should look somewhat like the one to the right. (You will probably only be able to obtain a measurement to one or two significant figures).

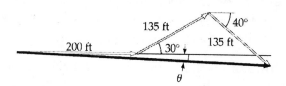

The distance R and the angle θ can be measured to give, upon use of your scale factor, the values of:

$R = 4.2 \times 10^2$ ft at about 3° below the horizontal. ◊

---

19. A vector has an x component of -25.0 units and a y component of 40.0 units. Find the magnitude and direction of this vector.

Solution

Conceptualize: First we should visualize the vector either in our mind or with a sketch. Since the hypotenuse of the right triangle must be greater than either the x or y components that form the legs, we can estimate the magnitude of the vector to be about 50 units. The direction of the vector (θ) appears to be about 120° from the +x axis.

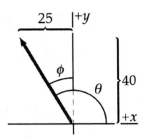

Categorize: The graphical analysis and visual estimates above may suffice for some situations, but we can use trigonometry to obtain a more precise result.

Analyze: The magnitude can be found by the Pythagorean theorem: $r = \sqrt{x^2 + y^2}$

$$r = \sqrt{(-25.0 \text{ units})^2 + (40.0 \text{ units})^2} = 47.2 \text{ units} \qquad \lozenge$$

We observe that $\tan\phi = x/y$ (if we consider x and y to both be positive).

$$\phi = \tan^{-1}\left(\frac{x}{y}\right) = \tan^{-1}\left(\frac{25.0 \text{ units}}{40.0 \text{ units}}\right) = \tan^{-1}(0.625) = 32.0° \qquad \lozenge$$

The angle from the +x axis can be found by adding 90° to ϕ.

$$\theta = \phi + 90° = 122° \qquad \lozenge$$

Finalize: Our calculated results agree with our graphical estimates. We should always remember to check that our answers are reasonable and make sense, especially for problems like this where it is easy to mistakenly calculate the wrong angle by confusing coordinates or overlooking a minus sign.

Quite often the direction angle of a vector can be specified in more than one way, and we must choose a notation that makes the most sense for the given problem. If compass directions were stated in this question, we could have reported the vector angle to be 32.0° west of north or a compass heading of 328°.

21. Obtain expressions in component form for the position vectors having polar coordinates: (a) 12.8 m, 150° (b) 3.30 cm, 60.0° (c) 22.0 in., 215°.

Solution   Find the x and y components of each vector using $x = R\cos\theta$ and $y = R\sin\theta$. In unit vector notation, $\mathbf{R} = R_x\hat{\mathbf{i}} + R_y\hat{\mathbf{j}}$

(a) $x = (12.8 \text{ m})\cos 150° = -11.1 \text{ m}$ $y = (12.8 \text{ m})\sin 150° = 6.40 \text{ m}$

$\mathbf{R} = (-11.1\hat{\mathbf{i}} + 6.40\hat{\mathbf{j}}) \text{ m}$ ◊

(b) $x = (3.30 \text{ cm})\cos 60.0° = 1.65 \text{ cm}$ $y = (3.30 \text{ cm})\sin 60.0° = 2.86 \text{ cm}$

$\mathbf{R} = (1.65\hat{\mathbf{i}} + 2.86\hat{\mathbf{j}}) \text{ cm}$ ◊

(c) $x = (22.0 \text{ in})\cos 215° = -18.0 \text{ in}$ $y = (22.0 \text{ in})\sin 215° = -12.6 \text{ in}$

$\mathbf{R} = (-18.0\hat{\mathbf{i}} - 12.6\hat{\mathbf{j}}) \text{ in}$ ◊

---

29. A man pushing a mop across a floor causes it to undergo two displacements. The first has a magnitude of 150 cm and makes an angle of 120° with the positive x axis. The resultant displacement has a magnitude of 140 cm and is directed at an angle of 35.0° to the positive x axis. Find the magnitude and direction of the second displacement.

Solution

The total displacement of the mop is $\mathbf{R} = \mathbf{d}_1 + \mathbf{d}_2$

Then $\mathbf{d}_2 = \mathbf{R} - \mathbf{d}_1$

$\mathbf{d}_2 = (140 \text{ cm at } 35°) - (150 \text{ cm at } 120°)$

$\mathbf{d}_2 = (140 \text{ cm at } 35°) + (150 \text{ cm at } 300°)$

$\mathbf{d}_2 = \left[140\cos 35°\hat{\mathbf{i}} + 140\sin 35°\hat{\mathbf{j}} + 150\cos 300°\hat{\mathbf{i}} + 150\sin 300°\hat{\mathbf{j}}\right] \text{ cm}$

$\mathbf{d}_2 = \left[(115 + 75)\hat{\mathbf{i}} + (80.3 - 130)\hat{\mathbf{j}}\right] \text{ cm}$

$\mathbf{d}_2 = \left[190\hat{\mathbf{i}} - 49.6\hat{\mathbf{j}}\right] \text{ cm}$

$\mathbf{d}_2 = \sqrt{(190 \text{ cm})^2 + (49.6 \text{ cm})^2}$ at $\tan^{-1}\left(\dfrac{49.6 \text{ cm}}{190 \text{ cm}}\right)$ from the x axis in quadrant IV

$\mathbf{d}_2 = 196 \text{ cm at } (360° - 14.7°) = 196 \text{ cm at } 345°$ ◊

---

31. Consider two vectors $\mathbf{A} = 3\hat{\mathbf{i}} - 2\hat{\mathbf{j}}$ and $\mathbf{B} = -\hat{\mathbf{i}} - 4\hat{\mathbf{j}}$. Calculate (a) $\mathbf{A} + \mathbf{B}$, (b) $\mathbf{A} - \mathbf{B}$, (c) $|\mathbf{A} + \mathbf{B}|$, (d) $|\mathbf{A} - \mathbf{B}|$, (e) the directions of $\mathbf{A} + \mathbf{B}$ and $\mathbf{A} - \mathbf{B}$.

Solution   Use the property of vector addition that states that

$$\text{if} \quad \mathbf{R} = \mathbf{A} + \mathbf{B}, \quad \text{then} \quad R_x = A_x + B_x \quad \text{and} \quad R_y = A_y + B_y$$

(a) $\mathbf{A} + \mathbf{B} = \left(3\hat{\mathbf{i}} - 2\hat{\mathbf{j}}\right) + \left(-\hat{\mathbf{i}} - 4\hat{\mathbf{j}}\right) = 2\hat{\mathbf{i}} - 6\hat{\mathbf{j}}$ ◊

(b) $\mathbf{A} - \mathbf{B} = \left(3\hat{\mathbf{i}} - 2\hat{\mathbf{j}}\right) - \left(-\hat{\mathbf{i}} - 4\hat{\mathbf{j}}\right) = 4\hat{\mathbf{i}} + 2\hat{\mathbf{j}}$ ◊

(c)   For a vector, if $\quad \mathbf{R} = R_x\hat{\mathbf{i}} + R_y\hat{\mathbf{j}} \qquad |\mathbf{R}| = \sqrt{R_x^2 + R_y^2}$

$|\mathbf{A} + \mathbf{B}| = \sqrt{2^2 + (-6)^2} = 6.32$ ◊

(d) $|\mathbf{A} - \mathbf{B}| = \sqrt{4^2 + 2^2} = 4.47$ ◊

The direction of a vector relative to the positive x axis is $\theta = \tan^{-1}\left(R_y / R_x\right)$

(e)   For $\mathbf{A} + \mathbf{B}$, $\theta = \tan^{-1}(-6/2) = -71.6° = 288°$ ◊

For $\mathbf{A} - \mathbf{B}$, $\theta = \tan^{-1}(2/4) = 26.6°$ ◊

---

33.   A particle undergoes the following consecutive displacements: 3.50 m south, 8.20 m northeast, and 15.0 m west. What is the resultant displacement?

Solution   Take the direction east to be along $+\hat{\mathbf{i}}$. The three displacements can be written as:

$$\mathbf{d}_1 = -3.50\hat{\mathbf{j}} \text{ m}$$

$$\mathbf{d}_2 = (8.20\cos 45.0°)\hat{\mathbf{i}} \text{ m} + (8.20\sin 45.0°)\hat{\mathbf{j}} \text{ m} = 5.80\hat{\mathbf{i}} \text{ m} + 5.80\hat{\mathbf{j}} \text{ m}$$

$$\mathbf{d}_3 = -15.0\hat{\mathbf{i}} \text{ m}$$

The resultant, $\mathbf{R} = \mathbf{d}_1 + \mathbf{d}_2 + \mathbf{d}_3 = (0 + 5.80 - 15.0)\hat{\mathbf{i}} \text{ m} + (-3.50 + 5.80 + 0)\hat{\mathbf{j}} \text{ m}$

$$\mathbf{R} = \left(-9.20\hat{\mathbf{i}} + 2.30\hat{\mathbf{j}}\right) \text{ m}$$ ◊

The magnitude of the resultant displacement is

$$|\mathbf{R}| = \sqrt{R_x^2 + R_y^2} = \sqrt{(-9.20 \text{ m})^2 + (2.30 \text{ m})^2} = 9.48 \text{ m}$$ ◊

The direction of the resultant vector is given by $\tan^{-1}\left(\dfrac{R_y}{R_x}\right) = \tan^{-1}\left(\dfrac{2.30}{-9.20}\right) = -14.0°$

or relative to the positive x axis, $\qquad \theta = 180° - 14.0° = 166°$ ◊

---

41. The vector **A** has x, y, and z components of 8.00, 12.0, and –4.00 units, respectively. (a) Write a vector expression for **A** in unit-vector notation. (b) Obtain a unit-vector expression for a vector **B** one fourth the length of **A** pointing in the same direction as **A**. (c) Obtain a unit-vector expression for a vector **C** three times the length of **A** pointing in the direction opposite the direction of **A**.

Solution

(a) $\mathbf{A} = A_x\hat{\mathbf{i}} + A_y\hat{\mathbf{j}} + A_z\hat{\mathbf{k}}$ $\qquad \mathbf{A} = 8.00\hat{\mathbf{i}} + 12.0\hat{\mathbf{j}} - 4.00\hat{\mathbf{k}}$ ◊

(b) $\mathbf{B} = \mathbf{A}/4$ $\qquad\qquad\qquad \mathbf{B} = 2.00\hat{\mathbf{i}} + 3.00\hat{\mathbf{j}} - 1.00\hat{\mathbf{k}}$ ◊

(c) $\mathbf{C} = -3\mathbf{A}$ $\qquad\qquad\qquad \mathbf{C} = -24.0\hat{\mathbf{i}} - 36.0\hat{\mathbf{j}} + 12.0\hat{\mathbf{k}}$ ◊

---

47. Vector **A** has a negative x component 3.00 units in length and a positive y component 2.00 units in length. (a) Determine an expression for **A** in unit-vector notation. (b) Determine the magnitude and direction of **A**. (c) What vector **B** when added to **A** gives a resultant vector with no x component and a negative y component 4.00 units in length?

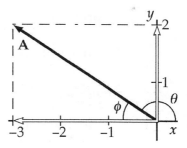

Solution

$\qquad A_x = -3.00 \text{ units}$ $\qquad\qquad\qquad A_y = 2.00 \text{ units}$

(a) $\mathbf{A} = A_x\hat{\mathbf{i}} + A_y\hat{\mathbf{j}} = -3.00\hat{\mathbf{i}} + 2.00\hat{\mathbf{j}} \text{ units}$ ◊

(b) $|\mathbf{A}| = \sqrt{A_x^2 + A_y^2} = \sqrt{(-3.00)^2 + (2.00)^2} = 3.61 \text{ units}$

$\qquad \tan\phi = \left|\dfrac{A_y}{A_x}\right| = \left|\dfrac{2.00}{-3.00}\right| = 0.667$ $\qquad$ so $\qquad \phi = 33.7°$ (relative to the $-x$ axis)

$\qquad$ **A** is in the **second quadrant**: $\qquad\qquad \theta = 180° - \phi = 146°$ ◊

(c) We are given that $R_x = 0$ and $R_y = -4.00$;

since $\mathbf{R} = \mathbf{A} + \mathbf{B}$, $\mathbf{B} = \mathbf{R} - \mathbf{A}$

$B_x = R_x - A_x = 0 - (-3.00) = 3.00$

$B_y = R_y - A_y = -4.00 - 2.00 = -6.00$

Therefore, $\mathbf{B} = B_x \hat{\mathbf{i}} + B_y \hat{\mathbf{j}} = \left(3.00\hat{\mathbf{i}} - 6.00\hat{\mathbf{j}}\right)$ units $\diamond$

---

49. Three displacement vectors of a croquet ball are shown in Figure P3.49, where $|\mathbf{A}| = 20.0$ units, $|\mathbf{B}| = 40.0$ units, and $|\mathbf{C}| = 30.0$ units. Find (a) the resultant in unit-vector notation and (b) the magnitude and direction of the resultant displacement.

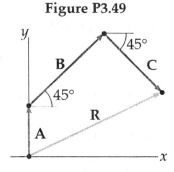

Figure P3.49

Solution

(a) $A_x = (20.0 \text{ units})\cos 90° = 0$

$A_y = (20.0 \text{ units})\sin 90° = 20.0$ units

$B_x = (40.0 \text{ units})\cos 45° = 28.3$ units

$B_y = (40.0 \text{ units})\sin 45° = 28.3$ units

$C_x = (30.0 \text{ units})\cos 315° = 21.2$ units

$C_y = (30.0 \text{ units})\sin 315° = -21.2$ units

$R_x = A_x + B_x + C_x = (0 + 28.3 + 21.2) \text{ units} = 49.5$ units

$R_y = A_y + B_y + C_y = (20.0 + 28.3 - 21.2) \text{ units} = 27.1$ units

$\mathbf{R} = 49.5\hat{\mathbf{i}} + 27.1\hat{\mathbf{j}}$ units $\diamond$

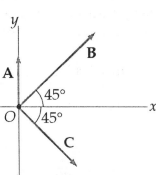

(b) $|\mathbf{R}| = \sqrt{R_x^2 + R_y^2} = \sqrt{(49.5 \text{ units})^2 + (27.1 \text{ units})^2} = 56.4$ units $\diamond$

$\theta = \tan^{-1}\left(\dfrac{R_y}{R_x}\right) = \tan^{-1}\left(\dfrac{27.1}{49.5}\right) = 28.7°$ $\diamond$

---

59. A person going for a walk follows the path shown in Figure P3.59. The total trip consists of four straight-line paths. At the end of the walk, what is the person's resultant displacement measured from the starting point?

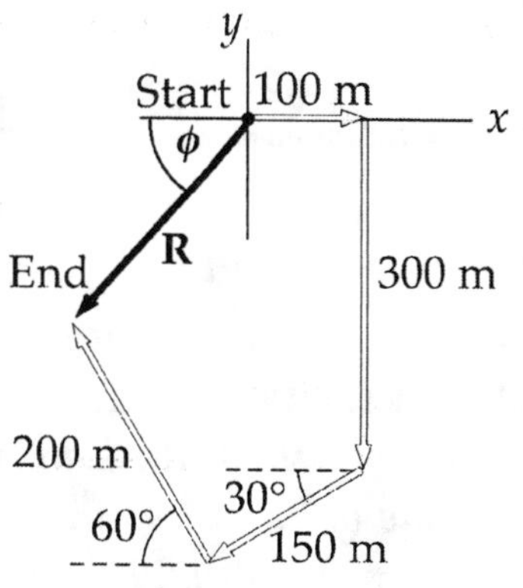

Figure P3.59

Solution The resultant displacement $\mathbf{R}$ is equal to the sum of the four individual displacements,

$$\mathbf{R} = \mathbf{d}_1 + \mathbf{d}_2 + \mathbf{d}_3 + \mathbf{d}_4$$

We translate from the pictorial representation to a mathematical representation by writing the individual displacements in unit-vector notation:

$$\mathbf{d}_1 = 100\hat{\mathbf{i}} \text{ m}$$

$$\mathbf{d}_2 = -300\hat{\mathbf{j}} \text{ m}$$

$$\mathbf{d}_3 = (-150\cos 30°)\hat{\mathbf{i}} \text{ m} + (-150\sin 30°)\hat{\mathbf{j}} \text{ m} = -130\hat{\mathbf{i}} \text{ m} - 75\hat{\mathbf{j}} \text{ m}$$

$$\mathbf{d}_4 = (-200\cos 60°)\hat{\mathbf{i}} \text{ m} + (200\sin 60°)\hat{\mathbf{j}} \text{ m} = -100\hat{\mathbf{i}} \text{ m} + 173\hat{\mathbf{j}} \text{ m}$$

Summing the components together,

$$R_x = d_{1x} + d_{2x} + d_{3x} + d_{4x} = (100 + 0 - 130 - 100) \text{ m} = -130 \text{ m}$$

$$R_y = d_{1y} + d_{2y} + d_{3y} + d_{4y} = (0 - 300 - 75 + 173) \text{ m} = -202 \text{ m}$$

$$\mathbf{R} = -130\hat{\mathbf{i}} \text{ m} - 202\hat{\mathbf{j}} \text{ m} \qquad \Diamond$$

$$|\mathbf{R}| = \sqrt{R_x{}^2 + R_y{}^2} = \sqrt{(-130 \text{ m})^2 + (-202 \text{ m})^2} = 240 \text{ m} \qquad \Diamond$$

$$\phi = \tan^{-1}\left(\frac{R_y}{R_x}\right) = \tan^{-1}\left(\frac{-202}{-130}\right) = 57.2°$$

Pitfall Prevention 3.4 in the text explains why this angle does not specify the resultant's direction in standard form. Instead, the angle counterclockwise from the +x axis is

$$\theta = 57.2° + 180° = 237° \qquad \Diamond$$

Chapter 4
MOTION IN TWO DIMENSIONS

EQUATIONS AND CONCEPTS

The **displacement** of a particle, $\Delta \mathbf{r} \equiv \mathbf{r}_f - \mathbf{r}_i$, is defined as the difference between the final and initial position vectors. *For the displacement illustrated in the figure, the magnitude of $\Delta \mathbf{r}$ is less than the actual path length from the initial to the final position.*

$$\Delta \mathbf{r} \equiv \mathbf{r}_f - \mathbf{r}_i \tag{4.1}$$

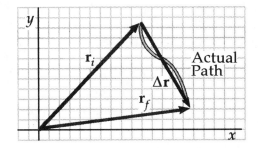

The **average velocity** of a particle which undergoes a displacement $\Delta \mathbf{r}$ in a time interval Δt equals the ratio $\Delta \mathbf{r}/\Delta t$. *The average velocity depends on the displacement vector and not on the path traveled.*

$$\overline{\mathbf{v}} \equiv \frac{\Delta \mathbf{r}}{\Delta t} \tag{4.2}$$

The **average speed** of a particle during any time interval is the ratio of the total distance of travel (length of path) to the total time. *If, during a time Δt a particle returns to the starting point, the displacement and average velocity over the interval will each be zero; the distance traveled and the average speed will not be zero.*

$$\text{average speed} = \frac{\text{total distance}}{\text{total time}}$$

The **instantaneous velocity** of a particle equals the limit of the average velocity as $\Delta t \to 0$. *The magnitude of the instantaneous velocity is called the speed.*

$$\mathbf{v} \equiv \lim_{\Delta t \to 0} \frac{\Delta \mathbf{r}}{\Delta t} = \frac{d\mathbf{r}}{dt} \tag{4.3}$$

The **average acceleration** of a particle which undergoes a change in velocity $\Delta \mathbf{v}$ in a time interval Δt equals the ratio $\Delta \mathbf{v}/\Delta t$.

$$\bar{\mathbf{a}} \equiv \frac{\mathbf{v}_f - \mathbf{v}_i}{t_f - t_i} = \frac{\Delta \mathbf{v}}{\Delta t} \qquad (4.4)$$

The **instantaneous acceleration** is defined as the limit of the average acceleration as $\Delta t \to 0$. *A particle experiences acceleration when the velocity vector undergoes a change in magnitude, direction or both.*

$$\mathbf{a} \equiv \lim_{\Delta t \to 0} \frac{\Delta \mathbf{v}}{\Delta t} = \frac{d\mathbf{v}}{dt} \qquad (4.5)$$

Motion in two dimensions with constant acceleration is described by equations for velocity and position which are vector versions of the one-dimensional kinematic equations.

$$\mathbf{v}_f = \mathbf{v}_i + \mathbf{a}t \qquad (4.8)$$

$$\mathbf{r}_f = \mathbf{r}_i + \mathbf{v}_i t + \frac{1}{2}\mathbf{a}t^2 \qquad (4.9)$$

A **projectile** moves with constant velocity along the horizontal direction and with constant acceleration (–g) along the vertical direction. *The path of a projectile is a parabola.*

$$v_x = v_i \cos \theta_i = \text{constant}$$

$$v_y = v_i \sin \theta_i - gt$$

$$a_x = 0$$

$$a_y = -g$$

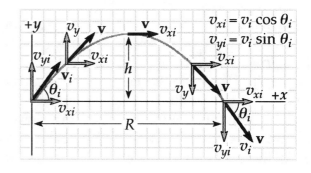

The **position coordinates of a projectile,** as a function of time, are given by Equations 4.11 and 4.12.

$$x_f = v_{xi}t = (v_i \cos \theta_i)t \qquad (4.11)$$

$$y_f = v_{yi}t + \frac{1}{2}a_y t^2 = (v_i \sin \theta_i)t - \frac{1}{2}gt^2 \qquad (4.12)$$

The **maximum height** (*h*) of a projectile can be written in terms of v_i and θ_i.

$$h = \frac{v_i^2 \sin^2 \theta_i}{2g} \qquad (4.13)$$

The **horizontal range** (R) of a projectile can also be stated in terms of v_i and θ_i. *The range is maximum when $\theta_i = 45°$.*

$$R = \frac{v_i^2 \sin 2\theta_i}{g} \qquad (4.14)$$

The **centripetal acceleration** vector for a particle in *uniform circular motion* is directed toward the center circular path.

$$a_c = \frac{v^2}{r} \qquad (4.15)$$

The **period** (T) of a particle moving with constant speed in circle of radius r is defined as the time required to complete one revolution.

$$T \equiv \frac{2\pi r}{v} \qquad (4.16)$$

The **total acceleration** of a particle moving on a curved path is the vector sum of the radial and the tangential components of acceleration (see figure below). The radial component, a_r, is directed toward the center of curvature and arises from the change in direction of the velocity vector. The tangential component, a_t, is perpendicular to the radius and causes the change in speed of the particle.

$$\mathbf{a} = \mathbf{a}_r + \mathbf{a}_t \qquad (4.17)$$

$$a_t = \frac{d|\mathbf{v}|}{dt} \qquad (4.18)$$

$$a_r = -a_c = -\frac{v^2}{r} \qquad (4.19)$$

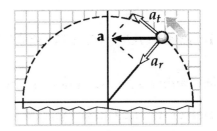

Galilean transformation equations relate the position and velocity, $\mathbf{r}'$ and $\mathbf{v}'$, of a particle as measured by an observer in a moving frame of reference to those values measured by an observer in a fixed frame (moving with the object).

$$\mathbf{r}' = \mathbf{r} - \mathbf{v}_0 t \qquad (4.21)$$

$$\mathbf{v}' = \mathbf{v} - \mathbf{v}_0 \qquad (4.22)$$

SUGGESTIONS, SKILLS, AND STRATEGIES

- A projectile moving under the influence of gravity has a parabolic trajectory and the equation of the path has the general form,

$$y = Ax - Bx^2 \quad \text{where} \quad A = \tan\theta_i \quad \text{and} \quad B = \frac{g}{2v_i^2 \cos^2\theta_i}$$

This expression for y assumes that the particle leaves the origin at $t = 0$, with a velocity $\mathbf{v}_i$ which makes an angle θ_i with the horizontal. A sketch of y versus x for this situation is shown at the right.

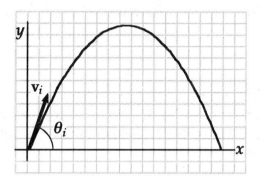

- Given the values for x and y at any time $t > 0$, an expression for the position vector $\mathbf{r}$ at that time can be written, using unit vector notation, in the form $\mathbf{r} = x\hat{\mathbf{i}} + y\hat{\mathbf{j}}$. If v_x and v_y are known at any time, the velocity vector $\mathbf{v}$ can be written in the form $\mathbf{v} = v_x\hat{\mathbf{i}} + v_y\hat{\mathbf{j}}$. From this, the **speed** $v = \sqrt{v_x^2 + v_y^2}$ and the angle $\theta = \tan^{-1}(v_y/v_x)$ that the vector makes with the x axis can be determined.

- Remember, at maximum height $v_y = 0$ and when $x = R$, $y = 0$ (for a projectile that lands at the same level from which it was launched).

PROBLEM-SOLVING STRATEGY: PROJECTILE MOTION

The following approach should be used in solving projectile motion problems:

- Select a coordinate system.

- Resolve the initial velocity vector into x and y components.

- Treat the horizontal motion and the vertical motion independently.

- Follow the techniques for solving problems with constant velocity to analyze the horizontal motion of the projectile.

- Follow the techniques for solving problems with constant acceleration to analyze the vertical motion of the projectile.

REVIEW CHECKLIST

You should be able to:

▷ Describe the displacement, velocity, and acceleration of a particle moving in the *xy* plane. (Sections 4.1 and 4.2)

▷ Find the velocity components and the coordinates of a projectile at any time t if the initial speed v_i and initial angle θ_i are known at a given point at $t = 0$. Also calculate the horizontal range R and maximum height h if v_i and θ_i are known.

▷ Calculate the acceleration of a particle moving in a circle with constant speed. In this situation, note that although $|\mathbf{v}| = \text{constant}$, the **direction** of $\mathbf{v}$ varies in time, the result of which is the radial, or centripetal acceleration.

▷ Calculate the components of acceleration for a particle moving on a curved path, where both the magnitude and direction of $\mathbf{v}$ are changing with time. In this case, the particle has a tangential component of acceleration and a radial component of acceleration.

▷ Use the Galilean transformation equations to relate the position and velocity of a particle as measured by observers in relative motion.

ANSWERS TO SELECTED CONCEPTUAL QUESTIONS

2. If you know the position vectors of a particle at two points along its path and also know the time it took to move from one point to the other, can you determine the particle's instantaneous velocity? Its average velocity? Explain.

Answer Its instantaneous velocity cannot be determined at any point from this information. However, the average velocity over the time interval can be determined from its definition and the given information.

□   □   □   □

6. A spacecraft drifts through space at a constant velocity. Suddenly a gas leak in the side of the spacecraft gives it a constant acceleration in a direction perpendicular to the initial velocity. The orientation of the spacecraft does not change, so that the acceleration remains perpendicular to the original direction of the velocity. What is the shape of the path followed by the spacecraft in this situation?

Answer   The spacecraft will follow a parabolic path. This is equivalent to a projectile thrown off a cliff with a horizontal velocity. For the projectile, gravity provides an acceleration which is always perpendicular to the initial velocity, resulting in a parabolic path. For the spacecraft, the initial velocity plays the role of the horizontal velocity of the projectile. The leaking gas provides an acceleration that plays the role of gravity for the projectile. If the orientation of the spacecraft were to change in response to the gas leak (which is by far the more likely result), then the acceleration would change direction and the motion could become quite complicated.

□   □   □   □

12.   A projectile is launched at some angle to the horizontal with some initial speed v_i, and air resistance is negligible. Is the projectile a freely falling body? What is its acceleration in the vertical direction? What is its acceleration in the horizontal direction?

Answer   Yes. The projectile is a freely falling body, because nothing counteracts the force of gravity. The vertical acceleration will be the local gravitational acceleration, g; the horizontal acceleration will be zero.

□   □   □   □

18.   Explain whether or not the following particles have an acceleration: (a) a particle moving in a straight line with constant speed and (b) a particle moving around a curve with constant speed.

Answer   (a) The acceleration is zero, since the magnitude and direction of **v** remain constant. (b) The particle has an acceleration since the direction of **v** changes.

□   □   □   □

SOLUTIONS TO SELECTED END-OF-CHAPTER PROBLEMS

1.   A motorist drives south at 20.0 m/s for 3.00 min, then turns west and travels at 25.0 m/s for 2.00 min, and finally travels northwest at 30.0 m/s for 1.00 min. For this 6.00-min trip, find (a) the total vector displacement, (b) the average speed, and (c) the average velocity. Let the positive x axis point east.

Solution

(a) For each segment of the motion we model the car as a particle under constant velocity. Her displacements are

$$\Delta \mathbf{r} = (20.0 \text{ m/s})(180 \text{ s}) \, \mathbf{south} + (25.0 \text{ m/s})(120 \text{ s}) \, \mathbf{west} + (30.0 \text{ m/s})(60.0 \text{ s}) \, \mathbf{northwest}$$

Choosing $\hat{\mathbf{i}}$ = east and $\hat{\mathbf{j}}$ = north, we have

$$\Delta \mathbf{r} = (3.60 \text{ km})(-\hat{\mathbf{j}}) + (3.00 \text{ km})(-\hat{\mathbf{i}}) + [(1.80 \text{ km})\cos 45.0°](-\hat{\mathbf{i}}) + [(1.80 \text{ km})\sin 45.0°](\hat{\mathbf{j}})$$

$$\Delta \mathbf{r} = [(3.00 + 1.27) \text{ km}](-\hat{\mathbf{i}}) + [(1.27 - 3.60) \text{ km}](\hat{\mathbf{j}}) = (-4.27\hat{\mathbf{i}} - 2.33\hat{\mathbf{j}}) \text{ km} \qquad \lozenge$$

The answer can also be written as

$$\Delta \mathbf{r} = \sqrt{(-4.27 \text{ km})^2 + (-2.33 \text{ km})^2} \quad \text{at} \quad \tan^{-1}\left(\frac{2.33}{4.27}\right) = 28.6° \text{ south of west}$$

or $\qquad \Delta \mathbf{r} = 4.87$ km at 209° from the east. $\qquad \lozenge$

(b) The total distance or path-length traveled is $(3.60 + 3.00 + 1.80) \text{ km} = 8.40 \text{ km}$

so $\qquad$ average speed $= \left(\dfrac{8.40 \text{ km}}{6.00 \text{ min}}\right)\left(\dfrac{1.00 \text{ min}}{60.0 \text{ s}}\right)\left(\dfrac{1000 \text{ m}}{\text{km}}\right) = 23.3 \text{ m/s} \qquad \lozenge$

(c) $\qquad \overline{\mathbf{v}} = \dfrac{\Delta \mathbf{r}}{t} = \dfrac{4.87 \text{ km}}{360 \text{ s}} = 13.5 \text{ m/s}$ at 209°

or $\qquad \overline{\mathbf{v}} = \dfrac{\Delta \mathbf{r}}{t} = \dfrac{(-4.27 \text{ east} - 2.33 \text{ north}) \text{ km}}{360 \text{ s}} = (11.9 \text{ west} + 6.46 \text{ south}) \text{ m/s} \qquad \lozenge$

---

7. A fish swimming in a horizontal plane has velocity $\mathbf{v}_i = (4.00\hat{\mathbf{i}} + 1.00\hat{\mathbf{j}}) \text{ m/s}$ at a point in the ocean where the position relative to a certain rock is $r_i = (10.0\hat{\mathbf{i}} - 4.00\hat{\mathbf{j}}) \text{ m}$. After the fish swims with constant acceleration for 20.0 s, its velocity is $\mathbf{v} = (20.0\hat{\mathbf{i}} - 5.00\hat{\mathbf{j}}) \text{ m/s}$. (a) What are the components of the acceleration? (b) What is the direction of the acceleration with respect to unit vector $\hat{\mathbf{i}}$? (c) If the fish maintains constant acceleration where is the fish at $t = 25.0$ s, and in what direction is it moving?

Solution Model the fish as a particle under constant acceleration.

At $t = 0$, $\qquad \mathbf{v}_i = (4.00\hat{\mathbf{i}} + 1.00\hat{\mathbf{j}}) \text{ m/s} \qquad$ and $\qquad r_i = (10.0\hat{\mathbf{i}} - 4.00\hat{\mathbf{j}}) \text{ m}$

$\qquad\qquad v_f = (20.0\hat{\mathbf{i}} - 5.00\hat{\mathbf{j}}) \text{ m/s}$

(a) $a_x = \dfrac{\Delta v_x}{\Delta t} = \dfrac{20.0 \text{ m / s} - 4.00 \text{ m / s}}{20.0 \text{ s}} = 0.800 \text{ m / s}^2$ ◊

$a_y = \dfrac{\Delta v_y}{\Delta t} = \dfrac{-5.00 \text{ m / s} - 1.00 \text{ m / s}}{20.0 \text{ s}} = -0.300 \text{ m / s}^2$ ◊

(b) $\theta = \tan^{-1}\left(\dfrac{a_y}{a_x}\right) = \tan^{-1}\left(\dfrac{-0.300 \text{ m / s}^2}{0.800 \text{ m / s}^2}\right) = -20.6°$ or $339°$ from the $+x$ axis ◊

(c) At $t = 25.0$ s, its coordinates are $x_f = x_i + v_{xi}t + \frac{1}{2}a_x t^2$ and $y_f = y_i + v_{yi}t + \frac{1}{2}a_y t^2$

$x_f = 10.0 \text{ m} + (4.00 \text{ m / s})(25.0 \text{ s}) + \frac{1}{2}(0.800 \text{ m / s}^2)(25.0 \text{ s})^2 = 360 \text{ m}$ ◊

$y_f = -4.00 \text{ m} + (1.00 \text{ m / s})(25.0 \text{ s}) + \frac{1}{2}(-0.300 \text{ m / s}^2)(25.0 \text{ s})^2 = -72.8 \text{ m}$ ◊

$v_{xf} = v_{xi} + a_y t = (4.00 \text{ m / s}) + (0.800 \text{ m / s}^2)(25.0 \text{ s}) = 24.0 \text{ m / s}$

$v_{yf} = v_{yi} + a_y t = (1.00 \text{ m / s}) - (0.300 \text{ m / s}^2)(25.0 \text{ s}) = -6.50 \text{ m / s}$

Therefore, $\mathbf{r} = (360\hat{\imath} - 72.8\hat{\jmath}) \text{ m}$

$\theta = \tan^{-1}(v_y/v_x) = \tan^{-1}\left(\dfrac{-6.50 \text{ m / s}}{24.0 \text{ m / s}}\right) = -15.2° = 345°$ from the $+x$ axis. ◊

---

11. In a local bar, a customer slides an empty beer mug down the counter for a refill. The bartender is momentarily distracted and does not see the mug, which slides off the counter and strikes the floor 1.40 m from the base of the counter. If the height of the counter is 0.860 m, (a) with what velocity did the mug leave the counter, and (b) what was the direction of the mug's velocity just before it hit the floor?

Solution

Conceptualize: Based on our everyday experiences and the description of the problem, a reasonable speed of the mug would be a few m/s and it will hit the floor at some angle between 0° and 90°, probably about 45°.

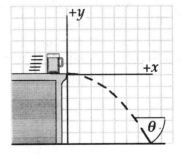

Categorize: We are looking for two different velocities, but we are only given two distances. Our approach will be to separate the vertical and horizontal motions. By using the height that the mug falls, we can find the time of the fall. Once we know the time, we can find the horizontal and vertical components of the velocity. For convenience, we will set the origin to be the point where the mug leaves the counter.

Analyze:

Vertical motion: $\quad\quad y = -0.860 \text{ m} \quad\quad v_{yi} = 0 \quad\quad v_y = ? \quad\quad a_y = -9.80 \text{ m} / \text{s}^2$

Horizontal motion: $\quad x = 1.40 \text{ m} \quad\quad\quad v_x = ? = \text{constant} \quad\quad a_x = 0$

(a) To find the time of fall, we use the free fall equation: $\quad y = v_{yi}t + \dfrac{1}{2}a_y t^2$

Solving, $\quad\quad\quad -0.860 \text{ m} = 0 - \dfrac{1}{2}\left(9.80 \text{ m} / \text{s}^2\right)t^2$

so $\quad\quad\quad\quad\quad t = 0.419 \text{ s}$

Then $\quad\quad\quad\quad v_x = \dfrac{x}{t} = \dfrac{1.40 \text{ m}}{0.419 \text{ s}} = 3.34 \text{ m} / \text{s}$

(b) The mug hits the floor with a vertical velocity of $\quad\quad\quad v_{yf} = v_{yi} + a_y t$

and an impact angle below the horizontal of $\quad\quad\quad \theta = \tan^{-1}\left(v_y / v_x\right)$

Solving for v_y, $\quad v_y = 0 - \left(9.80 \text{ m} / \text{s}^2\right)\left(0.419 \text{ s}\right) = -4.11 \text{ m} / \text{s}$

Thus, $\quad\quad\quad\quad \theta = \tan^{-1}\left(\dfrac{-4.11 \text{ m} / \text{s}}{3.34 \text{ m} / \text{s}}\right) = 50.9° \quad\quad\quad\quad\quad\quad\quad \Diamond$

Finalize: This was a multi-step problem that required several physics equations to solve; yet our answers do agree with our initial expectations. Since the problem did not ask for the time, we could have eliminated this variable by substitution, but then we would have had to substitute the algebraic expression $t = \sqrt{2y / g}$ into two other equations. So in this case it was easier to find a numerical value for the time as an intermediate step. Sometimes the most efficient method is not realized until each alternative solution is attempted.

15. A projectile is fired in such a way that its horizontal range is equal to three times its maximum height. What is the angle of projection?

Solution In this problem, we want to find θ_i such that $R = 3h$. We can use the equations for the range and height of a projectile path,

$$R = (v_i^2 \sin 2\theta_i) / g \qquad \text{and} \qquad h = (v_i^2 \sin^2 \theta_i) / 2g$$

Since we require that $R = 3h$,

$$\frac{v_i^2 \sin 2\theta_i}{g} = \frac{3v_i^2 \sin^2 \theta_i}{2g} \qquad \text{or} \qquad \frac{2}{3} = \frac{\sin^2 \theta_i}{\sin 2\theta_i}$$

But $\qquad \sin 2\theta_i = 2\sin\theta_i \cos\theta_i \qquad$ so $\qquad \dfrac{\sin^2 \theta_i}{\sin 2\theta_i} = \dfrac{\sin^2 \theta_i}{2\sin\theta_i \cos\theta_i} = \dfrac{\tan\theta_i}{2}$

Substituting and solving for θ_i, $\qquad \theta_i = \tan^{-1}\left(\dfrac{4}{3}\right) = 53.1° \qquad \Diamond$

---

19. A place kicker must kick a football from a point 36.0 m (about 40 yards) from the goal, and half the crowd hopes the ball will clear the crossbar, which is 3.05 m high. When kicked, the ball leaves the ground with a speed of 20.0 m/s at an angle of 53.0° to the horizontal. (a) By how much does the ball clear or fall short of clearing the crossbar? (b) Does the ball approach the crossbar while still rising or while falling?

Solution Model the football as a projectile, moving with constant horizontal velocity and with constant vertical acceleration.

(a) To find the actual height of the football when it reaches the goal line, we use the trajectory equation:

$$y_f = x_f \tan\theta_i - \frac{gx_f^2}{2v_i^2 \cos^2 \theta_i} \qquad \Diamond$$

where $\qquad x_f = 36.0 \text{ m}, \quad v_i = 20.0 \text{ m/s}, \quad \text{and} \quad \theta_i = 53.0°$

So, $\qquad y_f = (36.0 \text{ m})(\tan 53.0°) - \dfrac{(9.80 \text{ m/s}^2)(36.0 \text{ m})^2}{2(20.0 \text{ m/s})^2 \cos^2 53.0°}$

$$y_f = 47.774 - 43.834 = 3.939 \text{ m}$$

The ball clears the bar by $\quad 3.939 \text{ m} - 3.050 \text{ m} = 0.889 \text{ m} \qquad \Diamond$

(b)   The time the ball takes to reach the maximum height $(v_y = 0)$ is

$$t_1 = \frac{(v_i \sin\theta_i) - v_y}{g} = \frac{(20.0\text{ m / s})(\sin 53.0°) - 0}{9.8\text{ m / s}^2} = 1.63\text{ s}$$

The time to travel 36.0 m horizontally is

$$t_2 = \frac{x_f}{v_{xi}} = \frac{36.0\text{ m}}{(20.0\text{ m / s})(\cos 53.0°)} = 2.99\text{ s}$$

Since $t_2 > t_1$, the ball clears the goal on its way down.   ◊

---

27.   The athlete shown in Figure P4.27 of the textbook rotates a 1.00-kg discus along a circular path of radius 1.06 m. The maximum speed of the discus is 20.0 m/s. Determine the magnitude of the maximum radial acceleration of the discus.

Solution   The maximum radial acceleration occurs when maximum tangential speed is attained. Model the discus here as a particle in uniform circular motion.

Here, $a_c = \dfrac{v^2}{r} = \dfrac{(20.0\text{ m / s})^2}{1.06\text{ m}} = 377\text{ m / s}^2$ ◊

---

33.   A train slows down as it rounds a sharp horizontal turn, slowing from 90.0 km/h to 50.0 km/h in the 15.0 s that it takes to round the bend. The radius of the curve is 150 m. Compute the acceleration at the moment the train speed reaches 50.0 km/h. Assume that it continues to slow down at this time at the same rate.

Solution

Conceptualize: If the train is taking this turn at a safe speed, then its acceleration should be significantly less than g, perhaps a few m/s^2 (otherwise it might jump the tracks!), and it should be directed toward the center of the curve and backwards since the train is slowing.

Categorize: Since the train is changing both its speed and direction, the acceleration vector will be the vector sum of the tangential and radial acceleration components. The tangential acceleration can be found from the changing speed and elapsed time, while the radial acceleration can be found from the radius of curvature and the train's speed.

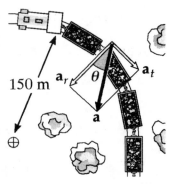

52

Analyze: First, let's convert the speed units from km/h to m/s:

$$v_i = 90.0 \text{ km / h} = (90.0 \text{ km / h})(10^3 \text{ m / km})(1 \text{ h/3 600 s}) = 25.0 \text{ m / s}$$

$$v_f = 50.0 \text{ km / h} = (50.0 \text{ km / h})(10^3 \text{ m / km})(1 \text{ h/3 600 s}) = 13.9 \text{ m / s}$$

The tangential acceleration and radial acceleration are:

$$a_t = \frac{\Delta v}{\Delta t} = \frac{13.9 \text{ m / s} - 25.0 \text{ m / s}}{15.0 \text{ s}} = -0.741 \text{ m / s}^2 \quad \text{(backward)}$$

$$a_r = \frac{v^2}{r} = \frac{(13.9 \text{ m / s})^2}{150 \text{ m}} = 1.29 \text{ m / s}^2 \quad \text{(inward)}$$

$$a = \sqrt{a_r^2 + a_t^2} = \sqrt{(1.29 \text{ m / s}^2)^2 + (-0.741 \text{ m / s}^2)^2} = 1.48 \text{ m / s}^2 \qquad \lozenge$$

$$\theta = \tan^{-1}\left(\frac{a_t}{a_r}\right) = \tan^{-1}\left(\frac{0.741 \text{ m / s}^2}{1.29 \text{ m / s}^2}\right) = 29.9° \qquad \lozenge$$

Finalize: The acceleration is clearly less than g, and it appears that most of the acceleration comes from the radial component, so it makes sense that the acceleration vector should point mostly inward toward the center of the curve and slightly backward due to the negative tangential acceleration.

---

35. Figure P4.35 represents the total acceleration of a particle moving clockwise in a circle of radius 2.50 m at a certain time. At this instant, find (a) the radial acceleration, (b) the speed of the particle, and (c) its tangential acceleration.

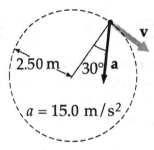

Solution

(a) The acceleration has an inward radial component:

$$a_c = a\cos 30.0° = (15.0 \text{ m / s}^2)\cos 30.0° = 13.0 \text{ m / s}^2 \qquad \lozenge$$

Figure P4.35

(b) The speed at the instant shown can be found by using

$$a_c = \frac{v^2}{r} \quad \text{or} \quad v = \sqrt{a_c r} = \sqrt{(13.0 \text{ m / s}^2)(2.50 \text{ m})} = 5.70 \text{ m / s} \qquad \lozenge$$

(c) The acceleration also has a tangential component:

$$a_t = a\sin 30.0° = (15.0 \text{ m / s}^2)\sin 30.0° = 7.50 \text{ m / s}^2 \qquad \lozenge$$

---

41. A river has a steady speed of 0.500 m/s. A student swims upstream a distance of 1.00 km and swims back to the starting point. If the student can swim at a speed of 1.20 m/s in still water, how long does the trip take? Compare this with the time the trip would take if the water were still.

Solution

Conceptualize: If we think about the time for a trip as a function of the stream's speed, we realize that if the stream is flowing at the same rate or faster than the student can swim, he will never reach the 1.00-km mark even after an infinite amount of time. Since the student can swim 1.20 km in 1000 s, we should expect that the trip will definitely take longer than in still water, maybe about 2000 s (~30 minutes).

Categorize: The total time in the river is the longer time upstream (against the current) plus the shorter time downstream (with the current). For each part, we will use the basic equation $t = d / v$, where v is the speed of the student relative to the shore.

Analyze:

$$t_{up} = \frac{d}{v_{student} - v_{stream}} = \frac{1\,000 \text{ m}}{1.20 \text{ m/s} - 0.500 \text{ m/s}} = 1\,430 \text{ s}$$

$$t_{dn} = \frac{d}{v_{student} + v_{stream}} = \frac{1\,000 \text{ m}}{1.20 \text{ m/s} + 0.500 \text{ m/s}} = 588 \text{ s}$$

Total time in the river,

$$t_{river} = t_{up} + t_{dn} = 2.02 \times 10^3 \text{ s}$$

In still water,

$$t_{still} = \frac{d}{v} = \frac{2\,000 \text{ m}}{1.20 \text{ m/s}} = 1.67 \times 10^3 \text{ s}$$

or

$$t_{still} = 0.827 t_{river} \qquad \qquad \Diamond$$

Finalize: As we predicted, it does take the student longer to swim up and back in the moving stream than in still water (21% longer in this case), and the amount of time agrees with our estimate.

---

45. A science student is riding on a flatcar of a train traveling along a straight horizontal track at a constant speed of 10.0 m/s. The student throws a ball into the air along a path that he judges to make an initial angle of 60.0° with the horizontal and to be in line with the track. The student's professor, who is standing on the ground nearby, observes the ball to rise vertically. How high does she see the ball rise?

Solution   Shown on the right, $\qquad \mathbf{v}_{be} = \mathbf{v}_{ce} + \mathbf{v}_{bc}$

with $\qquad \mathbf{v}_{bc}$ = velocity of the ball relative to the car

$\qquad \mathbf{v}_{be}$ = velocity of the ball relative to the Earth

and $\qquad \mathbf{v}_{ce}$ = velocity of the car relative to the Earth = 10.0 m/s

From the figure, we have $\qquad v_{ce} = v_{bc}\cos 60.0°$

So $\qquad v_{bc} = \dfrac{10.0 \text{ m / s}}{\cos 60.0°} = 20.0 \text{ m / s}$

Again from the figure, $\qquad v_{be} = v_{bc}\sin 60.0° + 0 = (20.0 \text{ m / s})(0.866) = 17.3 \text{ m / s}$

This is the initial velocity of the ball relative to the Earth. Now we can calculate the maximum height that the ball rises.

From $\quad v_{yf}^2 = v_{yi}^2 + 2ah \qquad\qquad 0 = (17.3 \text{ m / s})^2 + 2(-9.80 \text{ m / s}^2)h$

$$h = 15.3 \text{ m} \qquad\qquad\qquad ◊$$

---

51. Barry Bonds hits a home run so that the baseball just clears the top row of bleachers, 21.0 m high, located 130 m from home plate. The ball is hit at an angle of 35.0° to the horizontal, and air resistance is negligible. Find (a) the initial speed of the ball, (b) the time it takes the ball to reach the cheap seats, and (c) the velocity components and the speed of the ball when it passes over the top row. Assume the ball is hit at a height of 1.00 m above the ground.

Solution Let the initial speed of the ball be v_i, and the initial angle $\theta_i = 35.0°$. Set the starting point at the origin, $(x_i, y_i) = (0 \text{ m, } 0 \text{ m})$. When $x_f = 130 \text{ m}$, $y_f = 20.0 \text{ m}$.

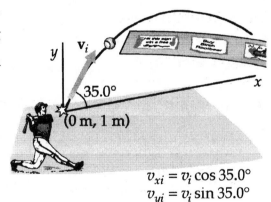

$$x_f = x_i + v_{xi}t = v_i t \cos 35.0°$$

$$v_i t \cos 35.0° = 130 \text{ m}$$

and $\qquad v_i t = \dfrac{130 \text{ m}}{\cos 35.0°} = 158.7 \text{ m}$

$v_{xi} = v_i \cos 35.0°$
$v_{yi} = v_i \sin 35.0°$

Next, $\qquad y_f = v_{yi}t - \frac{1}{2}gt^2 = (v_i \sin 35.0°)t - \frac{1}{2}(9.80 \text{ m / s}^2)t^2$

Substituting for $v_i t$, $\qquad 20.0 \text{ m} = (158.7 \text{ m})(\sin 35.0°) - \frac{1}{2}(9.80 \text{ m / s}^2)t^2$

(b) and $\qquad t = \sqrt{\dfrac{71.0 \text{ m}}{4.90 \text{ m / s}^2}} = 3.81 \text{ s} \qquad\qquad ◊$

(a) Therefore, $\qquad v_i = \dfrac{158.7 \text{ m}}{3.81 \text{ s}} = 41.7 \text{ m / s} \qquad\qquad ◊$

(c)
$$v_{yf} = v_{yi} - gt = v_i \sin 35.0° - gt = (23.9 \text{ m / s}) - (9.80 \text{ m / s}^2)t$$

At $t = 3.81$ s $\qquad v_y = -13.4$ m / s $\qquad\qquad\qquad ◊$

$$v_{xf} = v_{xi} = v_i \cos 35.0° = (41.7 \text{ m / s}) \cos 35.0° = 34.1 \text{ m / s} \qquad ◊$$

and $\qquad |\mathbf{v}| = \sqrt{v_x^2 + v_y^2} = \sqrt{34.1^2 + (-13.4)^2} = 36.7$ m / s $\qquad ◊$

---

59. Your grandfather is copilot of a bomber, flying horizontally over level terrain, with a speed of 275 m/s relative to the ground, at an altitude of 3 000 m. (a) The bombardier releases one bomb. How far will it travel horizontally between its release and its impact on the ground? Neglect the effects of air resistance. (b) Firing from the people on the ground suddenly incapacitates the bombardier before he can call "Bombs away!" Consequently, the pilot retains the plane's original course, altitude, and speed through a storm of flak. Where will the plane be when the bomb hits the ground? (c) The plane has a telescopic bomb sight set so that the bomb hits the target seen in the sight at the time of release. At what angle from the vertical was the bomb sight set?

Solution

(a) In horizontal flight, $v_{yi} = 0$

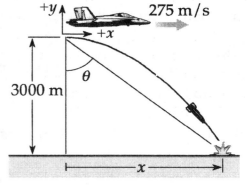

and $\qquad v_{xi} = 275$ m / s

Therefore, $\qquad y_f = y_i - \frac{1}{2}gt^2 = -\frac{1}{2}gt^2$

and $\qquad x = v_{xi}t$

Eliminate t between these two equations to find

$$x_f = v_{xi}\sqrt{\frac{-2y}{g}} = (275 \text{ m / s})\sqrt{\frac{(-2)(-3\,000 \text{ m})}{9.80 \text{ m / s}^2}} = 6\,800 \text{ m} \qquad ◊$$

(b) The plane and the bomb have the same constant **horizontal** velocity Therefore, the plane will be 3 000 m above the bomb at impact, and 6 800 m from the point of release.

(c) If θ is the angle between the vertical and the direct line of sight to the target at the time of release,

$$\theta = \tan^{-1}\left(\frac{x}{y}\right) = \tan^{-1}\left(\frac{6\,800}{3\,000}\right) = 66.2° \qquad ◊$$

---

61. A hawk is flying horizontally at 10.0 m/s in a straight line, 200 m above the ground. A mouse it has been carrying struggles free from its grasp. The hawk continues on its path at the same speed for 2.00 s before attempting to retrieve its prey. To accomplish the retrieval, it dives in a straight line at constant speed and recaptures the mouse 3.00 m above the ground. (a) Assuming no air resistance, find the diving speed of the hawk. (b) What angle did the hawk make with the horizontal during its descent? (c) For how long did the mouse "enjoy" free fall?

Solution

Conceptualize: We should first recognize the hawk cannot instantaneously change from slow horizontal motion to rapid downward motion. The hawk cannot move with infinite acceleration, but we assume that the time required for the hawk to accelerate is short compared to two seconds. Based on our everyday experiences, a reasonable diving speed for the hawk might be about 100 mph (~50 m/s) at some angle close to 90° and should last only a few seconds.

Categorize: We know the distance that the mouse and hawk fall, but to find the diving speed of the hawk, we must know the time of descent. If the hawk and mouse both maintain their original horizontal velocity of 10 m/s (as they should without air resistance), then the hawk only needs to think about diving straight down, but to a ground-based observer, the path will appear to be a straight line angled less than 90° below horizontal.

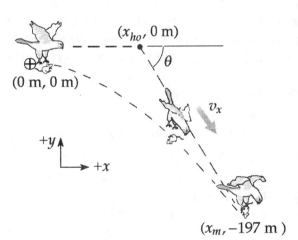

Analyze: The mouse falls a total vertical distance,

$$y = 200 \text{ m} - 3.00 \text{ m} = 197 \text{ m}$$

The time of fall is found from $y = v_{yi}t - \frac{1}{2}gt^2$:

$$t = \sqrt{\frac{2(197 \text{ m})}{9.80 \text{ m} / \text{s}^2}} = 6.34 \text{ s}$$

To find the diving speed of the hawk, we must first calculate the total distance covered from the vertical and horizontal components. We already know the vertical distance y; we just need the horizontal distance during the same time (minus the 2.00-s late start).

$$x = v_{xi}(t - 2.00 \text{ s}) = (10.0 \text{ m} / \text{s})(6.34 \text{ s} - 2.00 \text{ s}) = 43.4 \text{ m}$$

The total distance is $\quad d = \sqrt{x^2 + y^2} = \sqrt{(43.4 \text{ m})^2 + (197 \text{ m})^2} = 202 \text{ m}$

So the hawk's diving speed is

$$v_{hawk} = \frac{d}{t - 2.00 \text{ s}} = \frac{202 \text{ m}}{4.34 \text{ s}} = 46.5 \text{ m / s}$$

at an angle below the horizontal of

$$\theta = \tan^{-1}\left(\frac{y}{x}\right) = \tan^{-1}\left(\frac{197 \text{ m}}{43.4 \text{ m}}\right) = 77.6° \qquad \Diamond$$

Finalize: The answers appear to be consistent with our predictions, even though it is not possible for the hawk to reach its diving speed instantaneously. Sometimes we must make simplifying assumptions to solve complex physics problems, and sometimes these assumptions are not physically possible. Once the idealized problem is understood, we can attempt to analyze the more complex, real-world problem. For this problem, if we considered the realistic effects of air resistance and the maximum diving acceleration attainable by the hawk, we might find that the hawk could not catch the mouse before it hit the ground.

---

63. A car is parked on a steep incline overlooking the ocean. The incline makes an angle of 37.0° below the horizontal. The negligent driver leaves the car in neutral, and the parking brakes are defective. Starting from rest at $t = 0$, the car rolls down the incline with a constant acceleration of 4.00 m/s², traveling 50.0 m to the edge of a vertical cliff. The cliff is 30.0 m above the ocean. Find (a) the speed of the car when it reaches the edge of the cliff and the time it takes to get there, (b) the velocity of the car when it lands in the ocean, (c) the total time that the car is in motion, and (d) the position of the car when it lands in the ocean, relative to the base of the cliff.

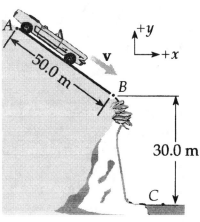

Solution

From point A to point B (along the incline), the car can be modeled as a particle under constant acceleration in one dimension, starting from rest $(v_i = 0)$. Therefore, taking s to be the position along the incline,

(a)

$$v_B^{\ 2} = v_i^{\ 2} + 2a\left(s_f - s_i\right) = 2a\left(s_B - s_A\right)$$

$$v_B = \sqrt{2(4.00 \text{ m / s}^2)(50.0 \text{ m})} = 20.0 \text{ m / s} \quad \Diamond$$

We can find the elapsed time from $v_B = v_i + at$:

$$t_{AB} = \frac{v_B - v_i}{a} = \frac{20.0 \text{ m / s} - 0}{4.00 \text{ m / s}^2} = 5.00 \text{ s} \qquad \Diamond$$

(b) After the car passes the top of the cliff, it becomes a projectile. At the edge of the cliff, the components of velocity v_B are:

$$v_{yB} = (-20.0 \text{ m} / \text{s}) \sin 37.0° = -12.0 \text{ m} / \text{s}$$

$$v_{xB} = (20.0 \text{ m} / \text{s}) \cos 37.0° = 16.0 \text{ m} / \text{s}$$

There is no further horizontal acceleration,

so
$$v_{xC} = v_{xB} = 16.0 \text{ m} / \text{s}$$

However, the downward (negative) vertical velocity is affected by free fall:

$$v_{yC} = \pm\sqrt{2a_y(\Delta y) + v_{yB}{}^2} = \pm\sqrt{2(-9.80 \text{ m} / \text{s}^2)(-30.0 \text{ m}) + (-12.0 \text{ m} / \text{s})^2} = -27.1 \text{ m} / \text{s}$$

$$\mathbf{v}_C = 16.0\hat{\mathbf{i}} - 27.1\hat{\mathbf{j}} \text{ m} / \text{s}$$ ◊

(c) From point B to C, the time

$$t_{BC} = \frac{v_{yC} - v_{yB}}{a_y} = \frac{(-27.1 \text{ m} / \text{s}) - (-12.0 \text{ m} / \text{s})}{(-9.80 \text{ m} / \text{s}^2)} = 1.53 \text{ s}$$

The total elapsed time is $\quad t_{AC} = t_{AB} + t_{BC} = 6.53 \text{ s}$ ◊

(d) The horizontal distance covered is

$$\Delta x = v_{xB} t_{BC} = (16.0 \text{ m} / \text{s})(1.54 \text{ s}) = 24.5 \text{ m}$$ ◊

---

67. A skier leaves the ramp of a ski jump with a velocity of 10.0 m/s, 15.0° above the horizontal, as in Figure P4.67. The slope is inclined at 50.0°, and air resistance is negligible. Find (a) the distance from the ramp to where the jumper lands and (b) the velocity components just before the landing. (How do you think the results might be affected if air resistance were included? Note that jumpers lean forward in the shape of an airfoil, with their hands at their sides, to increase their distance. Why does this work?)

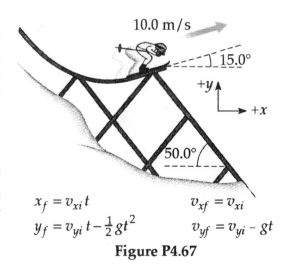

$$x_f = v_{xi} t$$
$$y_f = v_{yi} t - \tfrac{1}{2} g t^2$$
$$v_{xf} = v_{xi}$$
$$v_{yf} = v_{yi} - g t$$

Figure P4.67

Solution

Set point '0' where the skier takes off, and point '2' where the skier lands. Define the coordinate system as shown in the figure.

$$\mathbf{v}_i = 10.0 \text{ m / s at } 15.0°: \qquad v_{xi} = v_i \cos\theta_i = (10.0 \text{ m / s})\cos 15.0° = 9.66 \text{ m / s}$$

$$v_{yi} = v_i \sin\theta_i = (10.0 \text{ m / s})\sin 15.0° = 2.59 \text{ m / s}$$

(a)   The skier travels horizontally $x_2 - x_0 = v_{xi}t = (9.66 \text{ m / s})t$ **[1]**

      and vertically $\qquad\qquad y_2 - y_0 = v_{yi}t - \frac{1}{2}gt^2 = (2.59 \text{ m/s})t - (4.90 \text{ m/s}^2)t^2$ **[2]**

      The skier hits the slope when $\dfrac{y_2 - y_0}{x_2 - x_0} = \tan(-50.0°) = -1.19$ **[3]**

      Substituting [1] and [2] into [3], $\qquad \dfrac{(2.59 \text{ m / s})t - (4.90 \text{ m / s}^2)t^2}{(9.66 \text{ m / s})t} = -1.19$

      Since we ignore the solution $t = 0$, $-4.90t + 14.1 = 0$ and $t = 2.88 \text{ s}$

      Solving Equation [1], $\qquad\qquad\qquad x_2 - x_0 = (9.66 \text{ m / s})t = 27.8 \text{ m}$

      From the diagram $\qquad\qquad\qquad d = \dfrac{x_2 - x_0}{\cos 50.0°} = \dfrac{27.8 \text{ m}}{\cos 50.0°} = 43.2 \text{ m}$ ◊

(b)   The final horizontal velocity is $\qquad v_{xf} = v_{xi} = 9.66 \text{ m / s}$ ◊

      The vertical component is found from $v_{yf} = v_{yi} - gt = 2.59 \text{ m / s} - (9.80 \text{ m / s}^2)t$

      When $t = 2.88 \text{ s}$, $\qquad\qquad\qquad v_{yf} = -25.6 \text{ m / s}$ ◊

The 'drag' force of air resistance would necessarily decrease both components of the ski jumper's impact velocity. On the other hand, the lift force of the air could extend her time of flight and increase the distance of her jump. If the jumper has the profile of an airplane wing, she can deflect downward the air through which she passes, to make the air deflect her upward.

Chapter 5
THE LAWS OF MOTION

EQUATIONS AND CONCEPTS

A quantitative **measurement of mass** (the term used to measure inertia) can be made by comparing the accelerations that a given force will produce on different bodies. If a given force acting on a body of mass m_1 produces an acceleration a_1 and the same force acting on a body of mass m_2 produces an acceleration a_2, the ratio of the two masses is defined as the inverse ratio of the magnitudes of the accelerations. *Mass is an inherent property of an object and is independent of the surroundings and the method of measurement.*

$$\frac{m_1}{m_2} \equiv \frac{a_2}{a_1} \tag{5.1}$$

The **acceleration** of an object is proportional to the net force acting on it and inversely proportional to its mass. *This is a statement of Newton's second law.*

$$\sum \mathbf{F} = m\mathbf{a} \tag{5.2}$$

Three **component equations** are the equivalent of the vector equation expressing Newton's second law. *The orientation of the coordinate system can often be chosen so that the object has a nonzero acceleration along only one direction.*

$$\sum F_x = m a_x$$
$$\sum F_y = m a_y \tag{5.3}$$
$$\sum F_z = m a_z$$

The **SI unit of force** is the newton (N), defined as the force that, when acting on a 1-kg mass, produces an acceleration of 1 m/s². *Calculations with Equation 5.2 must be made using a consistent set of units for the quantities force, mass, and acceleration.*

$$1\,N \equiv 1\,kg \cdot m/s^2 \tag{5.4}$$

$$1\,lb \equiv 1\,slug \cdot ft/s^2 \tag{5.5}$$

$$1\,lb \equiv 4.448\,N$$

The **gravitational force** exerted by the earth on an object depends on the local value of **g** and varies with location. *Weight (the force of gravity on a mass) is the magnitude of the gravitational force and is not an inherent property of a body.*

$$\mathbf{F}_g = m\mathbf{g} \qquad (5.6)$$

$$\text{weight} = mg$$

Newton's third law states that the action force, $\mathbf{F}_{12}$, exerted by object 1 on object 2 is equal in magnitude and opposite in direction to the reaction force, $\mathbf{F}_{21}$, exerted by object 2 on object 1. *Remember, the two forces in an action-reaction pair always act on two different objects — they cannot add to give a net force of zero.*

$$\mathbf{F}_{12} = -\mathbf{F}_{21} \qquad (5.7)$$

The magnitude of the **force of static friction** between two surfaces in contact, but not in motion relative to each other, cannot be greater than $\mu_s n$, where n is the normal (perpendicular) force between the two surfaces. The coefficient of static friction, μ_s, is a dimensionless constant which depends on the nature of the pair of surfaces. *The equality sign holds when the two surfaces are on the verge of slipping (impending motion). The inequality sign holds when motion is not impending.*

$$f_s \leq \mu_s n \qquad (5.8)$$

The magnitude of the **force of kinetic friction** applies when two surfaces are in relative motion. *The friction force is parallel to the surface on which an object is in contact and is directed opposite the direction of actual or impending motion.*

$$f_k = \mu_k n \qquad (5.9)$$

SUGGESTIONS, SKILLS, AND STRATEGIES

The following procedure is recommended when dealing with problems involving the application of Newton's second law, including cases when objects are in static equilibrium:

- Draw a simple, neat diagram of the system.

- Isolate the object of interest whose motion is being analyzed. Draw a free-body diagram for this object; that is, a diagram showing **all external forces acting on the object**. For systems containing more than one object, draw **separate diagrams for each object**. Do not include forces that the object exerts on its surroundings.

- Establish convenient coordinate axes for each object and find the components of the forces along these axes.

- Apply Newton's second law, $\Sigma \mathbf{F} = m\mathbf{a}$, in the x and y directions for each object under consideration. Check to be sure that all terms in the resulting equation have units of force.

- Solve the component equations for the unknowns. Remember that you must have as many independent equations as you have unknowns in order to obtain a complete solution. Often in solving such problems, one must also use the equations of kinematics (motion with constant acceleration) to find all the unknowns.

- Make sure that your final results are consistent with the free-body diagram and check the predictions of your solution in the case of extreme values of the variables.

REVIEW CHECKLIST

You should be able to:

▷ State in your own words Newton's laws of motion, recall physical examples of each law, and identify the action-reaction force pairs in a multiple-body interaction problem as specified by Newton's third law. (Sections 5.2, 5.4, and 5.6)

▷ Apply Newton's laws of motion to various mechanical systems using the recommended procedure outlined in Suggestions, Skills and Strategies and discussed in Section 5.7 of your textbook. Identify all external forces acting on the system, draw the correct free-body diagram for each body of the system, and apply Newton's second law, $\Sigma \mathbf{F} = m\mathbf{a}$, in **component** form. (Section 5.7)

▷ Express the normal force in terms of other forces acting on an object and the acceleration of the object. Write out the equation which relates the coefficient of friction, force of friction and normal force between an object and surface on which it rests or moves. (Section 5.7)

▷ Apply the equations of kinematics (which involve the quantities displacement, velocity, time, and acceleration) as described in Chapter 2 along with those methods and equations of Chapter 5 (involving mass, force, and acceleration) to the solutions of problems using Newton's second law. (Section 5.7)

▷ Solve several linear equations simultaneously for the unknown quantities. Recall that you must have as many independent equations as you have unknowns.

ANSWERS TO SELECTED CONCEPTUAL QUESTIONS

4. In the motion picture *It Happened One Night* (Columbia Pictures, 1934), Clark Gable is standing inside a stationary bus in front of Claudette Colbert, who is seated. The bus suddenly starts moving forward and Clark falls into Claudette's lap. Why did this happen?

Answer When the bus starts moving, the mass of Claudette is accelerated by the force of the back of the seat on her body. Clark is standing, however, and the only force on him is the friction between his shoes and the floor of the bus. Thus, when the bus starts moving, his feet start accelerating forward, but the rest of his body experiences almost no accelerating force (only that due to his being attached to his accelerating feet!). As a consequence, his body tends to stay almost at rest, according to Newton's first law, relative to the ground. Relative to Claudette, however, he is moving toward her and falls into her lap. (Both performers won Academy Awards.)

□   □   □   □

7. A rubber ball is dropped onto the floor. What force causes the ball to bounce?

Answer When the ball hits the floor, it is compressed. As the ball returns to its original shape, it exerts a force on the floor, and the reaction to this thrusts it back into the air.

□   □   □   □

10. A weightlifter stands on a bathroom scale. He pumps a barbell up and down. What happens to the reading on the bathroom scale as this is done? **What if** he is strong enough to actually **throw** the barbell upward? How does the reading on the scale vary now?

Answer If the barbell is not moving, the reading on the bathroom scale is the combined weight of the weightlifter and the barbell. At the beginning of the lift of the barbell, the barbell accelerates upward. By Newton's third law, the barbell pushes downward on the hands of the weightlifter with more force than its weight, in order to accelerate. As a result, he is pushed with more force into the scale, increasing its reading. Near the top of the lift, the weightlifter reduces the upward force, so that the acceleration of the barbell is downward, causing it to come to rest. While the barbell is coming to rest, it pushes with less force on the weightlifter's hands, so the reading on the scale is below the combined stationary weight. If the barbell is held at rest for a moment at the top of the lift, the scale reading is simply the combined weight. As it begins to be brought down, the reading decreases, as the force of the weightlifter on the barbell is reduced. The reading increases as the barbell is slowed down at the bottom.

If we now consider the throwing of the barbell, we have the same behavior as before, except that the variations in scale reading will be larger, since more force must be applied to throw the barbell upward rather than just lift it. Once the barbell leaves the weightlifter's hands, the reading will suddenly drop to just the weight of the weightlifter, and will rise suddenly when the barbell is caught.

□   □   □   □

14. Identify the action–reaction pairs in the following situations: a man takes a step; a snowball hits a girl in the back; a baseball player catches a ball; a gust of wind strikes a window.

Answer   As a man takes a step, the action is the force his foot exerts on the Earth; the reaction is the force of the Earth on his foot. In the second case, the action is the force exerted on the girl's back by the snowball; the reaction is the force exerted on the snowball by the girl's back. The third action is the force of the glove on the ball; the reaction is the force of the ball on the glove. The fourth action is the force exerted on the window by the air molecules; the reaction is the force on the air molecules exerted by the window. We could equally well interchange the terms 'action' and 'reaction' in each case.

□   □   □   □

SOLUTIONS TO SELECTED END-OF-CHAPTER PROBLEMS

3.   A 3.00-kg object undergoes an acceleration given by $\mathbf{a} = (2.00\hat{\mathbf{i}} + 5.00\hat{\mathbf{j}})$ m/s^2. Find the resultant force $\mathbf{F}$ and its magnitude.

Solution $\sum \mathbf{F} = m\mathbf{a} = (3.00 \text{ kg})(2.00\hat{\mathbf{i}} + 5.00\hat{\mathbf{j}})$ m / s$^2 = (6.00\hat{\mathbf{i}} + 15.0\hat{\mathbf{j}})$ N   ◊

$|\mathbf{F}| = \sqrt{(F_x)^2 + (F_y)^2} = \sqrt{(6.00 \text{ N})^2 + (15.0 \text{ N})^2} = 16.2$ N   ◊

5. To model a spacecraft, a toy rocket engine is securely fastened to a large puck, which can glide with negligible friction over a horizontal surface, taken as the xy plane. The 4.00-kg puck has a velocity of $3.00\hat{i}$ m/s at one instant. Eight seconds later, its velocity is to be $(8.00\hat{i} + 10.0\hat{j})$ m/s. Assuming the rocket engine exerts a constant horizontal force, find (a) the components of the force and (b) its magnitude.

Solution We use the particle under constant acceleration and particle under net force models. We first calculate the acceleration of the puck:

$$\bar{a} = \frac{\Delta v}{\Delta t} = \frac{(8.00\hat{i} + 10.0\hat{j}) \text{ m/s} - 3.00\hat{i} \text{ m/s}}{8.00 \text{ s}} = 0.625\hat{i} \text{ m/s}^2 + 1.25\hat{j} \text{ m/s}^2$$

In $\Sigma F = ma$, the only horizontal force is the thrust F of the rocket:

(a) $\qquad F = (4.00 \text{ kg})(0.625\hat{i} \text{ m/s}^2 + 1.25\hat{j} \text{ m/s}^2) = 2.50\hat{i} \text{ N} + 5.00\hat{j} \text{ N}$ ◊

(b) Then, $\quad |F| = \sqrt{(2.50 \text{ N})^2 + (5.00 \text{ N})^2} = 5.59 \text{ N}$ ◊

---

7. An electron of mass 9.11×10^{-31} kg has an initial speed of 3.00×10^5 m/s. It travels in a straight line, and its speed increases to 7.00×10^5 m/s in a distance of 5.00 cm. Assuming its acceleration is constant, (a) determine the force on the electron and (b) compare this force with the weight of the electron, which we neglected.

Solution

Conceptualize: We should expect that only a very small force is required to accelerate an electron because of its small mass, but this force is probably much greater than the weight of the electron if the gravitational force can be neglected.

Categorize: Since this is simply a linear acceleration problem, we can use Newton's second law to find the force as long as the electron does not approach relativistic speeds (much less than 3×10^8 m/s), which is certainly the case for this problem. We know the initial and final velocities, and the distance involved, so from these we can find the acceleration needed to determine the force.

Analyze: From $\qquad\qquad\qquad\qquad\qquad v_f^2 = v_i^2 + 2ax \quad$ and $\quad \sum F = ma$

we can solve for the acceleration and the force $\quad a = \dfrac{v_f^2 - v_i^2}{2x}$

and so $\qquad\qquad\qquad\qquad\qquad\qquad \sum F = \dfrac{m(v_f^2 - v_i^2)}{2x}$

(a) $\sum F = \dfrac{\left(9.11\times10^{-31}\ \text{kg}\right)\left(\left(7.00\times10^{5}\ \text{m/s}\right)^{2} - \left(3.00\times10^{5}\ \text{m/s}\right)^{2}\right)}{2\left(0.050\ 0\ \text{m}\right)}$

$\sum F = 3.64\times10^{-18}\ \text{N}$ ◊

(b) The electron's weight is

$F_g = mg = \left(9.11\times10^{-31}\ \text{kg}\right)\left(9.80\ \text{m/s}^{2}\right) = 8.93\times10^{-30}\ \text{N}$ ◊

The ratio of the accelerating force to the weight is

$F / F_g = 4.08\times10^{11}$ ◊

Finalize: The force that causes the electron to accelerate is indeed a small fraction of a newton, but it is much greater than the gravitational force. For this reason, it is quite reasonable to ignore the weight of the electron in problems about electric forces.

---

11. Two forces F_1 and F_2 act on a 5.00-kg object. If $F_1 = 20.0\ \text{N}$ and $F_2 = 15.0\ \text{N}$, find the accelerations in (a) and (b) of Figure P5.11.

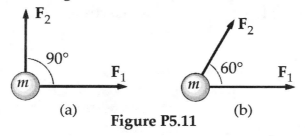

(a) (b)

Figure P5.11

Solution We use the particle under a net force model:

$m = 5.00\ \text{kg}$

(a) $\sum \mathbf{F} = \mathbf{F}_1 + \mathbf{F}_2 = \left(20.0\hat{\mathbf{i}} + 15.0\hat{\mathbf{j}}\right)\ \text{N}$

$\mathbf{a} = \dfrac{\sum \mathbf{F}}{m} = \left(4.00\hat{\mathbf{i}} + 3.00\hat{\mathbf{j}}\right)\ \text{m/s}^{2} = 5.00\ \text{m/s}^{2}\ \text{at } 36.9°$ ◊

(b) $\sum \mathbf{F} = \mathbf{F}_1 + \mathbf{F}_2 = \left[20.0\hat{\mathbf{i}} + \left(15.0\cos 60°\hat{\mathbf{i}} + 15.0\sin 60°\hat{\mathbf{j}}\right)\right]\ \text{N} = \left(27.5\hat{\mathbf{i}} + 13.0\hat{\mathbf{j}}\right)\ \text{N}$

$\mathbf{a} = \dfrac{\sum \mathbf{F}}{m} = \left(5.50\hat{\mathbf{i}} + 2.60\hat{\mathbf{j}}\right)\ \text{m/s}^{2} = 6.08\ \text{m/s}^{2}\ \text{at } 25.3°$ ◊

---

19. A bag of cement of weight F_g hangs from three wires as shown in Figure P5.18. Two of the wires make angles θ_1 and θ_2 with the horizontal. If the system is in equilibrium, show that the tension in the left-hand wire is

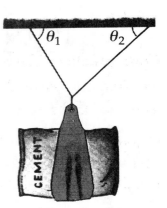

$$T_1 = \frac{F_g \cos\theta_2}{\sin(\theta_1 + \theta_2)}$$

Figure P5.18

Solution

We use the particle in equilibrium model. Draw a free-body diagram for the knot where the three ropes are joined. Choose the x axis to be horizontal and apply Newton's second law in component form.

$\Sigma F_x = 0:$ $T_2 \cos\theta_2 - T_1 \cos\theta_1 = 0$ **[1]**

$\Sigma F_y = 0:$ $T_2 \sin\theta_2 + T_1 \sin\theta_1 - F_g = 0$ **[2]**

Solve Equation [1] for $T_2 = \dfrac{T_1 \cos\theta_1}{\cos\theta_2}$

Substitute this expression for T_2 into Equation [2]:

$$\left(\frac{T_1 \cos\theta_1}{\cos\theta_2}\right)\sin\theta_2 + T_1 \sin\theta_1 = F_g$$

Solve for $T_1 = \dfrac{F_g \cos\theta_2}{\cos\theta_1 \sin\theta_2 + \sin\theta_1 \cos\theta_2}$

Use the trigonometric identity $\sin(\theta_1 + \theta_2) = \cos\theta_1 \sin\theta_2 + \sin\theta_1 \cos\theta_2$

to find $T_1 = \dfrac{F_g \cos\theta_2}{\sin(\theta_1 + \theta_2)}$ ◊

---

23. A 1.00-kg object is observed to have an acceleration of 10.0 m/s² in a direction 30.0° north of east (Fig. P5.23). The force $\mathbf{F}_2$ acting on the object has a magnitude of 5.00 N and is directed north. Determine the magnitude and direction of the force $\mathbf{F}_1$ acting on the mass.

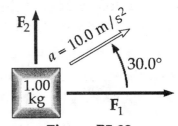

Figure P5.23

Solution

Conceptualize: The net force acting on the mass is $\Sigma F = ma = (1\ \text{kg})(10\ \text{m}/\text{s}^2) = 10\ \text{N}$. If we sketch a vector diagram of the forces drawn to scale ($F_2 = \Sigma F - F_1$), we see that $F_1 \approx 9\ \text{N}$, to the east.

Categorize: We can find a more precise result by examining the forces in terms of vector components. For convenience, we choose directions east and north along $\hat{i}$ and $\hat{j}$, respectively.

Analyze: $\quad \mathbf{a} = \left[(10.0\cos 30.0°)\hat{i} + (10.0\sin 30.0°)\hat{j}\right]\ \text{m}/\text{s}^2 = (8.66\hat{i} + 5.00\hat{j})\ \text{m}/\text{s}^2$

From Newton's second law,

$$\sum \mathbf{F} = m\mathbf{a} = (1.00\ \text{kg})(8.66\hat{i}\ \text{m}/\text{s}^2 + 5.00\hat{j}\ \text{m}/\text{s}^2) = (8.66\hat{i} + 5.00\hat{j})\ \text{N}$$

and $\qquad \sum \mathbf{F} = \mathbf{F}_1 + \mathbf{F}_2$

so $\qquad \mathbf{F}_1 = \sum \mathbf{F} - \mathbf{F}_2 = (8.66\hat{i} + 5.00\hat{j} - 5.00\hat{j})\ \text{N} = 8.66\hat{i}\ \text{N} = 8.66\ \text{N east} \qquad \Diamond$

Finalize: Our calculated answer agrees with the prediction from the force diagram.

---

25. A block is given an initial velocity of 5.00 m/s up a frictionless 20.0° incline (see Figure P5.22). How far up the incline does the block slide before coming to rest?

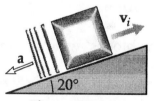

Figure P5.22

Solution Every successful physics student (this means you) learns to solve inclined-plane problems.

Hint one: Try to set one axis in the direction of motion. In this case, take the x axis along the incline, so that $a_y = 0$.

Hint two: Recognize that the 20.0° angle between the x axis and horizontal implies a 20.0° angle between the weight vector and the y axis. Why? Because "angles are equal if their sides are perpendicular, right side to right side and left side to left side." Either you learned this theorem in geometry class, or you learn it now, since it is a theorem used often in physics.

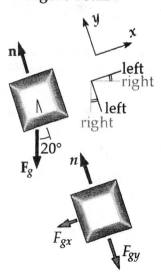

Hint three: The 20.0° angle lies between $m\mathbf{g}$ and the y axis, so split the weight vector into its x and y components:

$$mg_x = -mg\sin 20.0° \qquad\qquad mg_y = -mg\cos 20.0°$$

Now, Newton's law applies for **each** axis. Applying it to the x axis,

$\sum F_x = ma_x:$

$-mg \sin 20° = ma_x$

$a_x = -g \sin 20° = -\left(9.80 \text{ m} / \text{s}^2\right) \sin 20° = -3.35 \text{ m} / \text{s}^2$

From Equation 2.13,

$v_{xf}^2 = v_{xi}^2 + 2a_x\left(x_f - x_i\right)$

$0 = \left(5.00 \text{ m} / \text{s}\right)^2 + 2\left(-3.35 \text{ m} / \text{s}^2\right)\left(x_f - x_i\right)$

Solving,

$\left(x_f - x_i\right) = 3.73 \text{ m}$ ◊

---

31. In the system shown in Figure P5.31, a horizontal force F_x acts on the 8.00-kg object. The horizontal surface is frictionless. (a) For what values of F_x does the 2.00-kg object accelerate upward? (b) For what values of F_x is the tension in the cord zero? (c) Plot the acceleration of the 8.00-kg object versus F_x. Include values of F_x from -100 N to $+100$ N.

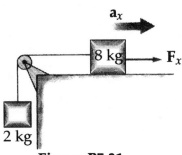

Figure P5.31

Solution

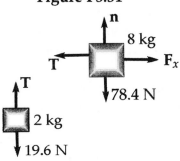

The blocks' weights are:

$F_{g1} = m_1 g = \left(8.00 \text{ kg}\right)\left(9.80 \text{ m} / \text{s}^2\right) = 78.4 \text{ N}$

$F_{g2} = m_2 g = \left(2.00 \text{ kg}\right)\left(9.80 \text{ m} / \text{s}^2\right) = 19.6 \text{ N}$

Let T be the tension in the connecting cord and draw a free-body diagram for each block.

(a) For the 2-kg mass, with the y axis directed upwards,

$\sum F_y = ma_y$ yields $T - 19.6 \text{ N} = \left(2.00 \text{ kg}\right)a_y$ [1]

Thus, we find that $a_y > 0$ when $T > 19.6 \text{ N}$.

For acceleration of the system of two blocks $F_x > T$,

so $F_x > 19.6 \text{ N}$ whenever the 2-kg mass accelerates upward. ◊

(b) Looking at the free-body diagram for the 8.00-kg mass, and taking the $+x$ direction to be directed to the right, we can apply Newton's law in the horizontal direction:

From $\sum F_x = ma_x$, $\qquad\qquad -T + F_x = (8.00 \text{ kg})a_x$ [2]

If $T = 0$, the cord goes slack; the 2-kg object is in free fall. The 8-kg object can have an acceleration to the left of larger magnitude:

$$a_x \leq -9.80 \text{ m} / \text{s}^2 \quad \text{with} \quad F_x \leq -78.4 \text{ N} \qquad \Diamond$$

(c) If $F_x \geq -78.4 \text{ N}$, then both Equations [1] and [2] apply. Substituting the value for T from Equation [1] into [2],

$$-(2.00 \text{ kg})a_y - 19.6 \text{ N} + F_x = (8.00 \text{ kg})a_x$$

In this case, $a_x = a_y$: $\qquad\qquad F_x = (8.00 \text{ kg} + 2.00 \text{ kg})a_x + 19.6 \text{ N}$

$$a_x = \frac{F_x}{10.0 \text{ kg}} - 1.96 \text{ m} / \text{s}^2 \qquad (F_x \geq -78.4 \text{ N}) \qquad [3]$$

From part (b), we find that if $F_x \leq -78.4 \text{ N}$, then $T = 0$ and Equation [2] becomes $F_x = (8.00 \text{ kg})a_x$:

$$a_x = F_x / 8.00 \text{ kg} \qquad (F_x \leq -78.4 \text{ N}) \qquad [4]$$

Observe that we have translated the pictorial representation into a simplified pictorial representation and then into a mathematical representation. We proceed to a tabular representation and a graphical representation of Equations (3) and (4):

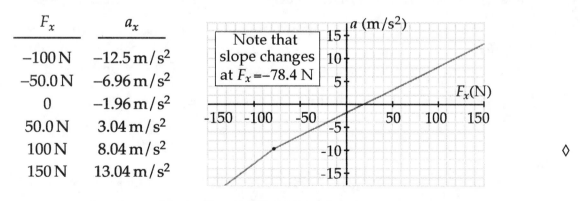

F_x	a_x
-100 N	$-12.5 \text{ m}/\text{s}^2$
-50.0 N	$-6.96 \text{ m}/\text{s}^2$
0	$-1.96 \text{ m}/\text{s}^2$
50.0 N	$3.04 \text{ m}/\text{s}^2$
100 N	$8.04 \text{ m}/\text{s}^2$
150 N	$13.04 \text{ m}/\text{s}^2$

Note that slope changes at $F_x = -78.4 \text{ N}$

$\Diamond$

33. A 72.0-kg man stands on a spring scale in an elevator. Starting from rest, the elevator ascends, attaining its maximum speed of 1.20 m/s in 0.800 s. It travels with this constant speed for the next 5.00 s. The elevator then undergoes a uniform acceleration in the negative y direction for 1.50 s and comes to rest. What does the spring scale register (a) before the elevator starts to move? (b) during the first 0.800 s? (c) while the elevator is traveling at constant speed? (d) during the time it is slowing down?

Solution

Conceptualize: Based on sensations experienced riding in an elevator, we expect that the man should feel slightly heavier when the elevator first starts to ascend, lighter when it comes to a stop, and his normal weight when the elevator is not accelerating. His apparent weight is registered by the spring scale beneath his feet, so the scale force should correspond to the force he feels through his legs (Newton's third law).

Categorize: We should draw free body diagrams for each part of the elevator trip and apply Newton's second law to find the scale force. The acceleration can be found from the change in speed divided by the elapsed time.

Analyze: Consider the free-body diagram of the man shown to the right. The force F is the upward force exerted on the man by the scale, and his weight is

$$F_g = mg = (72.0 \text{ kg})(9.80 \text{ m / s}^2) = 706 \text{ N}$$

With $+y$ defined to be upwards, Newton's 2nd law gives

$$\sum F_y = +F_s - F_g = ma$$

Thus, we calculate the upward scale force to be

$$F_s = 706 \text{ N} + (72.0 \text{ kg})a \qquad\qquad \text{[1]}$$

where a is the acceleration the man experiences as the elevator changes speed.

(a) Before the elevator starts moving, the elevator's acceleration is zero ($a = 0$). Therefore, Equation [1] gives the force exerted by the scale on the man as 706 N (upward), and the man exerts a downward force of 706 N on the scale. ◊

(b) During the first 0.800 s of motion, the man accelerates at a rate of

$$a = \frac{\Delta v}{\Delta t} = \frac{1.20 \text{ m / s} - 0}{0.800 \text{ s}} = 1.50 \text{ m / s}^2$$

Substituting a into Eq. [1] then gives: 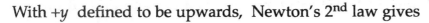 $F = 706 \text{ N} + (72.0 \text{ kg})(1.50 \text{ m / s}^2) = 814 \text{ N}$ ◊

(c) While the elevator is traveling upward at constant speed, the acceleration is zero and Equation [1] again gives a scale force $F = 706 \text{ N}$ ◊

(d) During the last 1.50 s, the elevator starts with an upward velocity of 1.20 m/s, and comes to rest with an acceleration of

$$a = \frac{\Delta v}{\Delta t} = \frac{0 - 1.20 \text{ m / s}}{1.50 \text{ s}} = -0.800 \text{ m / s}^2$$

Thus, the force of the man on the scale is:

$$F = 706 \text{ N} + (72.0 \text{ kg})(-0.800 \text{ m}/\text{s}^2) = 648 \text{ N}$$ ◊

Finalize: The calculated scale forces are consistent with our predictions. This problem could be extended to a couple of extreme cases. If the acceleration of the elevator were +9.80 m/s², then the man would feel twice as heavy, and if $a = -9.80$ m/s² (free fall), then he would feel "weightless", even though his true weight ($F_g = mg$) would remain the same.

---

41. A 3.00-kg block starts from rest at the top of a 30.0° incline and slides a distance of 2.00 m down the incline in 1.50 s. Find (a) the magnitude of the acceleration of the block, (b) the coefficient of kinetic friction between block and plane, (c) the friction force acting on the block, and (d) the speed of the block after it has slid 2.00 m.

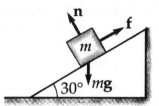

Solution We use the particle under constant acceleration and particle under net force models.

(a)   At constant acceleration, $x_f = v_i t + \dfrac{1}{2}at^2$

So, $$a = \frac{2(x_f - v_i t)}{t^2} = \frac{2(2.00 \text{ m} - 0 - 0)}{(1.50 \text{ s})^2} = 1.78 \text{ m}/\text{s}^2$$ ◊

From the acceleration, we can calculate the friction force, answer (c), next.

(c)   Choose the x axis parallel to the incline, take the positive direction down the incline (in the direction of the acceleration) and apply the second law.

$$\sum F_x = mg\sin\theta - f = ma: \quad f = m(g\sin\theta - a)$$

$$f = (3.00 \text{ kg})\left[(9.80 \text{ m}/\text{s}^2)\sin 30.0° - 1.78 \text{ m}/\text{s}^2\right] = 9.37 \text{ N} ◊$$

(b)   Applying Newton's law in the y direction (perpendicular to the incline),

$$\sum F_y = n - mg\cos\theta = 0: \quad n = mg\cos\theta$$

Because $f = \mu n$, $$\mu = \frac{f}{mg\cos\theta} = \frac{9.37 \text{ N}}{(3.00 \text{ kg})(9.80 \text{ m/s}^2)\cos 30.0°} = 0.368$$ ◊

(d) $v_f = v_i + at$, so $$v_f = 0 + (1.78 \text{ m}/\text{s}^2)(1.50 \text{ s}) = 2.67 \text{ m}/\text{s}$$ ◊

---

45. Two blocks connected by a rope of negligible mass are being dragged by a horizontal force **F** (Fig. P5.45). Suppose that $F = 68.0\,\text{N}$, $m_1 = 12.0\,\text{kg}$, $m_2 = 18.0\,\text{kg}$, and the coefficient of kinetic friction between each block and the surface is 0.100. (a) Draw a free-body diagram for each block. (b) Determine the tension, T, and the magnitude of the acceleration of the system.

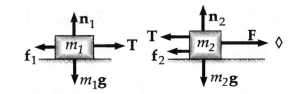

Figure P5.45

Solution

(a)   The free-body diagrams for m_1 and m_2 are:

(b)   Use the free-body diagrams to apply Newton's second law.

For m_1: $\sum F_x = T - f_1 = m_1 a$

or $T = m_1 a + f_1$ **[1]**

$\sum F_y = n_1 - m_1 g = 0$ or $n_1 = m_1 g$

Also, $f_1 = \mu_1 n_1 = (0.100)(12.0\ \text{kg})(9.80\ \text{m} / \text{s}^2) = 11.8\ \text{N}$

For m_2: $\sum F_x = F - T - f_2 = m_2 a$

or $T = F - m_2 a - f_2$ **[2]**

$\sum F_y = n_2 - m_2 g = 0$ or $n_2 = m_2 g$

Also, $f_2 = \mu n_2 = (0.100)(18.0\ \text{kg})(9.80\ \text{m} / \text{s}^2) = 17.6\ \text{N}$

Substituting T from Equation [1] into [2], we get $m_1 a + f_1 = F - m_2 a - f_2$

Solving for a, $a = \dfrac{F - f_1 - f_2}{m_1 + m_2} = \dfrac{(68.0 - 11.8 - 17.6)\ \text{N}}{(12.0 + 18.0)\ \text{kg}} = 1.29\ \text{m} / \text{s}^2$ ◊

From Eq. [1], $T = m_1 a + f_1 = (12.0\ \text{kg})(1.29\ \text{m} / \text{s}^2) + 11.8\ \text{N} = 27.2\ \text{N}$ ◊

51. An inventive child named Pat wants to reach an apple in a tree without climbing the tree. Sitting in a chair connected to a rope that passes over a frictionless pulley (Fig. P5.51), Pat pulls on the loose end of the rope with such a force that the spring scale reads 250 N. Pat's weight is 320 N, and the chair weighs 160 N. (a) Draw free-body diagrams for Pat and the chair considered as separate systems, and another diagram for Pat and the chair considered as one system. (b) Show that the acceleration of the system is upward and find its magnitude. (c) Find the force Pat exerts on the chair.

Solution

Figure P5.51

(a)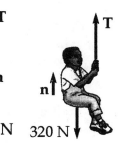

160 N   320 N   480 N

(b) First consider Pat and the chair as the system. Note that **two** ropes support the system, and $T = 250$ N in each rope.

Applying $\Sigma F = ma$, $2T - (160 \text{ N} + 320 \text{ N}) = ma$

where $m = \dfrac{480 \text{ N}}{9.80 \text{ m / s}^2} = 49.0 \text{ kg}$

Solving for a gives $a = \dfrac{(500 - 480) \text{ N}}{49.0 \text{ kg}} = 0.408 \text{ m / s}^2$

(c) On Pat, we apply $\Sigma F = ma$: $n + T - 320 \text{ N} = ma$

where $m = \dfrac{320 \text{ N}}{9.80 \text{ m / s}^2} = 32.7 \text{ kg}$

$n = ma + 320 \text{ N} - T$

$n = (32.7 \text{ kg})(0.408 \text{ m / s}^2) + 320 \text{ N} - 250 \text{ N}$

Therefore, $n = 83.3 \text{ N}$

55. An object of mass M is held in place by an applied force **F** and a pulley system as shown in Figure P5.55. The pulleys are massless and frictionless. Find (a) the tension in each section of rope, T_1, T_2, T_3, T_4, and T_5, and (b) the magnitude of **F**. (**Suggestion**: draw a free-body diagram for each pulley.)

Solution

Draw free-body diagrams and apply Newton's 2nd law. (All forces are along the y axis.)

For M,
$$\sum F = 0 = T_5 - Mg$$

$$T_5 = Mg$$

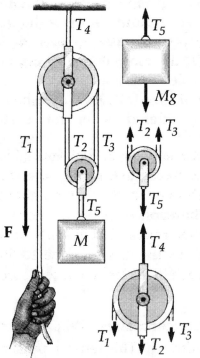

Figure P5.55

Assume frictionless pulleys. The tension is constant throughout a light, continuous rope.

Therefore,
$$T_1 = T_2 = T_3$$

For the bottom pulley
$$\sum F = 0 = T_2 + T_3 - T_5$$

So
$$2T_2 = T_5$$

and
$$T_1 = T_2 = T_3 = \frac{1}{2}Mg \qquad \Diamond$$

The applied force is
$$F = T_1 = \frac{1}{2}Mg \qquad \Diamond$$

For the top pulley,
$$\sum F = 0 = T_4 - T_1 - T_2 - T_3$$

Solving,
$$T_4 = T_1 + T_2 + T_3 = \frac{3}{2}Mg \qquad \Diamond$$

61. What horizontal force must be applied to the cart shown in Figure P5.61 so that the blocks remain stationary relative to the cart? Assume all surfaces, wheels, and pulley are frictionless. (**Hint:** Note that the force exerted by the string accelerates m_1.)

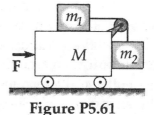

Figure P5.61

Solution Draw separate free-body diagrams for blocks m_1 and m_2.

Remembering that normal forces are always perpendicular to the contacting surface, and always **push** on a body, draw n_1 and n_2 as shown. Note that m_2 should be in **contact** with the cart, and therefore does have a normal force from the cart. Remembering that ropes always **pull** on bodies in the direction of the rope, draw the tension force **T**. Finally, draw the gravitational force on each block, which always points downwards.

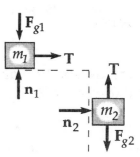

Conceptualize: What can keep m_2 from falling? Only tension in the cord connecting it with m_1. This tension pulls forward on m_1 to accelerate that mass. We might guess that the acceleration is proportional to both m_2 and g and inversely proportional to m_1, so perhaps $a = m_2 g / m_1$. If the entire system accelerates at this rate, then m_1 need not slide on M to achieve this acceleration. We should also expect the applied force to be proportional to the total mass of the system.

Categorize: Use $\Sigma F = ma$ and the free-body diagrams above.

Analyze: For m_2 $\qquad\qquad T - m_2 g = 0$ $\qquad$ or $\qquad T = m_2 g$

For m_1 $\qquad\qquad T = m_1 a$ $\qquad$ or $\qquad a = T / m_1$

Substituting for T, we have $\quad a = m_2 g / m_1$

For all 3 blocks $\qquad\qquad F = (M + m_1 + m_2)a$

Therefore $\qquad\qquad F = (M + m_1 + m_2)(m_2 g / m_1)$ $\qquad\qquad\qquad$ ◊

Finalize: Even though this problem did not have a numerical solution, we were still able to rationalize the algebraic form of the solution. This technique does not always work, especially for complex situations, but often we can think through a problem to see if an equation for the solution makes sense based on the physical principles we know.

69. A van accelerates down a hill (Fig. P5.69), going from rest to 30.0 m/s in 6.00 s. During the acceleration, a toy ($m = 0.100$ kg) hangs by a string from the van's ceiling. The acceleration is such that the string remains perpendicular to the ceiling. Determine (a) the angle θ and (b) the tension in the string.

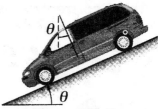

Figure P5.69

Solution   The acceleration is obtained from $v_f = v_i + at$:

$$30.0 \text{ m}/\text{s} = 0 + a(6.00 \text{ s})$$

$$a = 5.00 \text{ m}/\text{s}^2$$

The toy moves with the same acceleration as the van, $5.00 \text{ m}/\text{s}^2$ parallel to the hill.   We take the x axis in this direction, so

$$a_x = 5.00 \text{ m}/\text{s}^2 \quad \text{and} \quad a_y = 0$$

The only forces on the toy are the string tension in the y direction and its weight, as shown in the free-body diagram.

$$mg = (0.100 \text{ kg})(9.80 \text{ m}/\text{s}^2) = 0.980 \text{ N}$$

(a)   Using $\Sigma F_x = ma_x$: $(0.980 \text{ N})\sin\theta = (0.100 \text{ kg})(5.00 \text{ m}/\text{s}^2)$

$$\sin\theta = \frac{0.500}{0.980} \quad \text{and} \quad \theta = 30.7° \qquad \lozenge$$

(b)   Using $\Sigma F_y = ma_y$: $+T - (0.980 \text{ N})\cos\theta = 0$

$$T = (0.980 \text{ N})\cos 30.7° = 0.843 \text{ N} \qquad \lozenge$$

Chapter 6
CIRCULAR MOTION AND OTHER APPLICATIONS OF NEWTON'S LAWS

EQUATIONS AND CONCEPTS

The **net force** exerted on an object of mass m, moving uniformly in a circular path of radius r, is directed toward the center of the circle. This force causes the object's centripetal acceleration. *Equation 6.1 is a statement of Newton's second law along the radial direction. The F in the equation refers only to the radial components of the forces.*

$$\sum F = ma_c = m\frac{v^2}{r} \qquad (6.1)$$

A **resistive force R** will be exerted on an object moving through a medium (gas or liquid). The form of Equation 6.2 assumes that the resistive force is proportional to the speed of the object. *The constant b has a value that depends on the properties of the medium and the dimensions and shape of the object.*

$$\mathbf{R} = -b\mathbf{v} \qquad (6.2)$$

A **differential equation** is used to describe the motion of an object falling vertically in a viscous medium. *Equation 6.5 gives the speed as a function of time, when the object is released from rest at $t = 0$.*

$$\frac{dv}{dt} = g - \frac{b}{m}v \qquad (6.4)$$

$$v = \frac{mg}{b}\left(1 - e^{-bt/m}\right) = v_T\left(1 - e^{-t/\tau}\right) \qquad (6.5)$$

An object reaches **terminal speed** as the magnitude of the resistive force approaches the weight of the object.

$$v_T = \frac{mg}{b}$$

The **time constant** τ is the time at which an object, released from rest, will achieve a speed equal to 63.2% of the terminal speed.

$$\tau = \frac{m}{b}$$

The **Euler method** for solving a differential equation approximates derivatives as the ratio of finite differences. The manner in which the net force depends on position, velocity and time must be known. Equations 6.10 and 6.11 are used to calculate the velocity and position at the end of each of a series of small time increments. Following each calculation, the value of the calculated quantity is updated to be used in the next step. *The time increments must be sufficiently small that the acceleration can be considered constant during the increment.*

$$v(t + \Delta t) \approx v(t) + a(t)\Delta t \qquad (6.10)$$

$$x(t + \Delta t) \approx x(t) + v(t)\Delta t \qquad (6.11)$$

SUGGESTIONS, SKILLS, AND STRATEGIES

The following procedure is used in applying the analytical method to a dynamics problem in which the forces acting on a mass m are known.

1. Sum all of the forces acting on the particle to find the net force ΣF.

2. Use this net force to determine the acceleration from the relationship $a = \Sigma F / m$.

3. Use this acceleration to determine the velocity from the relationship $dv / dt = a$.

4. Use this velocity to determine the position from the relationship $dx / dt = v$.

Section 6.4 deals with the motion of an object through a viscous medium. The method of solution for Equation 6.4 (when the resistive force $\mathbf{R} = -b\mathbf{v}$) is shown below:

$$\frac{dv}{dt} = g - \frac{b}{m}v \qquad (6.4)$$

In order to solve this equation, it is convenient to change variables. If we let $y = g - (b/m)v$, it follows that $dy = -(b/m)dv$. With these substitutions, Equation 6.4 becomes

$$-\left(\frac{m}{b}\right)\frac{dy}{dt} = y \qquad \text{or} \qquad \frac{dy}{y} = -\frac{b}{m}dt$$

Integrating this expression (now that the variables are separated) gives

$$\int \frac{dy}{y} = -\frac{b}{m} \int dt \qquad \text{or} \qquad \ln y = -\frac{b}{m}t + \text{constant}$$

This is equivalent to

$$y = (\text{const}) \, e^{-bt/m} = g - (b/m)v \, .$$

Taking $v = 0$ and $t = 0$, we see that $\text{const} = g$.

so

$$v = \frac{mg}{b}(1 - e^{-bt/m}) = v_t(1 - e^{-t/\tau}) \qquad (6.5)$$

where $\tau = m/b$

REVIEW CHECKLIST

You should be able to:

▷ Apply Newton's second law to uniform and nonuniform circular motion. (Sections 6.1 and 6.2)

▷ Identify situations in which an observer is in an accelerating reference frame and explain the apparent violation of Newton's laws of motion in terms of "fictitious" forces acting on an object. (Section 6.3)

▷ Calculate the acceleration of an object moving in a medium in which there is a resistive force exerted on the object. Use the equations describing motion in a viscous medium to determine the speed of an object as a function of time and the value of the terminal speed. (Section 6.4)

▷ Use the procedure outlined in Suggestions, Skills and Strategies (and in Section 6.5 of the text) to apply the analytical method to a dynamics problem. (Section 6.5)

▷ Use Euler's method of numerical analysis to find speed and position at successive time intervals for an object falling in a resistive medium.

ANSWERS TO SELECTED CONCEPTUAL QUESTIONS

10. It has been suggested that rotating cylinders about 10 mi in length and 5 mi in diameter be placed in space and used as colonies. The purpose of the rotation is to simulate gravity for the inhabitants. Explain this concept for producing an effective imitation of gravity.

Answer   The centripetal force on the inhabitants is provided by the normal force exerted on them by the cylinder wall. If the rotation rate is adjusted to such a speed that this normal force is equal to their weight on Earth, then this artificial gravity would seem to the inhabitants to be the same as normal gravity.

□   □   □   □

14.   Why does a pilot tend to black out when pulling out of a steep dive?

Answer   When pulling out of a dive, blood leaves the pilot's head because the pilot's blood pressure is not great enough to compensate for both the gravitational force and the centripetal acceleration of the airplane. This loss of blood from the brain can cause the pilot to black out.

□   □   □   □

17.   A falling sky diver reaches terminal speed with her parachute closed. After the parachute is opened, what parameters change to decrease this terminal speed?

Answer   From the expression for the force of air resistance and Newton's law, we derive the equation that governs the motion of the skydiver:

$$m\frac{dv_y}{dt} = mg - \frac{D\rho A}{2}v_y^2$$

where D is the coefficient of drag of the parachutist, and A is the area of the parachutist's body. At terminal speed,

$$a_y = dv_y/dt = 0 \quad \text{and} \quad v_T = \sqrt{2mg/D\rho A}$$

When the parachute opens, the coefficient of drag D and the effective area A both increase, thus reducing the velocity of the skydiver.

Modern parachutes also add a third term, lift, to change the equation to

$$m\frac{dv_y}{dt} = mg - \frac{D\rho A}{2}v_y^2 - \frac{L\rho A}{2}v_x^2$$

where v_y is the vertical velocity, and v_x is the horizontal velocity. This lift is best seen in the "Paraplane," an ultralight airplane made from a fan, a chair, and a parachute.

□   □   □   □

SOLUTIONS TO SELECTED END-OF-CHAPTER PROBLEMS

1. A light string can support a stationary hanging load of 25.0 kg before breaking. A 3.00-kg object attached to the string rotates on a horizontal, frictionless table in a circle of radius 0.800 m, while the other end of the string is held fixed. What range of speeds can the object have before the string breaks?

Solution We use the models of a particle under a net force and a particle in uniform circular motion. The string will break if the tension T exceeds

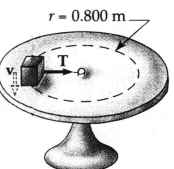

$r = 0.800$ m

$$T_{max} = mg = (25.0 \text{ kg})(9.80 \text{ m} / \text{s}^2) = 245 \text{ N}$$

As the 3.00-kg mass rotates in a horizontal circle, the tension provides the centripetal acceleration:

$$a = v^2/r$$

From $\Sigma F = ma$,

$$T = mv^2/r$$

Therefore,

$$v^2 = \frac{rT}{m} = \frac{(0.800 \text{ m})T}{3.00 \text{ kg}} \le \left(\frac{0.800 \text{ m}}{3.00 \text{ kg}} \right) T_{max}$$

Substituting $T_{max} = 245$ N, we find

$$v^2 \le 65.3 \text{ m}^2/\text{s}^2$$

and

$$0 < v < 8.08 \text{ m} / \text{s} \qquad \Diamond$$

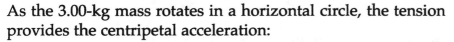

5. A coin placed 30.0 cm from the center of a rotating, horizontal turntable slips when its speed is 50.0 cm/s. (a) What force causes the centripetal acceleration when the coin is stationary relative to the turntable? (b) What is the coefficient of static friction between coin and turntable?

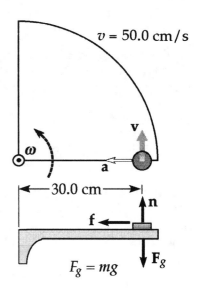

$v = 50.0$ cm/s

Solution

(a) The central force is provided by the force of static friction.

(b) The motion of the coin is shown in the upper diagram. The forces on the coin, shown in the free-body diagram below, are the normal force, the weight, and the force of static friction. The only force in the radial direction is **f**.

Therefore, $$f = m\frac{v^2}{r}$$

The normal force balances the weight, so $n - mg = 0$: $\quad n = mg$

The frictional force, $$f = \mu_s n = \mu_s mg$$

The coin slips when the frictional force is exactly enough to cause the centripetal acceleration.

$$\mu_s mg = mv^2/r$$

$$v^2/rg = \mu_s$$

Taking $r = 30$ cm, $v = 50$ cm/s, and $g = 980$ cm/s^2,

$$\mu_s = \frac{(50.0 \text{ cm/s})^2}{(30.0 \text{ cm})(980 \text{ cm/s}^2)} = 0.0850 \qquad \Diamond$$

---

7. A crate of eggs is located in the middle of the flat bed of a pickup truck as the truck negotiates an unbanked curve in the road. The curve may be regarded as an arc of a circle of radius 35.0 m. If the coefficient of static friction between crate and truck is 0.600, how fast can the truck be moving without the crate sliding?

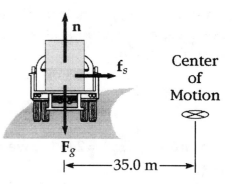

Solution

Call the mass of the egg crate m. The forces on it are its weight $\mathbf{F}_g = m\mathbf{g}$ vertically down, the normal force $\mathbf{n}$ of the truck bed vertically up, and static friction $\mathbf{f}_s$ directed to oppose relative sliding motion of the crate over the truck bed. The friction force is directed radially inward. It is the only horizontal force on the crate, so it must provide the centripetal acceleration. When the truck has maximum speed, friction f_s will have its maximum value with $f_s = \mu_s n$.

$\sum F_y = ma_y$ gives $\qquad n - mg = 0 \qquad$ or $\qquad n = mg$

$\sum F_x = ma_x$ gives $\qquad f_s = ma_r$

From these two equations, $\qquad \mu_s n = \frac{mv^2}{r} \qquad$ and $\qquad \mu_s mg = \frac{mv^2}{r}$

The mass divides out, leaving $\quad v = \sqrt{\mu_s gr} = \sqrt{(0.600)(9.80 \text{ m/s}^2)(35.0 \text{ m})} = 14.3 \text{ m/s} \, \Diamond$

---

13. A 40.0-kg child swings in a swing supported by two chains, each 3.00 m long. If the tension in each chain at the lowest point is 350 N, find (a) the child's speed at the lowest point and (b) the force exerted by the seat on the child at the lowest point. (Neglect the mass of the seat.)

Solution

Conceptualize: If the tension in each chain is 350 N at the lowest point, then the force of the seat on the child should be twice this force or 700 N. The child's speed is not as easy to determine, but somewhere between 0 and 10 m/s would be reasonable for the situation described.

Categorize: We should first draw a free body diagram that shows the forces acting on the seat and apply Newton's laws to solve the problem.

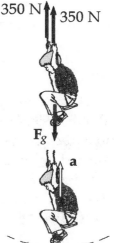

Analyze: We can see from the diagram that the only forces acting on the system of child + seat are the tension in the two chains and the weight of the boy:

$$\sum F = 2T - mg = ma$$

$$\sum F = F_{net} = 2(350 \text{ N}) - (40.0 \text{ kg})(9.80 \text{ m}/\text{s}^2) = 308 \text{ N (up)}$$

Because $a = v^2/r$ is the centripetal acceleration,

$$v = \sqrt{\frac{F_{net}r}{m}} = \sqrt{\frac{(308 \text{ N})(3.00 \text{ m})}{40.0 \text{ kg}}} = 4.81 \text{ m}/\text{s} \qquad \lozenge$$

The child feels a normal force exerted by the seat equal to the total tension in the chains:

$$\mathbf{n} = 2(350 \text{ N}) = 700 \text{ N} \quad \text{(upwards)} \qquad \lozenge$$

Finalize: Our answers agree with our predictions. It may seem strange that there is a net upward force on the boy yet he does not move upwards. We must remember that a net force causes an acceleration, but not necessarily a motion in the direction of the force. In this case, the acceleration is due to a change in the direction of the motion. It is also interesting to note that the boy feels about twice as heavy as normal, so his experience is equivalent to an acceleration of about 2 g's.

15. Tarzan ($m = 85.0$ kg) tries to cross a river by swinging from a vine. The vine is 10.0 m long, and his speed at the bottom of the swing (as he just clears the water) is 8.00 m/s. Tarzan doesn't know that the vine has a breaking strength of 1 000 N. Does he make it safely across the river?

Solution The forces acting on Tarzan are the force of gravity $m\mathbf{g}$ and the force from the rope, **T**. At the lowest point in his motion, **T** is upward and $m\mathbf{g}$ is downward as in the free-body diagram. Thus, Newton's second law gives

$$T - mg = \frac{mv^2}{r}$$

1011-7-23-50
© 1950, Edgar Rice Burroughs, Inc. –
™ Reg. U.S. Pat. Off.,
Distributed by United Feature Syndicate, Inc.

Solving for T, with $v = 8.00$ m/s, $r = 10.0$ m, and $m = 85.0$ kg,

gives $\qquad T = m\left(g + \frac{v^2}{r}\right) = (85.0 \text{ kg})\left(9.80 \text{ m/s}^2 + \frac{(8.00 \text{ m/s})^2}{10.0 \text{ m}}\right) = 1.38 \times 10^3$ N

Since T **exceeds** the breaking strength of the vine (1 000 N), Tarzan **doesn't make it!** ◊
The vine breaks before he reaches the bottom of the swing.

================

17. A pail of water is rotated in a vertical circle of radius 1.00 m. What is the minimum speed of the pail when it is upside down at the top of the circle if no water is to spill out?

Solution The normal force, **n**, will maintain exactly enough force to prevent the water from going **through** the bottom of the bucket. If water were to spill out, the force of gravity would exceed that required to provide the centripetal acceleration, and the normal force would be zero:

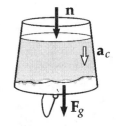

$$ma_c < mg \qquad \text{or} \qquad \frac{mv^2}{r} < mg$$

Since that is the only case in which water would spill out, all other cases must result in no water spilling out.

That is, $\qquad \dfrac{mv^2}{r} \geq mg \qquad$ or $\qquad v^2 \geq rg$

At the minimum speed, we have $\qquad v_{\min} = \sqrt{rg} = \sqrt{(1.00 \text{ m})(9.80 \text{ m/s}^2)} = 3.13$ m/s ◊

================

23. A 0.500-kg object is suspended from the ceiling of an accelerating boxcar as in Figure 6.13. If $a = 3.00\,\text{m/s}^2$, find (a) the angle that the string makes with the vertical and (b) the tension in the string.

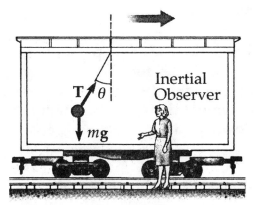

Figure 6.13

Solution

Conceptualize: If the horizontal acceleration were zero, then the angle would be 0°, and if $a = g$, then the angle would be 45°, but since the acceleration is $3.00\,\text{m/s}^2$, a reasonable estimate of the angle is about 20°. Similarly, the tension in the string should be slightly more than the weight of the object, which is about 5 N.

Categorize: We will apply Newton's second law to solve the problem.

Analyze: The only forces acting on the suspended object are the force of gravity $\mathbf{F}_g = m\mathbf{g}$ and the force of tension $\mathbf{T}$, as shown in the free-body diagram. Applying Newton's second law in the x and y directions,

$$\sum F_x = T\sin\theta = ma \qquad\qquad\qquad \textbf{[1]}$$

$$\sum F_y = T\cos\theta - mg = 0 \qquad \text{or} \qquad T\cos\theta = mg \qquad\qquad \textbf{[2]}$$

(a) Dividing Equation [1] by [2], $\qquad \tan\theta = \dfrac{a}{g} = \dfrac{3.00\ \text{m}/\text{s}^2}{9.80\ \text{m}/\text{s}^2} = 0.306$

   Solving for θ, $\qquad\qquad\qquad\qquad \theta = 17.0°$ ◊

(b) From Equation [1], $\qquad\qquad T = \dfrac{ma}{\sin\theta} = \dfrac{(0.500\ \text{kg})(3.00\ \text{m}/\text{s}^2)}{\sin(17.0°)} = 5.12\ \text{N}$ ◊

Finalize: Our answers agree with our original estimates. We used the same Equations [1] and [2] as in Example 6.8. We did not need the idea of a fictitious force.

---

25. A person stands on a scale in an elevator. As the elevator starts, the scale has a constant reading of 591 N. As the elevator later stops, the scale reading is 391 N. Assume the magnitude of the acceleration is the same during starting and stopping, and determine (a) the weight of the person, (b) the person's mass, and (c) the acceleration of the elevator.

Solution

The scale reads the upward normal force exerted by the floor on the passenger. The maximum force occurs during upward acceleration (when starting an upward trip or ending a downward trip). The minimum normal force occurs with downward acceleration. For each respective situation,

$\Sigma F_y = ma_y$ starting $+591\text{ N} - mg = +ma$

stopping $+391\text{ N} - mg = -ma$

(a)   These two simultaneous equations can be added to eliminate a and solve for mg.

$$+591\text{ N} - mg + 391\text{ N} - mg = 0$$

or $$982\text{ N} - 2mg = 0$$

$$F_g = mg = \frac{982\text{ N}}{2} = 491\text{ N} \qquad \Diamond$$

(b) $$m = \frac{F_g}{g} = \frac{491\text{ N}}{9.80\text{ m / s}^2} = 50.1\text{ kg} \qquad \Diamond$$

(c)   Substituting back gives $+591\text{ N} - 491\text{ N} = (50.1\text{ kg})a$

$$a = \frac{100\text{ N}}{50.1\text{ kg}} = 2.00\text{ m / s}^2 \qquad \Diamond$$

---

37.   A small, spherical bead of mass 3.00 g is released from rest at $t = 0$ in a bottle of liquid shampoo. The terminal speed is observed to be $v_T = 2.00\text{ cm/s}$. Find (a) the value of the constant b in Equation 6.4, (b) the time τ at which the bead reaches $0.630v_T$, and (c) the value of the resistive force when the bead reaches terminal speed.

Solution

(a)   The speed v varies with time according to Equation 6.5,

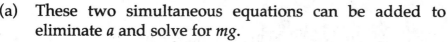

$$v = \frac{mg}{b}\left(1 - e^{-bt/m}\right) = v_T\left(1 - e^{-bt/m}\right)$$

where    $v_T = mg/b$ is the terminal speed.

Hence, $$b = \frac{mg}{v_T} = \frac{\left(3.00 \times 10^{-3}\text{ kg}\right)\left(9.80\text{ m / s}^2\right)}{2.00 \times 10^{-2}\text{ m / s}} = 1.47\text{ N} \cdot \text{s / m} \qquad \Diamond$$

88

(b) To find the time for v to reach $0.630v_T$, we substitute $v = 0.630v_T$ into Equation 6.5, giving

$$0.630v_T = v_T\left(1 - e^{-bt/m}\right) \quad \text{or} \quad 0.370 = e^{-(1.47t/0.00300)}$$

Solve for t by taking the natural logarithm of each side of the equation:

$$\ln(0.370) = -\frac{1.47t}{3.00 \times 10^{-3}} \quad \text{or} \quad t = 2.03 \times 10^{-3} \text{ s} \qquad \Diamond$$

(c) At terminal speed, $R = v_T b = mg$

Therefore, $\qquad R = \left(3.00 \times 10^{-3} \text{ kg}\right)\left(9.80 \text{ m}/\text{s}^2\right) = 2.94 \times 10^{-2} \text{ N} \qquad \Diamond$

---

39. A motor boat cuts its engine when its speed is 10.0 m/s and coasts to rest. The equation describing the motion of the motorboat during this period is $v = v_i e^{-ct}$, where v is the speed at time t, v_i is the initial speed, and c is a constant. At $t = 20.0$ s, the speed is 5.00 m/s. (a) Find the constant c. (b) What is the speed at $t = 40.0$ s? (c) Differentiate the expression for $v(t)$ and thus show that the acceleration of the boat is proportional to the speed at any time.

Solution

(a) We must fit the equation $v = v_i e^{-ct}$ to the two data points:

At $t = 0$, $v = 10.0 \text{ m}/\text{s}$: $\qquad v = v_i e^{-ct}$

$$10.0 \text{ m}/\text{s} = v_i e^0 = (v_i)(1)$$

$$v_i = 10.0 \text{ m}/\text{s}$$

At $t = 20.0$ s, $v = 5.00 \text{ m}/\text{s}$: $\quad 5.00 \text{ m}/\text{s} = (10.0 \text{ m}/\text{s})e^{-c(20.0 \text{ s})}$

$$0.500 = e^{-c(20.0 \text{ s})}$$

$$\ln(0.500) = (-c)(20.0 \text{ s})$$

$$c = \frac{-\ln(0.500)}{20.0 \text{ s}} = 0.0347 \text{ s}^{-1} \qquad \Diamond$$

(b) At all times, $\qquad v = (10.0 \text{ m}/\text{s})e^{-(0.0347)t}$

At $t = 40.0$ s, $\qquad v = (10.0 \text{ m}/\text{s})e^{-(0.0347)(40.0)} = 2.50 \text{ m}/\text{s} \qquad \Diamond$

(c) The acceleration is the rate of change of the velocity:

$$a = \frac{dv}{dt} = \frac{d}{dt} v_i e^{-ct} = v_i \left(e^{-ct} \right)(-c) = -c \left(v_i e^{-ct} \right) = -cv = \left(-0.0347 \text{ s}^{-1} \right) v$$

Thus, the acceleration is a negative constant times the speed. ◊

---

45. A hailstone of mass 4.80×10^{-4} kg falls through the air and experiences a net force given by $F = -mg + Cv^2$ where $C = 2.50 \times 10^{-5}$ kg/m. (a) Calculate the terminal speed of the hailstone. (b) Use Euler's method of numerical analysis to find the speed and position of the hailstone at 0.2-s intervals, taking the initial speed to be zero. Continue the calculation until the hailstone reaches 99% of terminal speed.

Solution

(a) At terminal speed the acceleration and total force are zero: $-mg + Cv^2 = 0$

We choose the negative root, because the velocity is downward.

$$v = \pm \sqrt{\frac{mg}{C}} = \pm \sqrt{\frac{\left(4.80 \times 10^{-4} \text{ kg} \right)\left(9.80 \text{ m/s}^2 \right)}{2.50 \times 10^{-5} \text{ kg/m}}} = -13.7 \text{ m/s}$$ ◊

(b) At $t = 0$, we take $x = 0$. The hailstone starts from rest, so $v = 0$; we then cycle through the following set of equations:

$$a = \frac{\sum F}{m} = \frac{-mg + Cv^2}{m} = -g + \frac{Cv^2}{m}$$ [1]

$$x_{new} = x_{old} + v_{old}\Delta t$$ [2]

$$v_{new} = v_{old} + a_{old}\Delta t$$ [3]

and we loop back to Equation [1] to find the next acceleration. The values we find are listed below, first for every step, and then (after 2.0 s), for every fifth step:

t (s)	0.00	0.200	0.400	0.600	0.800	1.00	1.20
a (m/s^2)	−9.80	−9.60	−9.02	−8.12	−7.02	−5.85	−4.72
x (m)	0.00	0.00	−0.392	−1.17	−2.30	−3.77	−5.51
v (m/s)	0.00	−1.96	−3.88	−5.68	−7.31	−8.71	−9.88

t (s)	1.40	1.60	1.80	2.00	3.00	4.00	5.00
a (m/s^2)	−3.70	−2.84	−2.14	−1.59	−0.321	−0.0606	−0.0113
x (m)	−7.48	−9.65	−12.0	−14.4	−27.4	−41.0	−54.7
v (m/s)	−10.8	−11.6	−12.1	−12.6	−13.5	−13.7	−13.7

The hailstone never attains terminal speed exactly, but it passes 99% of it at 3.4 seconds, and 99.99% at 6.2 seconds. ◊

---

57. Because the Earth rotates about its axis, a point on the equator experiences a centripetal acceleration of $0.0337\ \text{m/s}^2$, while a point at the poles experiences no centripetal acceleration. (a) Show that at the equator the gravitational force on an object must exceed the normal force required to support the object. (b) What is the apparent weight at the equator and at the poles of a person having a mass of 75.0 kg? (Assume the Earth is a uniform sphere and take $g = 9.800\ \text{m/s}^2$.)

Solution

Conceptualize: Since the centripetal acceleration is a small fraction (~0.3%) of g, we should expect that a person would have an apparent weight that is just slightly less at the equator than at the poles due to the rotation of the Earth.

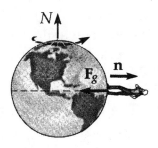

Categorize: We will apply Newton's second law and the equation for centripetal acceleration.

Analyze:

(a) Let **n** represent the force exerted on the person by a scale, which is the "apparent weight." The true weight is mg. Summing up forces on the object in the direction towards the Earth's center gives

$$mg - n = m\,a_c \qquad\qquad \textbf{[1]}$$

where
$$a_c = v^2/R_e = 0.033\ 7\ \text{m/s}^2$$

is the centripetal acceleration directed toward the center of the Earth.

Thus, we see that $\quad n = m(g - a_c) < mg$

or $\qquad\qquad\qquad mg = n + m\,a_c > n \qquad\qquad \textbf{[2]}\ \Diamond$

(b) If $\qquad\qquad m = 75.0\ \text{kg}\quad$ and $\quad g = 9.800\ \text{m/s}^2,$

at the Equator: $\quad n = m(g - a_c) = (75.0\ \text{kg})(9.800\ \text{m/s}^2 - 0.033\ 7\ \text{m/s}^2)$

$$n = 732\ \text{N} \qquad\qquad\qquad\qquad\qquad\qquad \Diamond$$

at the Poles, $a_c = 0$: $n = mg = (75.0\ \text{kg})(9.80\ \text{m/s}^2) = 735\ \text{N} \qquad\qquad \Diamond$

Finalize: As we expected, the person does appear to weigh about 0.3% less at the equator than the poles. We might extend this problem to consider the effect of the earth's bulge on a person's weight. Since the earth is fatter at the equator than the poles, would you expect g to be less than $9.80\ \text{m/s}^2$ at the equator and slightly more at the poles?

65. An amusement park ride consists of a large vertical cylinder that spins about its axis fast enough such that any person inside is held up against the wall when the floor drops away (Fig. P6.65). The coefficient of static friction between a person and the wall is μ_s, and the radius of the cylinder is R. (a) Show that the maximum period of revolution necessary to keep the person from falling is $T=(4\pi^2 R\mu_s/g)^{1/2}$. (b) Obtain a numerical value for T if $R=4.00\,\text{m}$ and $\mu_s=0.400$. How many revolutions per minute does the cylinder make?

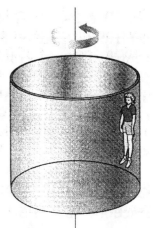

Solution

We model the person as a particle in uniform circular motion.

Figure P6.65

(a) The wall's normal force pushes inward:

$$n = \frac{mv^2}{R} = \frac{m}{R}\left(\frac{2\pi R}{T}\right)^2 = \frac{4\pi^2 Rm}{T^2}$$

The friction and weight balance: $f_s = \mu_s n = mg$

Therefore, with $\mu_s n = mg$,

$$\mu_s n = \mu_s \frac{4\pi^2 Rm}{T^2} = mg$$

Solving, $T^2 = \dfrac{4\pi^2 R\mu_s}{g}$ gives $T = \sqrt{\dfrac{4\pi^2 R\mu_s}{g}}$

(b)

$$T = \left(\frac{4\pi^2 (4.00\ \text{m})(0.400)}{9.80\ \text{m}/\text{s}^2}\right)^{1/2} = 2.54\ \text{s}$$

The angular speed is

$$\left(\frac{1\ \text{rev}}{2.54\ \text{s}}\right)\left(\frac{60\ \text{s}}{\text{min}}\right) = 23.6\ \text{rev}/\text{min} \qquad \diamond$$

Related Questions and Answers Why is the normal force horizontally inward? Because the wall is vertical, and on the outside.

Why is there no upward normal force? Because there is no floor.

Why is the frictional force directed upward? The friction opposes the possible relative motion of the person sliding down the wall.

Why is it not kinetic friction? Because person and wall are moving together, stationary with respect to each other.

Why is there no outward force on her? No other object pushes out on her. She pushes out on the wall as the wall pushes inward on her.

71. A model airplane of mass 0.750 kg flies in a horizontal circle at the end of a 60.0-m control wire, with a speed of 35.0 m/s. Compute the tension in the wire if it makes a constant angle of 20.0° with the horizontal. The forces exerted on the airplane are the pull of the control wire, its own weight, and aerodynamic lift, which acts at 20.0° inward from the vertical as shown in Figure P6.71.

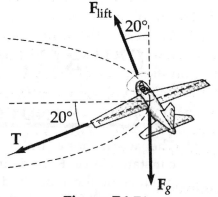

Figure P6.71

Solution The plane's acceleration is toward the center of the circle of motion, so it is horizontal.

The radius of the circle of motion is $(60.0 \text{ m})\cos 20.0° = 56.4 \text{ m}$

and the acceleration is $$a_c = \frac{v^2}{r} = \frac{(35 \text{ m / s})^2}{56.4 \text{ m}} = 21.7 \text{ m / s}^2$$

We can also calculate the weight of the airplane:

$$F_g = mg = (0.750 \text{ kg})(9.80 \text{ m / s}^2) = 7.35 \text{ N}$$

We define our axes for convenience. In this case, two of the forces — one of them our force of interest — are directed along the 20.0° lines.

We define the x axis to be directed in the (+**T**) direction, and the y axis to be directed in the direction of lift. With these definitions, the x component of the centripetal acceleration is

$$a_{cx} = a_c \cos(20°)$$

and $\Sigma F_x = m a_x$ yields $\qquad T + F_g \sin 20.0° = m a_{cx}$

Solving for T, $\qquad T = m a_{cx} - F_g \sin 20.0°$

$$T = (0.750 \text{ kg})(21.7 \text{ m / s}^2)\cos 20.0° - (7.35 \text{ N})\sin 20.0°$$

and $\qquad T = (15.3 \text{ N}) - (2.51 \text{ N}) = 12.8 \text{ N}$ ◊

Chapter 7
ENERGY AND ENERGY TRANSFER

EQUATIONS AND CONCEPTS

The **work done on a system by a constant force F** is defined to be the product of the magnitude of the force, magnitude of the displacement of the point of application of the force, and the cosine of the angle between the force and displacement vectors. *Work is a scalar quantity and represents a transfer of energy. The SI unit of work is the N·m. One N·m = 1 joule (J).*

$$W \equiv F\Delta r \cos\theta \tag{7.1}$$

The **scalar product (or dot product)** of any two vectors **A** and **B** is defined to be a scalar quantity whose magnitude is equal to the product of the magnitudes of the two vectors and the cosine of the angle between the directions of the two vectors.

$$\mathbf{A} \cdot \mathbf{B} \equiv AB \cos\theta \tag{7.2}$$

Work as a scalar product is shown in Equation 7.3. The force must be constant in magnitude and direction and $\mathbf{F} \cdot \Delta\mathbf{r}$ is read "F dot Δr".

$$W = F\Delta r \cos\theta = \mathbf{F} \cdot \Delta\mathbf{r} \tag{7.3}$$

The figure at right shows an object on a smooth incline and acted on by four forces. Illustrated below are cases for four different values of θ. Note that work done by a given force can be positive, negative, or zero depending on the value of θ.

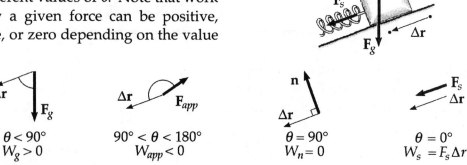

$$\begin{array}{cccc}
\theta < 90° & 90° < \theta < 180° & \theta = 90° & \theta = 0° \\
W_g > 0 & W_{app} < 0 & W_n = 0 & W_s = F_s \Delta r
\end{array}$$

94

The **scalar product of two vectors A and B** can be expressed in terms of the x, y and z components of the two vectors.

$$\mathbf{A} \cdot \mathbf{B} = A_x B_x + A_y B_y + A_z B_z \qquad (7.6)$$

The **work done by a variable force** on a particle which has been displaced along the x axis from x_i to x_f, is given by an integral expression. *Graphically, the work done equals the area under the F_x versus x curve.*

$$W = \int_{x_i}^{x_f} F_x \, dx \qquad (7.7)$$

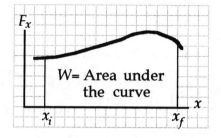

Hooke's law expresses the force exerted by a spring which is stretched or compressed from the equilibrium position. *The force exerted by the spring is always directed opposite the displacement from equilibrium.*

$$F_s = -kx \qquad (7.9)$$

The **work done by a spring force** $(-kx)$ as an object undergoes an arbitrary displacement from x_i to x_f is given by Equation 7.11. *Note: The work done **on a spring** by an applied force (external agent) while stretching a spring from x_i to x_f is equal to the negative of the work done **by the spring** between the same two initial and final points.*

$$W_s = \int_{x_i}^{x_f} (-kx) \, dx = \frac{1}{2}kx_i{}^2 - \frac{1}{2}kx_f{}^2 \qquad (7.11)$$

Kinetic energy, K, is energy associated with the motion of an object. *Kinetic energy is a scalar quantity and has the same units as work.*

$$K \equiv \frac{1}{2}mv^2 \qquad (7.15)$$

The **work-kinetic energy theorem** states that the work done by the net force on a particle equals the change in kinetic energy of the particle. If ΣW is positive, the kinetic energy and speed increase; if ΣW is negative, the kinetic energy and speed decrease. *The work-kinetic energy theorem is valid for a particle or for a system that can be modeled as a particle.*

$$\sum W = K_f - K_i = \Delta K \qquad (7.16)$$

The **principle of conservation of energy** is described mathematically by Equation 7.17. E represents the total energy of a system and T is the quantity of energy transferred across the system boundary by any transfer mechanism.

$$\Delta E_{\text{system}} = \sum T \qquad (7.17)$$

A **force of friction** results in a decrease in kinetic energy as an object slides across a surface.

$$-f_k d = \Delta K \qquad (7.20)$$

The **internal energy of a system** will change as a result of transfer of energy across the system boundary by frictional forces. *The increase in internal energy is equal to the decrease in kinetic energy.*

$$\Delta E_{\text{int}} = f_k d \qquad (7.22)$$

The **average power** supplied by a force is the ratio of the work done by that force to the time interval over which it acts.

$$\overline{\mathcal{P}} \equiv \frac{W}{\Delta t}$$

The **instantaneous power** is equal to the limit of the average power as the time interval approaches zero.

$$\mathcal{P} = \frac{dW}{dt} = \mathbf{F} \cdot \mathbf{v} \qquad (7.23)$$

The **general expression for power** defines power as any type of energy transfer. The rate at which energy crosses the boundary of a system by any transfer mechanism is dE/dt.

$$\mathcal{P} = \frac{dE}{dt} \qquad (7.24)$$

The **SI unit of power** is J/s, which is called a watt (W).

$$1\,W = 1\,J/s$$
$$1\,kW \cdot h = 3.60 \times 10^6 \, J$$

The **unit of power in the US customary system** is the horsepower (hp).

$$1\,hp = 746\,W$$

SUGGESTIONS, SKILLS, AND STRATEGIES

In order to apply the system model to problem solving, you must be able to identify the particular system boundary (an imaginary surface) that divides the Universe into two regions; the system of interest and the environment surrounding the system. A valid system may be a single object or particle, a collection of objects or particles, a region of space and may vary in size and shape.

The scalar (or dot) product is introduced as a new mathematical skill in this chapter.

$$\mathbf{A} \cdot \mathbf{B} = AB\cos\theta \qquad \text{where } \theta \text{ is the angle between } \mathbf{A} \text{ and } \mathbf{B}.$$

Also, $$\mathbf{A} \cdot \mathbf{B} = A_x B_x + A_y B_y + A_z B_z$$

Since $\mathbf{A} \cdot \mathbf{B}$ is a scalar, then the order of the product can be interchanged.

That is, $$\mathbf{A} \cdot \mathbf{B} = \mathbf{B} \cdot \mathbf{A}$$

Furthermore, $\mathbf{A} \cdot \mathbf{B}$ can be positive, negative, or zero depending on the value of θ. (That is, $\cos\theta$ varies from -1 to $+1$.) If vectors are expressed in unit vector form, then the dot product is conveniently carried out using the multiplication table for unit vectors:

$$\hat{\mathbf{i}} \cdot \hat{\mathbf{i}} = \hat{\mathbf{j}} \cdot \hat{\mathbf{j}} = \hat{\mathbf{k}} \cdot \hat{\mathbf{k}} = 1; \qquad \hat{\mathbf{i}} \cdot \hat{\mathbf{j}} = \hat{\mathbf{i}} \cdot \hat{\mathbf{k}} = \hat{\mathbf{k}} \cdot \hat{\mathbf{j}} = 0$$

The definite integral is introduced as a method to calculate the work done by a varying force. In Section 7.4, it is shown that the incremental bit of work done by a variable force F_x in displacing a particle a small distance Δx is given by

$$\Delta W \approx F_x \Delta x$$

(ΔW equals the area of the shaded rectangle in the figure at the right.) The total work done by F_x as the particle is displaced from x_i to x_f is given approximately by the **sum** of such terms.

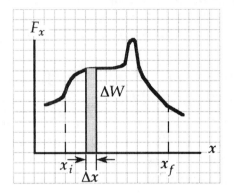

By letting the widths of the displacements (dx) approach zero, the number of terms in the sum increases without limit and we get the actual work done:

$$W = \lim_{\Delta x \to 0} \sum_{x_i}^{x_f} F_x \, \Delta x = \int_{x_i}^{x_f} F_x \, dx$$

The quantity on the right is a definite integral, which graphically represents the **area under the F_x versus x curve**, as in the figure. Appendix B6 of the text represents a brief review of integration operations with some examples.

REVIEW CHECKLIST

You should be able to:

▷ Calculate the work done by a constant force using Equations 7.1 or 7.3. (Section 7.2)

▷ Calculate the scalar or dot product of any two vectors A and B using the definition $\mathbf{A} \cdot \mathbf{B} = AB \cos \theta$, or by writing $\mathbf{A}$ and $\mathbf{B}$ in unit vector notation and using Equation 7.6. Also, find the value of the angle between two vectors. (Section 7.3)

▷ Calculate the work done by a variable force from the force vs. distance curve and by evaluating the integral, $\int F(x)dx$. (Section 7.4)

▷ Determine the kinetic energy of an object of mass m given either speed v or velocity $\mathbf{v}$. (Section 7.5)

▷ Make calculations using the work-kinetic energy theorem. (Section 7.5)

▷ Determine the force of kinetic friction acting on an object and calculate the decrease in kinetic energy (and increase in internal energy) due to frictional forces. (Section 7.7)

▷ Make calculations of average and instantaneous power. (Section 7.8)

ANSWERS TO SELECTED CONCEPTUAL QUESTIONS

5. As a simple pendulum swings back and forth, the forces acting on the suspended mass are the gravitational force, the tension in the supporting cord, and air resistance. (a) Which of these forces, if any, does no work on the pendulum? (b) Which of these forces does negative work at all times during its motion? (c) Describe the work done by the gravitational force while the pendulum is swinging.

Answer (a) The tension in the supporting cord does no work, because the motion of the pendulum is always perpendicular to the cord, and therefore to the force exerted by the string. (b) The air resistance does negative work at all times, since the air resistance is always acting in a direction opposite to the motion. (c) The weight always acts downward; therefore, the work done by the gravitational force is positive on the downswing, and negative on the upswing.

□ □ □ □

10. Can kinetic energy be negative? Explain.

Answer No. Kinetic energy = $mv^2/2$. Since v^2 is always positive, K is always positive.

□ □ □ □

12. One bullet has twice the mass of a second bullet. If both are fired so that they have the same speed, which has more kinetic energy? What is the ratio of the kinetic energies of the two bullets?

Answer The kinetic energy of the more massive bullet is twice that of the lower mass bullet.

□ □ □ □

14. (a) If the speed of a particle is doubled, what happens to its kinetic energy? (b) What can be said about the speed of a particle if the net work done on it is zero?

Answer (a) Kinetic energy, K, depends on the square of the velocity. Therefore if the speed is doubled, the kinetic energy will increase by a factor of four. (b) If net work is zero, the speed of a particle does not change.

□ □ □ □

Chapter 7

SOLUTIONS TO SELECTED END-OF-CHAPTER PROBLEMS

1.   A block of mass 2.50 kg is pushed 2.20 m along a frictionless horizontal table by by a constant 16.0-N force directed 25.0° below the horizontal. Determine the work done on the block by (a) the applied force, (b) the normal force exerted by the table, and (c) the gravitational force? (d) Determine the total work done on the block.

Solution $W = F\Delta r \cos\theta$

(a) By the applied force, $W_{app} = (16.0 \text{ N})(2.20 \text{ m})\cos(25.0°) = 31.9 \text{ J}$ ◊

(b) By normal force, $W_n = n\Delta r \cos\theta = n\Delta r \cos(90°) = 0$ ◊

(c) By force of gravity, $W_g = F_g\Delta r \cos\theta = mg\Delta r \cos(90°) = 0$ ◊

(d) Net work done on the block: $W_{net} = W_{app} + W_n + W_g = 31.9 \text{ J}$ ◊

---

3.   A certain superhero, whose mass is 80.0 kg, is dangling on the free end of a 12.0-m rope, the other end of which is fixed to a tree limb above. He is able to get the rope in motion as only he knows how, eventually getting it to swing enough that he can reach a ledge when the rope makes a 60.0° angle with the vertical. How much work was done by the gravitational force on the superhero in this maneuver?

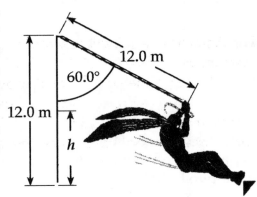

Solution

The work done is $W = \mathbf{F} \cdot \Delta\mathbf{r}$, where the gravitational force is

$$F_g = -mg\hat{\jmath} = (80.0 \text{ kg})(-9.80 \text{ m} / \text{s}^2\hat{\jmath}) = -784\hat{\jmath} \text{ N}$$

The superhero travels $\Delta\mathbf{r} = L\sin 60°\hat{\imath} + L\cos 60°\hat{\jmath}$

$\Delta\mathbf{r} = (12.0 \text{ m})\sin 60°\hat{\imath} + (12.0 \text{ m})\cos 60°\hat{\jmath}$

Thus, $W = \mathbf{F} \cdot \Delta\mathbf{r} = (784\hat{\jmath} \text{ N})(10.39\hat{\imath} \text{ m} + 6.00\hat{\jmath} \text{ m}) = -4.70 \times 10^3 \text{ J}$ ◊

7. A force $\mathbf{F} = (6\hat{\mathbf{i}} - 2\hat{\mathbf{j}})$ N acts on a particle that undergoes a displacement $\Delta\mathbf{r} = (3\hat{\mathbf{i}} + \hat{\mathbf{j}})$ m. Find (a) the work done by the force on the particle and (b) the angle between $\mathbf{F}$ and $\Delta\mathbf{r}$.

Solution We use the mathematical representation of the definition of work.

(a) $W = \mathbf{F} \cdot \Delta\mathbf{r} = (6\hat{\mathbf{i}} + 2\hat{\mathbf{j}}) \cdot (3\hat{\mathbf{i}} + 1\hat{\mathbf{j}}) = (6\ \text{N})(3\ \text{m}) + (-2\ \text{N})(1\ \text{m}) = 18\ \text{J} - 2\ \text{J} = 16.0\ \text{J}$ ◊

(b) $|\mathbf{F}| = \sqrt{F_x^2 + F_y^2} = \sqrt{(6\ \text{N})^2 + (-2\ \text{N})^2} = 6.32\ \text{N}$

$|\Delta\mathbf{r}| = \sqrt{\Delta r_x^2 + \Delta r_y^2} = \sqrt{(3\ \text{m})^2 + (1\ \text{m})^2} = 3.16\ \text{m}$

$W = F\Delta r\cos\theta:$ $\cos\theta = \dfrac{W}{F\Delta r} = \dfrac{16.0\ \text{J}}{(6.32\ \text{N})(3.16\ \text{m})} = 0.800$

and $\theta = \cos^{-1}(0.800) = 36.9°$ ◊

---

13. A particle is subject to a force F_x that varies with position as in Figure P7.13. Find the work done by the force on the particle as it moves (a) from $x = 0$ to $x = 5.00$ m, (b) from $x = 5.00$ m to $x = 10.0$ m, and (c) from $x = 10.0$ m to $x = 15.0$ m. (d) What is the total work done by the force over the distance $x = 0$ to $x = 15.0$ m?

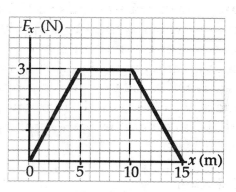

Figure P7.13

Solution $W = \int F_x\,dx$

We use the graphical representation of the definition of work. W equals the area under the force-displacement curve.

(a) For the region $0 \le x \le 5.00$ m, $W = \dfrac{(3.00\ \text{N})(5.00\ \text{m})}{2} = 7.50\ \text{J}$ ◊

(b) For the region $5.00\ \text{m} \le x \le 10.0$ m, $W = (3.00\ \text{N})(5.00\ \text{m}) = 15.0\ \text{J}$ ◊

(c) For the region $10.0\ \text{m} \le x \le 15.0$ m, $W = \dfrac{(3.00\ \text{N})(5.00\ \text{m})}{2} = 7.50\ \text{J}$ ◊

(d) For the region $0 \le x \le 15.0$ m, $W = (7.50\ \text{J} + 7.50\ \text{J} + 15.0\ \text{J}) = 30.0\ \text{J}$ ◊

19. If it takes 4.00 J of work to stretch a Hooke's-law spring 10.0 cm from its unstressed length, determine the extra work required to stretch it an additional 10.0 cm.

Solution

Conceptualize: We know that the force required to stretch a spring is proportional to the distance the spring is stretched, and since the work required is proportional to the force **and** to the distance, then $W \propto x^2$. This means if the extension of the spring is doubled, the work will increase by a factor of 4, so that for $x = 20$ cm, $W = 16$ J, requiring 12 J of additional work.

Categorize: Let's confirm our answer using Hooke's law and the definition of work.

Analyze: The linear spring force relation is given by Hooke's law: $F_s = -kx$

Integrating with respect to x, we find the work done by the spring is:

$$W_s = \int_{x_i}^{x_f} F \, dx = \int_{x_i}^{x_f} (-kx) \, dx = \frac{1}{2} k \left(x_f^2 - x_i^2 \right)$$

However, we want the work done **on** the spring, which is

$$W = -W_s = \frac{1}{2} k \left(x_f^2 - x_i^2 \right)$$

We know the work for the first 10 cm, so we can find the force constant:

$$k = \frac{2W}{x_f^2 - x_i^2} = \frac{2(4.00 \text{ J})}{(0.100 \text{ m})^2 - 0} = 800 \text{ N / m}$$

Substituting for k, x_i and x_f, the extra work for the next step of extension is

$$W = \frac{1}{2}(800 \text{ N / m})\left[(0.200 \text{ m})^2 - (0.100 \text{ m})^2 \right] = 12.0 \text{ J}$$ ◊

Finalize: Our calculated answer agrees with our prediction. It is helpful to remember that the force required to stretch a spring is proportional to the distance the spring is extended, but the work is proportional to the square of the extension.

27. A 2 100-kg pile driver is used to drive a steel I-beam into the ground. The pile driver falls 5.00 m before coming into contact with the top of the beam, and it drives the beam 12.0 cm further into the ground before coming to rest. Using energy considerations, calculate the average force the beam exerts on the pile driver while the pile driver is brought to rest.

Solution

Conceptualize: Anyone who has hit their thumb with a hammer knows that the resulting force is greater than just the weight of the hammer, so we should also expect the force of the pile driver to be significantly greater than its weight: $F \gg mg \sim 20\,\text{kN}$. The force **on** the pile driver will be directed upwards.

Categorize: The average force stopping the driver can be found from the work that results from the gravitational force starting its motion. The initial and final kinetic energies are zero.

Analyze: Choose the initial point when the mass is elevated and the final point when it comes to rest again 5.12 m below. Two forces do work on the pile driver: gravity (weight) and the normal force exerted by the beam on the pile driver.

$$\Sigma W = K_f - K_i \qquad\qquad \text{so that} \quad mgd_w \cos(0) + nd_n \cos(180°) = 0 - 0$$

where $\qquad d_w = 5.12\,\text{m}, \ d_n = 0.120\,\text{m}, \quad$ and $\qquad m = 2\,100\,\text{kg}$

In this situation, the weight vector is in the direction of motion and the beam exerts a force on the pile driver opposite the direction of motion.

$$(2\,100\,\text{kg})(9.80\,\text{m / s}^2)(5.12\,\text{m}) + n(0.120\,\text{m})(-1) = 0$$

Solve for n. $\quad n = \dfrac{1.05 \times 10^5\,\text{J}}{0.120\,\text{m}} = 878\,\text{kN}$ ◊

Finalize: The normal force is larger than 20 kN as we expected, and is actually about 43 times greater than the weight of the pile driver, which is why this machine is so effective.

Additional Calculation: Show that the work done by gravity on an object can be represented by mgh, where h is the vertical height that the object falls. Apply your results to the problem above.

By the figure to the right, where **d** is the path of the object, and h is the height that the object falls,

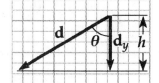

$$h = |\mathbf{d}_y| = d\cos\theta$$

Since $F = mg$, $\quad mgh = Fd\cos\theta = \mathbf{F} \cdot \mathbf{d}$ ◊

In this problem, $mgh = n(d_n)$ so $n = mgh / d_n$

and $\qquad\qquad n = (2\,100\,\text{kg})(9.80\,\text{m / s}^2)(5.12\,\text{m})/(0.120\,\text{m}) = 878\,\text{kN}$ ◊

31. A 40.0-kg box initially at rest is pushed 5.00 m along a rough, horizontal floor with a constant applied horizontal force of 130 N. If the coefficient of friction between box and floor is 0.300, find (a) the work done by the applied force, (b) the increase in internal energy in the box-floor system due to friction, (c) the work done by the normal force, (d) the work done by the gravitational force, (e) the change in kinetic energy of the box, and (f) the final speed of the box.

Solution $\mu_k = 0.300$

$\Delta r = 5.00$ m

$v_i = 0$

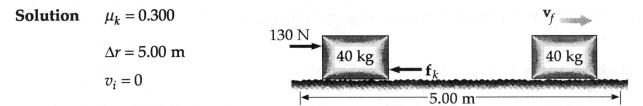

(a)   The applied force and the motion are both **horizontal.**

$W_F = \mathbf{F} \cdot \Delta \mathbf{r} = F \Delta r \cos(0°) = (130 \text{ N})(5.00 \text{ m})(1) = 650 \text{ J}$ ◊

(b) $f_k = \mu_k n = \mu_k mg = 0.300(40.0 \text{ kg})(9.80 \text{ m}/\text{s}^2) = 117.6 \text{ N}$

$\Delta E_{int} = f_k \Delta x = (117.6 \text{ N})(5.00 \text{ m}) = 588 \text{ J}$ ◊

(c)   Since the normal force is perpendicular to the motion,

$W_n = F \Delta r \cos(90°) = (130 \text{ N})(5.00 \text{ m})(0) = 0$ ◊

(d)   The force of gravity is also perpendicular to the motion, so $W_g = 0$.   ◊

We use the energy version of the nonisolated system model.

(e) $\Delta K = W_{\text{other forces}} - f_k \Delta r = 650 \text{ J} - 588 \text{ J} = 62.0 \text{ J}$ ◊

(f) $\frac{1}{2} m v_i^2 + W_{\text{other forces}} - f_k \Delta r = \frac{1}{2} m v_f^2$

$v_f = \sqrt{\frac{2}{m}\left[\Delta K + \frac{1}{2} m v_i^2\right]} = \sqrt{\left(\frac{2}{40.0 \text{ kg}}\right)\left[62.0 \text{ J} + \frac{1}{2}(40.0 \text{ kg})(0)\right]} = 1.76 \text{ m}/\text{s}$ ◊

33. A crate of mass 10.0 kg is pulled up a rough incline with an initial speed of 1.50 m/s. The pulling force is 100 N parallel to the incline, which makes an angle of 20.0° with the horizontal. The coefficient of kinetic friction is 0.400, and the crate is pulled 5.00 m. (a) How much work is done by the gravitational force on the crate? (b) Determine the increase in internal energy of the crate-incline system due to friction (c) How much work is done by the 100-N force on the crate? (d) What is the change in kinetic energy of the crate? (e) What is the speed of the crate after being pulled 5.00 m?

Solution   The force of gravity is $(10.0\,\text{kg})(9.80\,\text{m}/\text{s}^2) = 98.0\,\text{N}$ straight down, at an angle of $(90.0° + 20.0°) = 110.0°$ with the motion. The work done by gravity on the crate is

(a) $W_g = \mathbf{F} \cdot \Delta\mathbf{r} = (98.0\ \text{N})(5.00\ \text{m})\cos 110.0° = -168\ \text{J}$ ◇

(b)   Setting the *xy* axes parallel and perpendicular to the incline,

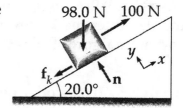

From $\Sigma F_y = ma_y$, $+n - (98.0\ \text{N})\cos 20.0° = 0$

$$n = 92.1\ \text{N}$$

and $f_k = \mu_k n = 0.400(92.1\ \text{N}) = 36.8\ \text{N}$

Therefore, $\Delta E_{\text{int}} = f_k \Delta x = (36.8\ \text{N})(5.00\ \text{m}) = 184\ \text{J}$ ◇

(c) $W = \mathbf{F} \cdot \Delta\mathbf{r} = 100\ \text{N}(5.00\ \text{m})\cos(0°) = +500\ \text{J}$ ◇

(d)   We use the energy version of the nonisolated system model.

By Equation 7.21, $\Delta K = -f_k \Delta x + \sum W_{\text{other forces}}$

$$\Delta K = -f_k \Delta x + W_g + W_{\text{applied force}} + W_n$$

$$\Delta K = -168\ \text{J} - 184\ \text{J} + 500\ \text{J} + 0 = 148\ \text{J}$$ ◇

The normal force does zero work, because it is at 90° to the motion.

(e)   Since $K_f = K_i + \Delta K$, $\frac{1}{2}(10.0\ \text{kg})v_f^2 = \frac{1}{2}(10.0\ \text{kg})(1.50\ \text{m}/\text{s})^2 + 148\ \text{J} = 159\ \text{J}$

Thus, $v_f = \sqrt{\dfrac{2(159\ \text{kg}\cdot\text{m}^2/\text{s}^2)}{10.0\ \text{kg}}} = 5.65\ \text{m}/\text{s}$ ◇

35.   A sled of mass *m* is given a kick on a frozen pond. The kick imparts to it an initial speed of 2.00 m/s. The coefficient of kinetic friction between sled and ice is 0.100. Use energy considerations to find the distance the sled moves before it stops.

Solution

Conceptualize: Since the sled's initial speed of $2\,\text{m/s}$ ($\sim 4\,\text{mph}$) is reasonable for a moderate kick, we might expect the sled to travel several meters before coming to rest.

Categorize: We could solve this problem using Newton's second law, but we are asked to use the work-kinetic energy theorem: $W = \Delta K = K_f - K_i$, where the only work done on the sled after the kick results from the friction between the sled and ice. (The weight and normal force both act at 90° to the motion, and therefore do no work on the sled.)

Analyze: The work due to friction is $W = -f_k d$

where $$f_k = -\mu_k n = -\mu_k mg$$

Since the final kinetic energy is zero, $$W = \Delta K = 0 - K_i = -\frac{1}{2}mv_i^2$$

So $$-\mu_k mgd = -\frac{1}{2}mv_i^2$$

Thus, $$d = \frac{mv_i^2}{2f_k} = \frac{mv_i^2}{2\mu_k mg} = \frac{v_i^2}{2\mu_k g} = \frac{(2.00 \text{ m / s})^2}{2(0.100)(9.80 \text{ m / s}^2)} = 2.04 \text{ m}$$ ◊

Finalize: The distance agrees with the prediction. It is interesting that the distance does not depend on the mass and is proportional to the square of the initial velocity. This means that a small car and a massive truck should be able to stop within the same distance if they both skid to a stop from the same initial speed. Also, doubling the speed requires 4 times as much stopping distance, which is consistent with advice given by transportation safety officers who suggest at least a 2 second gap between vehicles (as opposed to a fixed distance like 100 feet).

---

37. A 700-N Marine in basic training climbs a 10.0-m vertical rope at a constant speed in 8.00 s. What is his power output?

Solution   The marine must exert a 700 N upward force opposite the force of gravity to lift his body at constant speed. Then his muscles do work:

$$W = \mathbf{F} \cdot \Delta \mathbf{r} = (700\hat{\mathbf{j}} \text{ N}) \cdot (10.0\hat{\mathbf{j}} \text{ m}) = 7\,000 \text{ J}$$

The power he puts out is $$\overline{\mathcal{P}} = \frac{W}{\Delta t} = \frac{7\,000 \text{ J}}{8.00 \text{ s}} = 875 \text{ W}$$ ◊

---

45. A compact car of mass 900 kg has an overall motor efficiency of 15.0%. (That is, 15.0% of the energy supplied by the fuel is delivered to the wheels of the car.) (a) If burning one gallon of gasoline supplies 1.34×10^8 J of energy, find the amount of gasoline used in accelerating the car from rest to 55.0 mi/h. Here you may ignore the effects of air resistance and rolling friction. (b) How many such accelerations will one gallon provide? (c) The mileage claimed for the car is 38.0 mi/gal at 55.0 mi/h. What power is delivered to the wheels (to overcome frictional effects) when the car is driven at this speed?

Solution   We first must convert units from miles per hour to SI units:

$$(55.0 \text{ mi} / \text{h})\left(\frac{1\,609 \text{ m}}{1 \text{ mi}}\right)\left(\frac{1 \text{ h}}{3\,600 \text{ s}}\right) = 24.6 \text{ m} / \text{s}$$

(a) The engine must do work to provide kinetic energy to the chassis, according to

$$\sum W = K_f - K_i = \frac{1}{2}mv_f^2 - \frac{1}{2}mv_i^2$$

$$W_{\text{engine output}} = \frac{1}{2}(900 \text{ kg})(24.6 \text{ m} / \text{s})^2 - 0 = 2.72 \times 10^5 \text{ J}$$

This output is only 15.0% of the chemical energy the engine takes in; the other 85.0% is lost as heat.

Since   efficiency $= \dfrac{\text{useful energy output}}{\text{total energy input}} = \dfrac{2.72 \times 10^5 \text{ J}}{\text{total energy input}}$

total energy input $= \dfrac{2.72 \times 10^5 \text{ J}}{0.150} = 1.82 \times 10^6 \text{ J}$

and   amt of gas $= (1.82 \times 10^6 \text{ J})\dfrac{1 \text{ gal}}{1.34 \times 10^8 \text{ J}} = 0.013\,5 \text{ gal}$

(b) $1 \text{ gal} = (n \text{ accelerations})(0.013\,5 \text{ gal} / \text{acceleration})$

$n = 73.8 \text{ accelerations}$

(c)   Consider driving 38.0 miles with energy input 1.34×10^8 J

and output work equal to $(0.150)(1.34 \times 10^8 \text{ J}) = 2.01 \times 10^7 \text{ J}$

The time for this trip is $\dfrac{38.0 \text{ mi}}{55.0 \text{ mi} / \text{h}} = 0.691 \text{ h} = 2\,490 \text{ s}$

So, the output power is $\overline{\mathcal{P}} = \dfrac{W}{\Delta t} = \dfrac{2.01 \times 10^7 \text{ J}}{2\,490 \text{ s}} = 8.08 \text{ kW}$ ◊

49. A 4.00-kg particle moves along the x axis. Its position varies with time according to $x = t + 2.0t^3$, where x is in meters and t is in seconds. Find (a) the kinetic energy at any time t, (b) the acceleration of the particle and the force acting on it at time t, (c) the power being delivered to the particle at time t, and (d) the work done on the particle in the interval $t = 0$ to $t = 2.00$ s.

Solution Given $m = 4.00$ kg and $x = t + 2.0t^3$, we find

(a) $v = \dfrac{dx}{dt} = \dfrac{d}{dt}\left(t + 2.00t^3\right) = 1 + 6.00t^2$

$K = \tfrac{1}{2}mv^2 = \tfrac{1}{2}(4.00 \text{ kg})\left(1 + 6.0t^2\right)^2 = \left(2.00 + 24.0t^2 + 72.0t^4\right) \text{ J}$ ◊

(b) $a = \dfrac{dv}{dt} = \dfrac{d}{dt}\left(1 + 6.0t^2\right) = 12t \text{ m/s}^2$ ◊

$F = ma = (4.00 \text{ kg})(12t) = 48t \text{ N}$ ◊

(c) $\mathcal{P} = \dfrac{dW}{dt} = \dfrac{dK}{dt} = \dfrac{d}{dt}\left(2.00 + 24t^2 + 72t^4\right) = \left(48t + 288t^3\right) \text{ W}$ ◊

[or use $\mathcal{P} = Fv = 48t\left(1.00 + 6.0t^2\right)$]

(d) $W = K_f - K_i$ where $t_i = 0$ and $t_f = 2.00$ s.

At $t_i = 0$, $K_i = 2.00$ J

At $t_f = 2.00$ s, $K_i = \left[2.00 + 24(2.00 \text{ s})^2 + 72(2.00 \text{ s})^4\right] = 1250$ J

Therefore, $W = 1248 \text{ J} = 1.25 \times 10^3 \text{ J}$ ◊

[or use $W = \int_{t_i}^{t_f} \mathcal{P}\, dt = \int_0^2 \left(48t + 288t^3\right)dt$, etc.]

---

61. A 200-g block is pressed against a spring of force constant 1.40 kN/m until the block compresses the spring 10.0 cm. The spring rests at the bottom of a ramp inclined at 60.0° to the horizontal. Using energy considerations, determine how far up the incline the block moves before it stops (a) if there is no friction between the block and the ramp and (b) if the coefficient of kinetic friction is 0.400.

Solution   We have two forms of energy involved: work done by gravity, and work done by the spring.

$$W_g = \mathbf{F} \cdot \mathbf{d} = mgd \cos 150° \qquad\qquad W_s = \frac{1}{2}kx_m{}^2$$

(a)   Apply the work-energy theorem between the starting point and the point of maximum travel up the incline. The initial and final states of the block are stationary, so $\Delta K = 0$, and $\Sigma W = \Delta K$ becomes $W_g + W_s = 0$. Substituting known values into this equation,

$$mgd \cos 150° + \frac{1}{2}kx_m{}^2 = 0$$

and $$d = \frac{-kx_m{}^2}{2mg \cos 150°} = \frac{(-1\,400 \text{ N / m})(0.100 \text{ m})^2}{2(0.200 \text{ kg})(9.80 \text{ m / s}^2)(-0.866)} = 4.12 \text{ m} \qquad\qquad \lozenge$$

(b)   We add a term for the friction's energy loss:

$$W_f = f_k d \cos 180° = -f_k d$$

Taking our vertical axis to be perpendicular to the incline, $\Sigma F_y = ma_y$:

$$+n - (1.96 \text{ N})(\cos 60.0°) = 0 \qquad \text{and} \qquad n = 0.980 \text{ N}$$

Then the force of friction is $f_k = \mu_k n$: $f_k = (0.400)(0.980 \text{ N}) = 0.392 \text{ N}$

We again apply our energy equation $K_i + \sum W = K_f$

this time adding a friction term: $K_i + W_g + W_f + W_s = K_f$

Since K_i and K_f are both zero,

$$mgd \cos 150° + f_k d \cos 180° + \frac{1}{2}kx_m{}^2 = 0$$

$$0 + (1.96 \text{ N})(d)(-0.866) + (0.392 \text{ N})(d)(-1.00) + 7.00 \text{ J} = 0$$

$$d = \frac{7.00 \text{ J}}{1.70 \text{ N} + 0.392 \text{ N}} = 3.35 \text{ m} \qquad\qquad \lozenge$$

63. The ball launcher in a pinball machine has a spring that has a force constant of 1.20 N/cm (Fig. P7.63). The surface on which the ball moves is inclined 10.0° with respect to the horizontal. If the spring is initially compressed 5.00 cm, find the launching speed of a 100-g ball when the plunger is released. Friction and the mass of the plunger are negligible.

Figure P7.63

Solution Use the work-kinetic energy theorem.

$W_{net} = \Delta K:$ $\qquad$ $W_s + W_g = \Delta K$

$$\tfrac{1}{2}kx^2 - mg\Delta r \sin 10.0° = \tfrac{1}{2}m\left(v_f^2 - v_i^2\right)$$

Since $v_i = 0$: $\qquad$ $v_f = \sqrt{\dfrac{2}{m}\left[\tfrac{1}{2}kx^2 - mg\Delta r \sin 10.0°\right]}$

In this case, $\qquad$ $x = \Delta r = 5.00 \text{ cm} = 5.00 \times 10^{-2} \text{ m}$

and $\qquad$ $k = 1.20 \text{ N / cm} = 120 \text{ N / m}$

$$v_f = \sqrt{\dfrac{2\left[\tfrac{1}{2}(120 \text{ N / m})(5.00 \times 10^{-2} \text{ m})^2 - (0.100 \text{ kg})(9.80 \text{ m / s}^2)(5.00 \times 10^{-2} \text{ m})\sin 10.0°\right]}{0.100 \text{ kg}}}$$

and $v_f = 1.68 \text{ m / s}$ $\qquad\qquad\qquad\qquad\qquad\qquad\qquad\qquad$ ◊

Chapter 8
POTENTIAL ENERGY

EQUATIONS AND CONCEPTS

The **gravitational potential energy** associated with an object at any point in space is the product of the object's weight and the vertical coordinate relative to an arbitrary reference level.

$$U_g \equiv mgy \tag{8.2}$$

The work done on any object by the gravitational force is equal to the negative of the change in the gravitational potential energy.

$$W_g = -\left(U_f - U_i\right) = -\Delta U_g$$

For **an isolated system** (e.g. mass-Earth system), there is no transfer of energy across the boundary of the system and the sum of the potential and kinetic energies of the system remains constant. *When gravitational potential energy decreases, there is a corresponding increase in kinetic energy.*

$$\Delta K = -\Delta U_g \tag{8.6}$$

The **units of energy** (kinetic and potential) are the same as the units of work — joules.

$$1\,J = 1\,N \cdot m$$

The **total mechanical energy** of a system is the sum of the kinetic and potential energies. *In Equation 8.8, U represents the total of all types of potential energy.*

$$E_{\text{mech}} = K + U \tag{8.8}$$

111

Conservation of mechanical energy requires that if no external forces do work on an isolated system and if no nonconservative forces act on objects inside the system, the sum of the kinetic and potential energies of the system remains constant — or is conserved. *According to this important conservation law, if the kinetic energy of the system increases by some amount, the potential energy must decrease by the same amount — and vice versa.*

$$K_f + U_f = K_i + U_i \qquad (8.9)$$

An **elastic potential energy function** is associated with a mass-spring system which has been deformed a distance x from the equilibrium position. The force constant k has units of N/m and is characteristic of a particular spring. *The elastic potential energy of a deformed spring is always positive.*

$$U_s \equiv \frac{1}{2}kx^2 \qquad (8.11)$$

The **work done by a conservative force** on an object that is a member of a system as the object moves from one position to another is equal to the negative of the change in the potential energy of the system.

$$W_c = U_i - U_f = -\Delta U \qquad (8.12)$$

If a **friction force** acts within a system, the total mechanical energy $(U+K)$ will be decreased. In Equation 8.14, ΔU is the change in all forms of potential energy. *Note that Equation 8.14 is equivalent to Equation 8.9 when the friction force is zero.*

$$\Delta E_{\text{mech}} = \Delta K + \Delta U = -f_k d \qquad (8.14)$$

The **continuity equation for energy** is a starting point for solving a wide class of mechanical problems. The first three terms on the left-hand side of the equation refer to the energy of a system at a chosen time and location (referred to as the "initial" point, designated by "i"). The work and friction terms on the left-hand side of the equation account for the changes in the mechanical energy of a nonisolated system. The terms on the right-hand side of the equation add to give the total energy at a "final" time and location designated by "f". In many situations (and depending on the arbitrary choice of the initial point) several terms in this general equation may be equal to zero. *Also, if the system includes more than one object, each object can make a contribution to the energy terms.*

$$K_i + U_{gi} + U_{si} + W_{external} - f_k d = K_f + U_{gf} + U_{sf}$$

A **potential energy function** U can be defined for a conservative force such that the work done by a conservative force equals the decrease in the potential energy of the system. *The work done on an object by a conservative force depends only on the initial and final positions of the object and equals zero around a closed path. If more than one conservative force acts, then a potential energy function is associated with each force.*

$$\Delta U = U_f - U_i = -\int_{x_i}^{x_f} F_x \, dx \qquad (8.16)$$

If the only conservative force is the **gravitational force**, the equation for conservation of mechanical energy takes a special form. *The coordinate y is the position of the mass relative to an arbitrary reference level.*

$$\frac{1}{2}mv_i^2 + mgy_i = \frac{1}{2}mv_f^2 + mgy_f$$

The x **component of a conservative force** equals the negative derivative of the system's potential energy with respect to x.

$$F_x = -\frac{dU}{dx} \qquad (8.18)$$

SUGGESTIONS, SKILLS, AND STRATEGIES

CHOOSING A ZERO LEVEL

In working problems involving gravitational potential energy, it is always necessary to choose a convenient reference level for gravitational potential energy — a position in the gravitational where the gravitational potential energy is taken to be zero. This choice is completely arbitrary because the important quantity is the difference in potential energy, and that difference is independent of the location of zero. It is often convenient, but not essential, to choose the surface of the Earth as the reference position for zero potential energy. In most cases, the statement of the problem suggests a convenient level to use.

CONSERVATION OF ENERGY

Take the following steps in applying the principle of conservation of energy:

• Identify the system of interest, which may consist of more than one object.

• Select a reference position for the zero point of gravitational potential energy.

• Determine whether or not nonconservative forces are present.

• If mechanical energy is conserved (that is, if only conservative forces are present), you can write the total initial energy at some point as the sum of the kinetic and potential energies at that point, $K_i + U_i$. Then, write an expression for the total final energy, $K_f + U_f$, at the final point of interest. Since mechanical energy is conserved, you can use Equation 8.9, $K_f + U_f = K_i + U_i$.

• If nonconservative forces such as friction are present (and thus mechanical energy is not conserved), first write expressions for the total initial and total final energies. In this case, the difference between the two total energies is equal to the decrease in mechanical energy due to the presence of nonconservative force(s); and Equation 8.14 applies: $\Delta E_{mech} = \Delta K + \Delta U = -fd$.

REVIEW CHECKLIST

You should be able to:

▷ Identify a mechanical system of interest, select an appropriate reference level for gravitational or elastic potential energy, determine whether or not nonconservative forces are present and apply Equation 8.9 or Equation 8.14 as appropriate. (Sections 8.1, 8.2, 8.3, and 8.4)

▷ Calculate the potential energy function for a system when the conservative force acting on the system is given. (Section 8.5)

▷ Calculate the force components (rectangular or radial) as the negative derivative of the potential function. (Section 8.5)

▷ Plot the potential as a function of coordinate and determine points of equilibrium. (Section 8.6)

ANSWERS TO SELECTED CONCEPTUAL QUESTIONS

3. One person drops a ball from the top of a building while another person at the bottom observes its motion. Will these two people agree on the value of the gravitational potential energy of the ball-Earth system? On the change in potential energy? On the kinetic energy?

Answer The two will not necessarily agree on the potential energy, since this depends on the origin — which may be chosen differently for the two observers. However, the two **must** agree on the value of the **change** in potential energy, which is independent of the choice of the reference frames. The two will also agree on the kinetic energy of the ball, assuming both observers are at rest with respect to each other, and hence measure the same v.

□   □   □   □

9. You ride a bicycle. In what sense is your bicycle solar-powered?

Answer The energy to ride the bicycle comes from your body. The source of that energy is the food that you ate at some previous time. The energy in the food, assuming that we focus on vegetables, came from the growth of the plant, for which photosynthesis is a major factor. The light for the photosynthesis comes from the Sun. The argument for meats has a couple of extra steps, but also goes through the process of photosynthesis in the plants eaten by animals. Thus, the source of the energy to ride the bicycle is the Sun, and your bicycle is solar-powered!

□   □   □   □

11. A bowling ball is suspended from the ceiling of a lecture hall by a strong cord. The ball is drawn away from its equilibrium position and released from rest at the tip of the demonstrator's nose as in Figure Q8.11. If the demonstrator remains stationary, explain why she is not struck by the ball on its return swing. Would this demonstrator be safe if the ball were given a push from its starting position at her nose?

Answer The total energy of the system (the bowling ball and the Earth) must be conserved. Since the system initially has a potential energy mgh, and the ball has no kinetic energy, it cannot have any kinetic energy when returning to its initial position. Of course, air resistance will cause the ball to return to a point slightly below its initial position. On the other hand, if the ball is given a push, the demonstrator's nose will be in big trouble.

Figure Q8.11

□   □   □   □

SOLUTIONS TO SELECTED END-OF-CHAPTER PROBLEMS

5. A bead slides without friction around a loop-the-loop (Fig. P8.5). The bead is released from a height $h = 3.50R$. (a) What is its speed at point Ⓐ? (b) How large is the normal force on it if its mass is 5.00 g?

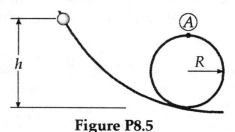

Figure P8.5

Solution

Conceptualize: Since the bead is released above the top of the loop, it will have enough potential energy to reach point Ⓐ and still have excess kinetic energy. The energy of the bead at point Ⓐ will be proportional to h and g. If it is moving relatively slowly, the track will exert an upward force on the bead, but if it is whipping around fast, the normal force will push it toward the center of the loop.

Categorize: The speed at the top can be found from the conservation of energy, and the normal force can be found from Newton's second law.

Analyze:

(a)   We define the bottom of the loop as the zero level for the gravitational potential energy.

Since $v_i = 0$,
$$E_i = K_i + U_i = 0 + mgh = mg(3.50R)$$

The total energy of   the bead at point Ⓐ can be written as

$$E_A = K_A + U_A = \frac{1}{2}mv_A{}^2 + mg(2R)$$

Since mechanical energy is conserved, $E_i = E_A$

and we get
$$\frac{1}{2}mv_A{}^2 + mg(2R) = mg(3.50R)$$

$$v_A{}^2 = 3.00gR$$

or
$$v_A = \sqrt{3.00gR} \qquad \lozenge$$

(b) To find the normal force at the top, we may construct a free-body diagram as shown, where we assume that **n** is downward, like *mg*. Newton's second law gives $\Sigma F = ma_c$, where a_c is the centripetal acceleration.

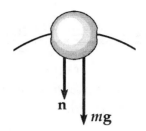

$$n + mg = \frac{mv_A{}^2}{R} = \frac{m(3.00gR)}{R} = 3.00mg$$

$$n = 3.00mg - mg = 2.00mg$$

$$n = 2.00(5.00 \times 10^{-3}\,\text{kg})(9.80\ \text{m/s}^2) = 0.0980\ \text{N downward}$$ ◊

Finalize: Our answer represents the speed at point Ⓐ as proportional to the square root of the product of *g* and *R*, but we must not think that simply increasing the diameter of the loop will increase the speed of the bead at the top. In general, the speed will increase with increasing release height, which for this problem was defined in terms of the radius. The normal force may seem small, but it is twice the weight of the bead.

---

11.   A block of mass 0.250 kg is placed on top of a light vertical spring of force constant 5 000 N/m and pushed downward so that the spring is compressed 0.100 m. After the block is released from rest, it travels upward and then leaves the spring. To what maximum height above the point of release does it rise?

Solution   In both the initial and final states, the block is not moving. Therefore, the initial and final energies of the block-spring-Earth system are:

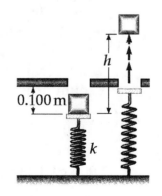

$$E_i = K_i + U_i = 0 + (U_g + U_s)_i = 0 + \left(0 + \frac{1}{2}kx^2\right)$$

$$E_f = K_f + U_f = 0 + (U_g + U_s)_f = 0 + (mgh + 0)$$

Since $E_i = E_f$, $mgh = \frac{1}{2}kx^2$

and $$h = \frac{kx^2}{2mg} = \frac{(5\,000\ \text{N/m})(0.100\ \text{m})^2}{2(0.250\ \text{kg})(9.80\ \text{m/s}^2)} = 10.2\ \text{m}$$ ◊

---

13. Two objects are connected by a light string passing over a light frictionless pulley as shown in Figure P8.13. The 5.00-kg object is released from rest. Using the principle of conservation of energy, (a) determine the speed of the 3.00-kg object just as the 5.00-kg object hits the ground. (b) Find the maximum height to which the 3.00-kg object rises.

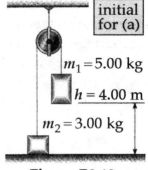

Solution

Figure P8.13

As the system choose the two blocks A and B, string, pulley, and Earth.

(a) Choose the initial point before release and the final point just before the larger object hits the floor. The total energy of the system remains constant and the energy version of the isolated system model gives

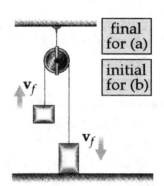

$$\left(K_A + K_B + U_g\right)_i = \left(K_A + K_B + U_g\right)_f$$

At the initial point K_{Ai} and K_{Bi} are zero and we define the gravitational potential energy of the system as zero. No external forces do work on the system and no friction acts within the system. Thus,

$$0 = \frac{1}{2}\left(5.00 \text{ kg} + 3.00 \text{ kg}\right)v_f^2 + \left(3.00 \text{ kg}\right)\left(9.80\,\frac{\text{m}}{\text{s}^2}\right)\left(4.00 \text{ m}\right) + \left(5.00 \text{ kg}\right)\left(9.80\,\frac{\text{m}}{\text{s}^2}\right)\left(-4.00 \text{ m}\right)$$

$$\left(2.00 \text{ kg}\right)\left(9.80 \text{ m / s}^2\right)\left(4.00 \text{ m}\right) = \frac{1}{2}\left(8.00 \text{ kg}\right)v_f^2$$

$$v_f = 4.43 \text{ m / s} \qquad \lozenge$$

(b) Now the string goes slack. The 3.00-kg object becomes a projectile. We focus now on the system of the 3.00-kg object and the Earth. Take the initial point at the previous final point, and the new final point at its maximum height:

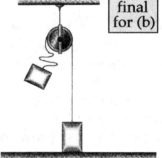

$$\left(K + U_g\right)_i = \left(K + U_g\right)_f$$

$$\frac{1}{2}\left(3.00 \text{ kg}\right)\left(4.43 \text{ m / s}\right)^2 + \left(3.00 \text{ kg}\right)\left(9.80 \text{ m / s}^2\right)\left(4.00 \text{ m}\right) = 0 + \left(3.00 \text{ kg}\right)\left(9.80 \text{ m / s}^2\right)y_f$$

$$y_f = 5.00 \text{ m} \qquad \lozenge$$

or 1.00 m higher than the height of the 5.00-kg mass when it was released.

21. A 4.00-kg particle moves from the origin to position C, having coordinates $x = 5.00$ m and $y = 5.00$ m. One force on the particle is the gravitational force acting in the negative y direction (Fig. P8.21). Using Equation 7.3, calculate the work done by the gravitational force in going from O to C along (a) OAC. (b) OBC. (c) OC. Your results should all be identical. Why?

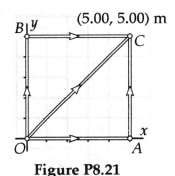

Figure P8.21

Solution

$$F_g = mg = (4.00 \text{ kg})(9.80 \text{ m / s}^2) = 39.2 \text{ N}$$

(a) $\quad W_{OAC} = W_{OA} + W_{AC}$

$$W_{OAC} = F_g d_{OA} \cos(270°) + F_g d_{AC} \cos(180°)$$

$$W_{OAC} = (39.2 \text{ N})(5.00 \text{ m})(0) + (39.2 \text{ N})(5.00 \text{ m})(-1) = -196 \text{ J} \qquad \Diamond$$

(b) $\quad W_{OBC} = W_{OB} + W_{BC}$

$$W_{OBC} = (39.2 \text{ N})(5.00 \text{ m}) \cos 180° + (39.2 \text{ N})(5.00 \text{ m}) \cos 90° = -196 \text{ J} \qquad \Diamond$$

(c) $\quad W_{OC} = F_g d_{OC} \cos 135° = (39.2 \text{ N})\left(5\sqrt{2} \text{ m}\right)\left(-\frac{\sqrt{2}}{2}\right) = -196 \text{ J} \qquad \Diamond$

The results should all be the same since the gravitational force is conservative, and the work done by a conservative force is independent of the path. $\qquad \Diamond$

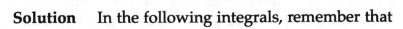

23. A force acting on a particle moving in the xy plane is given by $\mathbf{F} = (2y\hat{\mathbf{i}} + x^2\hat{\mathbf{j}})$ N, where x and y are in meters. The particle moves from the origin to a final position having coordinates $x = 5.00$ m and $y = 5.00$ m, as in Figure P8.21. Calculate the work done by $\mathbf{F}$ along (a) OAC, (b) OBC, (c) OC. (d) Is $\mathbf{F}$ conservative or nonconservative? Explain.

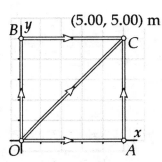

Figure P8.21

Solution   In the following integrals, remember that

$$\hat{\mathbf{i}} \cdot \hat{\mathbf{i}} = \hat{\mathbf{j}} \cdot \hat{\mathbf{j}} = 1 \qquad \text{and} \qquad \hat{\mathbf{i}} \cdot \hat{\mathbf{j}} = 0$$

(a) $\quad W_{OA} = \int_0^{5.00} \left(2y\hat{i} + x^2\hat{j}\right) \cdot \left(\hat{i}\, dx\right) = \int_0^{5.00} 2y\, dx = 2y \int_0^{5.00} dx = 2yx \Big]_{x=0,y=0}^{x=5.00,y=0} = 0$

$\quad W_{AC} = \int_0^{5.00} \left(2y\hat{i} + x^2\hat{j}\right) \cdot \left(\hat{j}\, dy\right) = \int_0^{5.00} x^2 dy = x^2 \int_0^{5.00} dy = x^2 y \Big]_{x=5.00,y=0}^{x=5.00,y=5.00} = 125\text{ J}$

$\quad W_{AC} = 0 + 125\text{ J} = 125\text{ J}$ ◊

(b) $\quad W_{OB} = \int_0^{5.00} \left(2y\hat{i} + x^2\hat{j}\right) \cdot \left(\hat{j}\, dy\right) = \int_0^{5.00} x^2 dy = x^2 \int_0^{5.00} dy = x^2 y \Big]_{x=0...}^{x=0...} = 0$

$\quad W_{BC} = \int_0^{5.00} \left(2y\hat{i} + x^2\hat{j}\right) \cdot \left(\hat{i}\, dx\right) = \int_0^{5.00} 2y\, dx = 2y \int_0^{5.00} dx = 2(5.00)x \Big]_{x=0}^{x=5.00} = 50.0\text{ J}$

$\quad W_{OBC} = 0 + 50.0\text{ J} = 50.0\text{ J}$ ◊

(c) $\quad W_{OC} = \int \left(2y\hat{i} + x^2\hat{j}\right) \cdot \left(\hat{i}\, dx + \hat{j}\, dy\right) = \int_{x=0,y=0}^{x=5.00,y=5.00} \left(2y\,dx + x^2 dy\right)$

$\quad$ Since $x = y$ along OC, $dx = dy$ and $\quad W_{OC} = \int_0^{5.00} \left(2x + x^2\right) dx = 66.7\text{ J}$ ◊

(d) **F** is non-conservative since the work done is path dependent. ◊

---

31. The coefficient of friction between the 3.00-kg block and the surface in Figure P8.31 is 0.400. The system starts from rest. What is the speed of the 5.00-kg ball when it has fallen 1.50 m?

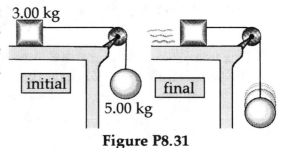

Figure P8.31

Solution

Conceptualize: Assuming that the block does not reach the pulley within the 1.50 m distance, a reasonable speed for the ball might be somewhere between 1 and 10 m/s based on common experience.

Categorize: We could solve this problem by using $\Sigma F = ma$ to give a pair of simultaneous equations in the unknown acceleration and tension; then we would have to solve a motion problem to find the final speed. We may find it easier to solve using the work-energy theorem.

Analyze:

For the Earth plus objects A (block) and B (ball), the work-kinetic energy theorem is

$$\left(K_A + K_A + U_A + U_B\right)_i + W_{app} - f_k d = \left(K_A + K_A + U_A + U_B\right)_f$$

Choose the initial point before release and the final point after each block has moved 1.50 m. Choose $U_g = 0$ with the 3.00-kg block on the tabletop and the 5.00-kg block in its final position.

So
$$K_{Ai} = K_{Bi} = U_{Ai} = U_{Af} = U_{Bf} = 0$$

Also, since the only external forces are gravity and friction,

$$W_{app} = 0$$

We now have
$$0 + 0 + 0 + m_B g y_{Bi} + 0 - f_k d = \frac{1}{2} m_A v_f^2 + \frac{1}{2} m_B v_f^2 + 0 + 0$$

where the frictional force is $f_k = \mu_k n = \mu_k m_A g$

and causes a negative change in mechanical energy since the force opposes the motion. Since all of the variables are known except for v_f, we can substitute and solve for the final speed.

$$(5.00 \text{ kg})(9.80 \text{ m / s}^2)(1.50 \text{ m}) - (0.400)(3.00 \text{ kg})(9.80 \text{ m / s}^2)(1.50 \text{ m}) = \ldots$$

$$\ldots = \frac{1}{2}(3.00 \text{ kg})v_f^2 + \frac{1}{2}(5.00 \text{ kg})v_f^2$$

$$73.5 \text{ J} - 17.6 \text{ J} = \frac{1}{2}(8.00 \text{ kg})v_f^2$$

or $\quad v_f = \sqrt{\dfrac{2(55.9 \text{ J})}{8.00 \text{ kg}}} = 3.74 \text{ m / s}$ ◊

Finalize: The final speed seems reasonable based on our expectation. This speed must also be less than if the rope were cut and the ball simply fell, in which case its final speed would be

$$v_f' = \sqrt{2gy} = \sqrt{2(9.80 \text{ m / s}^2)(1.50 \text{ m})} = 5.42 \text{ m / s}$$

33. A 5.00-kg block is set into motion up an inclined plane with an initial speed of 8.00 m/s (Fig. P8.33). The block comes to rest after traveling 3.00 m along the plane, which is inclined at an angle of 30.0° to the horizontal. For this motion determine (a) the change in the block's kinetic energy, (b) the change in the potential energy of the block-Earth system, (c) the friction force exerted on the block (assumed to be constant). (d) What is the coefficient of kinetic friction?

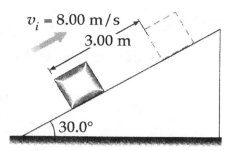

Figure P8.33

Solution

(a) $\quad \Delta K = K_f - K_i = \frac{1}{2}mv_f^2 - \frac{1}{2}mv_i^2 = 0 - \frac{1}{2}(5.00 \text{ kg})(8.00 \text{ m/s})^2 = -160 \text{ J}$ ◊

(b) $\quad \Delta U = U_{gf} - U_{gi} = mgy_f - 0 = (5.00 \text{ kg})(9.80 \text{ m/s}^2)(3.00 \text{ m}) \sin 30.0° = 73.5 \text{ J}$ ◊

(c) $\quad (K + U_g)_i + W_{external} - f_k \Delta x = (K + U_g)_f$

$\quad \frac{1}{2}(5.00 \text{ kg})(8.00 \text{ m/s})^2 + 0 - f(3.00 \text{ m}) = 0 + (5.00 \text{ kg})(9.80 \text{ m/s}^2)(1.50 \text{ m})$

$\quad 160 \text{ J} - f(3.00 \text{ m}) = 73.5 \text{ J} \quad$ and $\quad f = \dfrac{86.5 \text{ J}}{3.00 \text{ m}} = 28.8 \text{ N}$ ◊

(d) The forces perpendicular to the incline must add to zero.

$\quad \sum F_y = 0: \qquad\qquad\qquad +n - mg \cos 30.0° = 0$

Substituting, $\qquad\qquad\qquad n = (5.00 \text{ kg})(9.80 \text{ m/s}^2) \cos 30.0° = 42.4 \text{ N}$

Now, $\quad f_k = \mu_k n$ gives $\qquad\qquad \mu_k = \dfrac{f_k}{n} = \dfrac{28.8 \text{ N}}{42.4 \text{ N}} = 0.679$ ◊

---

41. A single conservative force acts on a 5.00-kg particle. The equation $F_x = (2x + 4)$ N describes the force, where x is in meters. As the particle moves along the x axis from $x = 1.00$ m to $x = 5.00$ m, calculate (a) the work done by this force, (b) the change in the potential energy of the system, and (c) the kinetic energy of the particle at $x = 5.00$ m if its speed is 3.00 m/s at $x = 1.00$ m.

Solution

(a)
$$W_F = \int_{x_i}^{x_f} F_x \, dx$$

where
$$F_x = (2x + 4) \text{ N}, \quad x_i = 1.00 \text{ m}, \quad \text{and} \quad x_f = 5.00 \text{ m}$$

therefore,
$$W_f = \int_{1.00 \text{ m}}^{5.00 \text{ m}} (2x + 4) dx \text{ N} \cdot \text{m} = x^2 + 4x \Big|_{1.00 \text{ m}}^{5.00 \text{ m}} \text{ N} \cdot \text{m} = 40.0 \text{ J} \ \lozenge$$

(b) The change in potential energy equals the negative of the work done by the conservative force.

$$\Delta U = -W_F = -40.0 \text{ J} \qquad\qquad \lozenge$$

(c) When only conservative forces act, conservation of energy gives

$$K_i + U_i = K_f + U_f$$

Rearranging,
$$K_f = K_i - \left(U_f - U_i\right) = \tfrac{1}{2}mv_i^2 - \Delta U$$

so
$$K_f = \tfrac{1}{2}(5.00 \text{ kg})(3.00 \text{ m}/\text{s})^2 - (-40.0 \text{ J}) = 62.5 \text{ J} \qquad \lozenge$$

---

43. The potential energy of a system of two particles separated by a distance r is given by $U(r) = A/r$, where A is a constant. Find the radial force $\mathbf{F}_r$ that each particle exerts on the other.

Solution

The force is the negative derivative of the potential energy with respect to distance:

$$F_r = -\frac{dU}{dr} = -\frac{d}{dr}\left(Ar^{-1}\right) = -A(-1)r^{-2} = \frac{A}{r^2} \qquad \lozenge$$

This describes an inverse-square-law force of repulsion, as between two negative point electric charges.

---

53. What if? The particle described in Problem 52 (Fig. P8.52) is released from rest at Ⓐ, and the surface of the bowl is rough. The speed of the particle at Ⓑ is 1.50 m/s. (a) What is its kinetic energy at Ⓑ? (b) How much mechanical energy is transformed into internal energy as the particle moves from Ⓐ to Ⓑ? (c) Is it possible to determine the coefficient of friction from these results in any simple manner? Explain.

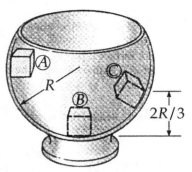

Figure P8.52

Solution   Let us take $U = 0$ for the particle-bowl-Earth system when the particle is at Ⓑ.

Since $v_i = 0$ at Ⓐ, $K_A = 0$ and $U_A = mgR$

(a)   Since $v_B = 1.50$ m/s   and $m = 200$ g,

$$K_B = \tfrac{1}{2}mv_B^{\,2} = \tfrac{1}{2}(0.200 \text{ kg})(1.50 \text{ m/s})^2 = 0.225 \text{ J} \qquad \Diamond$$

(b)   At Ⓐ, $E_i = K_A + U_A = 0 + mgR = (0.200 \text{ kg})(9.80 \text{ m/s}^2)(0.300 \text{ m}) = 0.588 \text{ J}$

At Ⓑ, $E_f = K_B + U_B = 0.225 \text{ J} + 0$

The decrease in mechanical energy is equal to the increase in internal energy.

$$E_{\text{mech},i} - \Delta E_{\text{int}} = E_{\text{mech},f}$$

The energy transformed is

$$\Delta E_{\text{int}} = -\Delta E_{\text{mech}} = E_i - E_f = 0.588 \text{ J} - 0.225 \text{ J} = 0.363 \text{ J} \qquad \Diamond$$

(c)   Even though the energy transformed is known, both the normal force and the friction force change with position as the block slides on the inside of the bowl. Therefore, there is no easy way to find the coefficient of friction. $\qquad \Diamond$

---

57. A 10.0-kg block is released from point Ⓐ in Figure P8.57. The track is frictionless except for the portion between points Ⓑ and Ⓒ, which has a length of 6.00 m. The block travels down the track, hits a spring of force constant 2 250 N/m, and compresses the spring 0.300 m from its equilibrium position before coming to rest momentarily. Determine the coefficient of kinetic friction between the block and the rough surface between Ⓑ and Ⓒ.

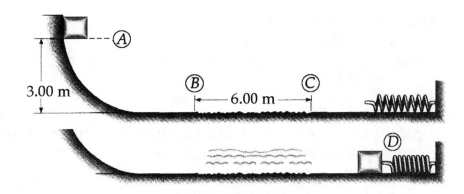

Figure P8.57

Solution

Conceptualize: We should expect the coefficient of friction to be somewhere between 0 and 1 since this is the range of typical μ_k values. It is possible that μ_k could be greater than 1, but it can never be less than 0.

Categorize: The easiest way to solve this problem is by considering the energy changes experienced by the block between the point of release (initial) and the point of full compression of the spring (final). Recall that the change in potential energy (gravitational and elastic) plus the change in kinetic energy must equal the work done on the block by non-conservative forces. Choose the gravitational energy to be zero along the flat portion of the track.

Analyze: Putting the energy equation into symbols: $K_A + U_{gA} - f_k d_{BC} = K_D + U_{sD}$

Expanding into specific variables: $0 + mgy_A - f_k d_{BC} = 0 + \frac{1}{2} k x_s^2$

The friction is $f_k = \mu_k mg$ $\qquad mgy_A - \frac{1}{2} k x^2 = \mu_k mgd$

Solving for the unknown variable μ_k: $\qquad \mu_k = \frac{y_A}{d} - \frac{kx^2}{2mgd}$

Substituting: $\mu_k = \dfrac{3.00 \text{ m}}{6.00 \text{ m}} - \dfrac{(2250 \text{ N / m})(0.300 \text{ m})^2}{2(10.0 \text{ kg})(9.80 \text{ m / s}^2)(6.00 \text{ m})} = 0.328$ $\qquad \diamond$

Finalize: Our calculated value seems reasonable based on the text's tabulation of coefficients of friction. The most important aspect to solving these energy problems is considering how the energy is transferred from the initial to final energy states and remembering to subtract the energy resulting from any non-conservative forces (like friction).

59. A 20.0-kg block is connected to a 30.0-kg block by a string that passes over a light frictionless pulley. The 30.0-kg block is connected to a spring that has negligible mass and a force constant of 250 N/m, as in Figure P8.59. The spring is unstretched when the system is as shown in the figure, and the incline is frictionless. The 20.0-kg block is pulled 20.0 cm down the incline (so that the 30.0-kg block is 40.0 cm above the floor) and released from rest. Find the speed of each block when the 30.0-kg block is 20.0 cm above the floor (when the spring is unstretched).

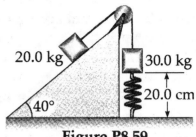

Figure P8.59

Solution

Let x be the distance the spring is stretched from equilibrium ($x = 0.200$ m), which corresponds to the upward displacement of the 30.0-kg mass. Also let $U_g = 0$ be measured with respect to the lowest position of the 20.0-kg mass when the system is released from rest. Finally, define v as the speed of both blocks at the moment the spring passes through its unstretched position. Since all forces are conservative, conservation of energy yields

$$\Delta K + \Delta U_s + \Delta U_g = 0$$

Solving for each variable,

$$\Delta K = \frac{1}{2}(m_1 + m_2)v^2 - 0 = \frac{1}{2}(50.0 \text{ kg})v^2 = (25.0 \text{ kg})v^2$$

$$\Delta U_s = 0 - \frac{1}{2}kx^2 = -\frac{1}{2}(250 \text{ N}/\text{m})(0.200 \text{ m})^2 = -5.00 \text{ N} \cdot \text{m}$$

$$\Delta U_g = (m_2 \sin\theta - m_1)gx$$
$$\Delta U_g = \left[(20.0 \text{ kg})\sin 40.0° - 30.0 \text{ kg}\right](9.80 \text{ m}/\text{s}^2)(0.200 \text{ m}) = -33.6 \text{ N} \cdot \text{m}$$

Substituting into our equation representing the energy version of the isolated system model,

$$(25.0 \text{ kg})v^2 - 5.00 \text{ N} \cdot \text{m} - 33.6 \text{ N} \cdot \text{m} = 0$$

Solving for v gives $\qquad v = 1.24 \text{ m}/\text{s}$ $\qquad\qquad$ ◊

61. A block of mass 0.500 kg is pushed against a horizontal spring of negligible mass until the spring is compressed a distance x (Fig. P8.61). The force constant of the spring is 450 N/m. When it is released, the block travels along a frictionless, horizontal surface to point B, the bottom of a vertical circular track of radius $R = 1.00$ m, and continues to move up the track. The speed of the block at the bottom of the track is $v_B = 12.0$ m/s, and the block experiences an average friction force of 7.00 N while sliding up the track. (a) What is x? (b) What speed do you predict for the block at the top of the track? (c) Does the block actually reach the top of the track, or does it fall off before reaching the top?

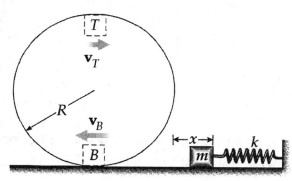

Figure P8.61

Solution The energy of the block-spring system is conserved in the firing of the block. Therefore,

(a) $\frac{1}{2}kx^2 = \frac{1}{2}mv^2$ or $\frac{1}{2}(450 \text{ N / m})x^2 = \frac{1}{2}(0.500 \text{ kg})(12.0 \text{ m / s})^2$

Thus, $x = 0.400$ m   ◊

(b) To find speed of block at the top, we consider the block-Earth system.

$$\left(K + U_g\right)_B - f_k x = \left(K + U_g\right)_T : \qquad \left(mgh_B + \frac{1}{2}mv_B^2\right) - f(\pi R) = \left(mgh_T + \frac{1}{2}mv_T^2\right)$$

Substituting, $mgh_T = (0.500 \text{ kg})(9.80 \text{ m / s}^2)(2.00 \text{ m}) = 9.80$ J

We have $0 + \frac{1}{2}(0.500 \text{ kg})(12.0 \text{ m / s})^2 - (7.00 \text{ N})(\pi)(1.00 \text{ m}) = 9.80 \text{ J} + \frac{1}{2}(0.500 \text{ kg})v_T^2$

and $0.250v_T^2 = 4.21$

Thus, $v_T = 4.10$ m / s   ◊

(c) The block falls if $a_c < g$ $a_c = \frac{v_T^2}{R} = \frac{(4.10 \text{ m / s})^2}{1.00 \text{ m}} = 16.8 \text{ m / s}^2$

Therefore, $a_c > g$. Some downward normal force is required along with the block's weight to provide the centripetal acceleration, and the block stays on the track.   ◊

71. A ball whirls around in a vertical circle at the end of a string. If the total energy of the ball-Earth system remains constant, show that the tension in the string at the bottom is greater than the tension at the top by six times the weight of the ball.

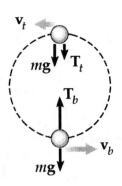

Solution

Applying Newton's second law at the bottom (*b*) and top (*t*) of the circular path gives

$$T_b - mg = \frac{mv_b^2}{R} \qquad \qquad [1]$$

$$-T_t - mg = -\frac{mv_t^2}{R} \qquad \qquad [2]$$

Adding Equations [2] and [1] gives $\quad T_b = T_t + 2mg + \dfrac{m\left(v_b^2 - v_t^2\right)}{R} \qquad [3]$

Also, energy must be conserved; that is, $\Delta K + \Delta U = 0$

So, $\qquad\qquad\qquad \frac{1}{2}m\left(v_b^2 - v_t^2\right) + \left(0 - 2mgR\right) = 0$

or $\qquad\qquad\qquad \dfrac{v_b^2 - v_t^2}{R} = 4g \qquad\qquad\qquad\qquad [4]$

Substituting [4] into [3] gives $\qquad T_b = T_t + 6mg \qquad\qquad\qquad \lozenge$

Chapter 9
LINEAR MOMENTUM AND COLLISIONS

EQUATIONS AND CONCEPTS

The **linear momentum p** of a particle is defined as the product of its mass m and its velocity **v**. *This vector equation is equivalent to three component scalar equations, one along each of the coordinate axes.*

$$\mathbf{p} \equiv m\mathbf{v} \tag{9.2}$$

The **net force** acting on a particle is equal to the time rate of change of the linear momentum of the particle. *This is an alternative form of Newton's second law.*

$$\sum \mathbf{F} = \frac{d(m\mathbf{v})}{dt} = \frac{d\mathbf{p}}{dt} \tag{9.3}$$

The **total momentum of an isolated pair of particles** remains constant. This is true because force is the time rate of change of momentum and the force of particle 1 on particle 2 is equal and opposite to the force of particle 2 on particle 1. *Momentum is a vector quantity and the momentum along each of the coordinate directions is independently conserved.*

$$\mathbf{p}_{tot} = \mathbf{p}_1 + \mathbf{p}_2 = \text{constant} \tag{9.4}$$

In general, the **law of conservation of momentum states**: when the **net external force** acting on a system of particles is zero, the total linear momentum of the system is conserved. There may be **internal** forces acting between particles within the system. *This fundamental law is especially useful in treating problems involving collisions between two bodies.*

$$\mathbf{p}_{1i} + \mathbf{p}_{2i} = \mathbf{p}_{1f} + \mathbf{p}_{2f} \tag{9.5}$$

The **impulse-momentum theorem** (Equation 9.8) states that the impulse of a force **F** acting on a particle equals the change in the momentum of the particle.

$$\Delta \mathbf{p} = \mathbf{p}_f - \mathbf{p}_i = \int_{t_i}^{t_f} \mathbf{F}\, dt \qquad (9.8)$$

Impulse is a vector quantity, defined by Equation 9.9. The magnitude of an impulse caused by a force is equal to the area under the force-time curve. Impulsive forces are generally time-varying forces and act for short time; therefore the time-averaged force can be used as in Equation 9.11 below.

$$\mathbf{I} \equiv \int_{t_i}^{t_f} \mathbf{F}\, dt \qquad (9.9)$$

The **time-averaged force** $\overline{\mathbf{F}}$ is defined as that constant force which would give the same impulse to a particle as an actual time-varying force over the same time interval Δt. See the shaded area in the figure above. *The impulse approximation assumes that one of the forces acting on a particle acts for a short time and is much larger than any other force present. This approximation is usually made in collision problems, where the force is the contact force between the particles during the collision.*

$$\overline{\mathbf{F}} \equiv \frac{1}{\Delta t} \int_{t_i}^{t_f} \mathbf{F}\, dt \qquad (9.10)$$

and

$$\mathbf{I} = \overline{\mathbf{F}}\, \Delta t \qquad (9.11)$$

An **elastic collision** is one in which both linear momentum and kinetic energy are conserved.

$$\left. \begin{array}{l} \mathbf{p}_1 + \mathbf{p}_2 = \text{const} \\ \\ K_1 + K_2 = \text{const} \end{array} \right\} \quad \text{(Elastic collision)}$$

An **inelastic collision** is one in which only linear momentum is conserved. *A perfectly inelastic collision is an inelastic collision in which the two bodies stick together after the collision.*

$$\mathbf{p}_1 + \mathbf{p}_2 = \text{const} \quad \text{(Inelastic collision)}$$

The **common velocity** following a perfectly inelastic collision between two bodies can be calculated in terms of the two mass values and the two initial velocities.

$$\mathbf{v}_f = \frac{m_1\mathbf{v}_{1i} + m_2\mathbf{v}_{2i}}{m_1 + m_2} \tag{9.14}$$

The **relative velocity** before a perfectly elastic collision between two bodies equals the negative of the relative velocity of the two bodies following the collision.

$$v_{1i} - v_{2i} = -\left(v_{1f} - v_{2f}\right) \tag{9.19}$$

The **final velocities following a one-dimensional elastic collision** between two particles can be calculated when the masses and initial velocities of both particles are known.

$$v_{1f} = \left(\frac{m_1 - m_2}{m_1 + m_2}\right)v_{1i} + \left(\frac{2m_2}{m_1 + m_2}\right)v_{2i} \tag{9.20}$$

$$v_{2f} = \left(\frac{2m_1}{m_1 + m_2}\right)v_{1i} + \left(\frac{m_2 - m_1}{m_1 + m_2}\right)v_{2i} \tag{9.21}$$

An **important special case** occurs when the second particle (m_2, the "target") is initially at rest. *Remember the appropriate algebraic signs (designating direction) must be included for v_{1i} and v_{2i}.*

$$v_{1f} = \left(\frac{m_1 - m_2}{m_1 + m_2}\right)v_{1i} \tag{9.22}$$

$$v_{2f} = \left(\frac{2m_1}{m_1 + m_2}\right)v_{1i} \tag{9.23}$$

The **coordinates of the center of mass** of n particles with individual coordinates of (x_1, y_1, z_1), (x_2, y_2, z_2), $(x_3, y_3, z_3) \ldots$ and masses of $m_1, m_2, m_3 \ldots$ are given by Eq. 9.28 and 9.29. The total mass $M = \Sigma m_i$ where the sum runs over all n particles. *The center of mass of a homogeneous, symmetric body must lie on an axis of symmetry and on any plane of symmetry.*

$$x_{CM} \equiv \frac{\sum_i m_i x_i}{\sum_i m_i} = \frac{\sum_i m_i x_i}{M} \tag{9.28}$$

$$y_{CM} \equiv \frac{\sum_i m_i y_i}{M} \tag{9.29}$$

$$z_{CM} \equiv \frac{\sum_i m_i z_i}{M}$$

The **position vector** can also be used to locate the center of mass of a collection of particles.

$$\mathbf{r}_{CM} \equiv \frac{\sum_i m_i \mathbf{r}_i}{M} \tag{9.30}$$

The **center of mass of an extended object** can be calculated by integrating over the total length, area, or volume which includes the total mass M.

$$\mathbf{r}_{CM} = \frac{1}{M}\int \mathbf{r}\, dm \tag{9.33}$$

In this expression for the **velocity of the center of mass of a system of particles**, v_i is the velocity of the ith particle and M is the total mass of the system.

$$\mathbf{v}_{CM} = \frac{d\mathbf{r}_{CM}}{dt} = \frac{\sum_i m_i \mathbf{v}_i}{M} \tag{9.34}$$

The **acceleration of the center of mass** of a system of particles depends on the value of the acceleration for each of the individual particles.

$$\mathbf{a}_{CM} = \frac{d\mathbf{v}_{CM}}{dt} = \frac{1}{M}\sum_i m_i \mathbf{a}_i \tag{9.36}$$

The **net external force** acting on a system of particles equals the product of the total mass of the system and the acceleration of the center of mass. *The center of mass of a system of particles of combined mass M moves like an equivalent single particle of mass M would move if acted on by the same external force.*

$$\sum \mathbf{F}_{ext} = M\mathbf{a}_{CM} \tag{9.38}$$

The **total momentum of a system of particles** is equal to the total mass M multiplied by the velocity of the center of mass.

$$M\mathbf{v}_{CM} = \mathbf{p}_{tot} = \text{constant} \tag{9.39}$$

$$(\text{when } \Sigma \mathbf{F}_{ext} = 0)$$

The basic **expression for rocket propulsion** states that the change in speed of the rocket, as the mass decreases from M_i to M_f, is proportional to the exhaust speed of the ejected gases.

$$v_f - v_i = v_e \ln\left(\frac{M_i}{M_f}\right) \tag{9.41}$$

The **thrust on a rocket** is the force exerted on the rocket by the ejected exhaust gases. *The thrust increases as the exhaust speed increases and as the burn rate increases.*

$$\text{Thrust} = M\frac{dv}{dt} = \left| v_e \frac{dM}{dt} \right| \qquad (9.42)$$

SUGGESTIONS, SKILLS, AND STRATEGIES

The following procedure is recommended when dealing with problems involving collisions between two objects:

- Set up a coordinate system and show the velocities with respect to that system. That is, objects moving in the direction selected as the positive direction of the x axis are considered as having a positive velocity and negative if moving in the negative x direction. It is convenient to have the x axis coincide with one of the initial velocities.

- In your sketch of the coordinate system, draw all velocity vectors with labels and include all the given information.

- Write expressions for the momentum of each object before and after the collision. **(In two-dimensional collision problems, write expressions for the x and y components of momentum before and after the collision.)** Remember to include the appropriate signs for the velocity vectors.

- Now write expressions for the **total momentum before and after the collision and equate the two.** (For two-dimensional collisions, this expression should be written for the momentum in both the x and y directions.) It is important to emphasize that it is the momentum of the **system** (the two colliding objects) that is conserved, not the momentum of the individual objects.

- If the collision is **inelastic**, you should then proceed to solve the momentum equations for the unknown quantities.

- If the collision is **elastic**, kinetic energy is also conserved, so you can equate the total kinetic energy before the collision to the total kinetic energy after the collision. This gives an additional relationship between the various velocities. The conservation of kinetic energy for elastic collisions leads to the expression $v_{1i} - v_{2i} = -(v_{1f} - v_{2f})$, which is often easier to use in solving elastic collision problems than is an expression for conservation of kinetic energy.

REVIEW CHECKLIST

You should be able to:

▷ Apply the law of conservation of linear momentum to a two-body system. (Section 9.1)

▷ Calculate linear momentum and change in momentum (impulse) due to a time-averaged force. (Sections 9.1 and 9.2)

▷ Determine the impulse delivered by a time-varying force from the area under the Force vs. Distance curve. (Section 9.2)

▷ Calculate the common velocity following an inelastic collision and the individual velocities following an elastic collision between two objects. (Section 9.3)

▷ Calculate the final velocities of two objects following a two-dimensional inelastic collision. (Section 9.4)

▷ Determine the center of mass of a system of discrete point-masses and, by integration, for a homogenous solid of sufficient symmetry. Determine also the motion (coordinate, velocity, and acceleration) of the center of mass of a system of objects. (Sections 9.5 and 9.6)

▷ Make calculations using the basic equations of rocket propulsion. (Section 9.7)

Important points to remember:

▷ The impulse of a force acting on a particle during some time interval equals the **change** in momentum of the particle, and the impulse equals the area under the force-time graph.

▷ The momentum of any isolated system (one for which the net external force is zero) is conserved, regardless of the nature of the forces between the masses which comprise the system.

▷ There are two types of collisions that can occur between two particles, namely elastic and inelastic collisions. A **perfectly** inelastic collision is an inelastic collision in which the colliding particles stick together after the collision, and hence move as a composite particle.

▷ The conservation of linear momentum applies not only to head-on collisions (one-dimensional), but also to glancing collisions (two- or three-dimensional). For example, in a two-dimensional collision, the total momentum in the x direction is conserved and the total momentum in the y direction is conserved.

▷ The equations for momentum and kinetic energy can be used to calculate the final velocities in a two-body head-on elastic collision, and to calculate the final velocity and the change of kinetic energy in a two-body system for a completely inelastic collision.

ANSWERS TO SELECTED CONCEPTUAL QUESTIONS

3. If two particles have equal kinetic energies, are their momenta necessarily equal? Explain.

Answer No, their momenta need not be equal. Equal kinetic energies means that $mv^2/2$ is the same for both. If they are moving and their masses are different, then their speeds must also be different. Let the speed of the lighter particle be larger by the factor α. Then v^2 is larger by the factor α^2, and the mass of the lighter particle must be smaller by the factor $1/\alpha^2$. The momentum takes the increased speed into account only once, not twice like the kinetic energy. The less massive particle will have a momentum smaller by the factor $(1/\alpha^2)\alpha = 1/\alpha$.

□   □   □   □

7. Explain how linear momentum is conserved when a ball bounces from a floor.

Answer The ball's downward momentum increases as it accelerates downward. A larger upward momentum change occurs when it touches the floor and rebounds. The outside forces of gravity and the normal force inject impulses to change its momentum.

If we think of ball-and-Earth-together as our system, these forces are internal and do not change the total momentum. It is conserved, if we neglect the curvature of the Earth's orbit. As the ball falls down, the Earth lurches up to meet it, on the order of 10^{25} times more slowly. Then, ball and Earth bounce off each other and separate. In other words, while you dribble a ball, you also dribble the Earth.

□   □   □   □

13. A sharpshooter fires a rifle while standing with the butt of the gun against his shoulder. If the forward momentum of a bullet is the same as the backward momentum of the gun, why isn't it as dangerous to be hit by the gun as by the bullet?

Answer It is the product mv which is the same for both the bullet and the gun. The bullet has a large velocity and a small mass, while the gun has a small velocity and a large mass. Furthermore, the bullet carries much more kinetic energy than the gun.

□   □   □   □

26. Does the center of mass of a rocket in free space accelerate? Explain. Can the speed of a rocket exceed the exhaust speed of the fuel? Explain.

Answer The center of mass of a rocket, plus its exhaust does not accelerate: momentum must be conserved. However, if you consider the "rocket" to be the mechanical system plus the unexpended fuel, then it becomes obvious that the center of mass of that "rocket" does accelerate.

Basically, a rocket is no more complicated than two rubber balls that are pressed together, and released. One springs in one direction with one velocity; the other springs in the other direction with a velocity equal to

$$v_2 = v_1(m_1 / m_2)$$

In a similar manner, the mass of the rocket times its speed must equal the mass of all parts of the exhaust, times their speed. After a sufficient quantity of fuel has been exhausted the ratio m_2/m_2 will be greater than 1; then the speed of the rocket **can** exceed the speed of the exhaust.

□   □   □   □

SOLUTIONS TO SELECTED END-OF-CHAPTER PROBLEMS

7. An estimated force-time curve for a baseball struck by a bat is shown in Figure P9.7. From this curve, determine (a) the impulse delivered to the ball, (b) the average force exerted on the ball, and (c) the peak force exerted on the ball.

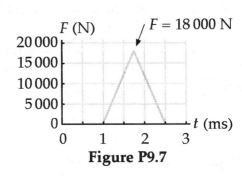

Figure P9.7

Solution The impulse delivered to the ball is equal to the area under the F-t graph. Thus,

(a) $I = \left(\dfrac{0 + 18\,000\ \text{N}}{2}\right)\left(2.5 \times 10^{-3}\ \text{s} - 1.0 \times 10^{-3}\ \text{s}\right) = 13.5\ \text{N} \cdot \text{s}$ ◊

(b) $\overline{F} = \dfrac{\int F\,dt}{\Delta t} = \dfrac{13.5\ \text{N} \cdot \text{s}}{(2.5 - 1.0)10^{-3}\ \text{s}} = 9.00\ \text{kN}$ ◊

(c)   From the graph, $F_{max} = 18\,000\ \text{N}$ ◊

9. A 3.00-kg steel ball strikes a wall with a speed of 10.0 m/s at an angle of 60.0° with the surface. It bounces off with the same speed and angle (Fig. P9.9). If the ball is in contact with the wall for 0.200 s, what is the average force exerted by the wall on the ball?

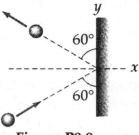

Figure P9.9

Solution

Conceptualize: If we think about the angle as a variable and consider the limiting cases, then the force should be zero when the angle is 0° (no contact between the ball and the wall). When the angle is 90° the force will be its maximum and can be found from the momentum-impulse equation, so that $F = \Delta p / \Delta t < 300$ N, and the force on the ball must be directed to the left.

Categorize: Use the momentum-impulse equation to find the force, and carefully consider the direction of the velocity vectors by defining up and to the right as positive.

Analyze: $\Delta \mathbf{p} = \mathbf{F} \Delta t$

$$\Delta p_y = m\left(v_{fy} - v_{iy}\right) = m(v\cos 60.0° - v\cos 60.0°) = 0$$

So the wall does not exert a force on the ball in the y direction.

$$\Delta p_x = m\left(v_{fx} - v_{ix}\right) = m(-v\sin 60.0° - v\sin 60.0°) = -2mv\sin 60.0°$$

$$\Delta p_x = -2(3.00 \text{ kg})(10.0 \text{ m / s})(0.866) = -52.0 \text{ kg} \cdot \text{m / s}$$

$$\bar{\mathbf{F}} = \frac{\Delta \mathbf{p}}{\Delta t} = \frac{\Delta p_x \hat{\mathbf{i}}}{\Delta t} = \frac{-52.0\hat{\mathbf{i}} \text{ kg} \cdot \text{m / s}}{0.200 \text{ s}} = -260\hat{\mathbf{i}} \text{ N}$$ ◊

Finalize: The force is to the left and has a magnitude less than 300 N as expected.

_____

17. A 10.0-g bullet is fired into a stationary block of wood ($m=5.00$ kg). The relative motion of the bullet stops inside the block. The speed of the bullet-plus-wood combination immediately after the collision is 0.600 m/s. What was the original speed of the bullet?

Solution

Conceptualize: A reasonable speed of a bullet should be somewhere between 100 and 1 000 m/s.

Categorize: We can find the initial speed of the bullet from conservation of momentum. We are told that the block of wood was originally stationary.

Analyze: Since there is no external force on the block and bullet system, the total momentum of the system is constant so that $\Delta\mathbf{p}=0$.

$$\mathbf{p}_{1i}+\mathbf{p}_{2i}=\mathbf{p}_{1f}+\mathbf{p}_{2f}$$

$$(0.010\ 0\ \text{kg})\mathbf{v}_{1i}+0=(0.010\ 0\ \text{kg})(0.600\ \text{m}/\text{s})\hat{\mathbf{i}}+(5.00\ \text{kg})(0.600\ \text{m}/\text{s})\hat{\mathbf{i}}$$

$$\mathbf{v}_{1i}=\frac{(5.01\ \text{kg})(0.600\ \text{m}/\text{s})\hat{\mathbf{i}}}{0.010\ 0\ \text{kg}}=301\hat{\mathbf{i}}\ \text{m}/\text{s} \qquad \lozenge$$

Finalize: The speed seems reasonable, and is in fact just under the speed of sound in air (343 m/s at 20° C).

---

21. A 45.0-kg girl is standing on a plank that has a mass of 150 kg. The plank, originally at rest, is free to slide on a frozen lake, which is a flat, frictionless supporting surface. The girl begins to walk along the plank at a constant speed of 1.50 m/s relative to the plank. (a) What is her speed relative to the ice surface? (b) What is the speed of the plank relative to the ice surface?

Solution $\mathbf{v}_g$ = velocity of the girl relative to the ice

$\mathbf{v}_{gp}$ = velocity of the girl relative to the plank

$\mathbf{v}_p$ = velocity of the plank relative to the ice

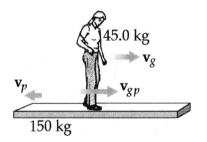

The girl and the plank exert forces on each other, but the ice isolates them from outside horizontal forces. Therefore, the net momentum is zero for the combined girl plus plank system.

$$0=m_g\mathbf{v}_g+m_p\mathbf{v}_p$$

Further, the relation among relative speeds can be written $\mathbf{v}_g=\mathbf{v}_{gp}+\mathbf{v}_p$

$$\mathbf{v}_g=1.50\hat{\mathbf{i}}\ \text{m}/\text{s}+\mathbf{v}_p$$

We substitute: $\qquad 0=(45.0\ \text{kg})(1.50\hat{\mathbf{i}}\ \text{m}/\text{s}+\mathbf{v}_p)+(150\ \text{kg})\mathbf{v}_p$

$$(195\ \text{kg})\mathbf{v}_p=-(45.0\ \text{kg})(1.50\hat{\mathbf{i}}\ \text{m}/\text{s})$$

(b) $$\mathbf{v}_p = -0.346\hat{\mathbf{i}} \ \text{m}/\text{s}$$ ◊

(a) $$\mathbf{v}_g = 1.50\hat{\mathbf{i}} - 0.346\hat{\mathbf{i}} \ \text{m}/\text{s} = 1.15\hat{\mathbf{i}} \ \text{m}/\text{s}$$ ◊

---

23. A neutron in a nuclear reactor makes an elastic head-on collision with the nucleus of a carbon atom initially at rest. (a) What fraction of the neutron's kinetic energy is transferred to the carbon nucleus? (b) If the initial kinetic energy of the neutron is 1.60×10^{-13} J, find its final kinetic energy and the kinetic energy of the carbon nucleus after the collision. (The mass of the carbon nucleus is nearly 12.0 times the mass of the neutron.)

Solution

(a) This is a perfectly elastic head-on collision, so we use Eq. 8.19:

$$v_{1i} - v_{2i} = -\left(v_{1f} - v_{2f}\right)$$

Let object 1 be the neutron, and object 2 be the carbon nucleus, with $m_2 = 12m_1$.

Since $v_{2i} = 0$, $\qquad\qquad v_{2f} = v_{1i} + v_{1f}$

Now, by conservation of momentum, $\quad m_1 v_{1i} + m_2 v_{2i} = m_1 v_{1f} + m_2 v_{2f}$

or $\qquad\qquad m_1 v_{1i} = m_1 v_{1f} + 12 m_1 v_{2f}$

Substituting our velocity equation, $\quad v_{1i} = v_{1f} + 12(v_{1i} + v_{1f})$

We solve $-11 v_{1i} = 13 v_{1f}$: $\qquad v_{1f} = -\dfrac{11}{13} v_{1i}$

and $\qquad\qquad v_{2f} = v_{1i} - \dfrac{11}{13} v_{1i} = \dfrac{2}{13} v_{1i}$

The neutron's original kinetic energy is $\frac{1}{2} m_1 v_{1i}^{\ 2}$

The carbon's final kinetic energy is $\quad \frac{1}{2} m_2 v_{2f}^{\ 2} = \frac{1}{2}(12 m_1)\left(\dfrac{2}{13}\right)^2 v_{1i}^{\ 2} = \left(\dfrac{48}{169}\right)\left(\dfrac{1}{2}\right)m_1 v_{1i}^{\ 2}$

So, $\qquad \dfrac{48}{169} = 0.284 \qquad$ or $\qquad$ 28.4% of the total energy is transferred. ◊

(b) For the carbon nucleus, $\qquad\qquad K_{2f} = (0.284)\left(1.60 \times 10^{-13} \ \text{J}\right) = 45.4 \ \text{fJ}$ ◊

The collision is perfectly elastic, so the neutron retains the rest of the energy,

$$K_{1f} = (1.60 - 0.454) \times 10^{-13} \ \text{J} = 1.15 \times 10^{-13} \ \text{J} = 115 \ \text{fJ}$$ ◊

---

25. A 12.0-g wad of sticky clay is hurled horizontally at a 100-g wooden block initially at rest on a horizontal surface. The clay sticks to the block. After impact, the block slides 7.50 m before coming to rest. If the coefficient of friction between the block and the surface is 0.650, what was the speed of the clay immediately before impact?

Solution   We use the momentum version of the isolated system model to analyze the collision, and the energy version of the isolated system model to analyze the subsequent sliding process. The collision, for which figures (1) and (2) are before and after pictures, is **totally inelastic**, and momentum is conserved for the system of clay and block:

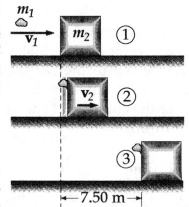

$$m_1 v_1 = (m_1 + m_2) v_2$$

In the sliding process occurring between figures (2) and (3), the original kinetic energy of the surface, block, and clay is equal to the increase in internal energy of the system due to friction:

$$\frac{1}{2}(m_1 + m_2)v_2^2 = f_f L$$

$$\frac{1}{2}(m_1 + m_2)v_2^2 = \mu(m_1 + m_2)gL$$

Solving for v_2, $v_2 = \sqrt{2\mu L g} = \sqrt{2(0.650)(7.50 \text{ m})(9.80 \text{ m}/\text{s}^2)} = 9.77 \text{ m}/\text{s}$

From the momentum conservation equation,

$$v_1 = \left(\frac{m_1 + m_2}{m_1}\right)v_2 = \left(\frac{112 \text{ g}}{12.0 \text{ g}}\right)(9.77 \text{ m}/\text{s}) = 91.2 \text{ m}/\text{s} \qquad \Diamond$$

---

33. A billiard ball moving at 5.00 m/s strikes a stationary ball of the same mass. After the collision, the first ball moves, at 4.33 m/s, at an angle of 30.0° with respect to the original line of motion. Assuming an elastic collision (and ignoring friction and rotational motion), find the struck ball's velocity after the collision.

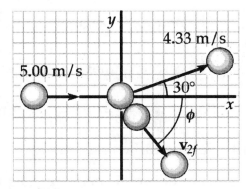

Solution   Call each mass m, and call $\mathbf{v}_{2f}$ the velocity of the second ball after the collision, as in the figure. Then apply conservation of momentum to the two-ball system.

In the x direction, $\qquad m(5.00 \text{ m / s}) = m(4.33 \text{ m / s})\cos 30° + mv_{2fx}$

$$v_{2fx} = 1.25 \text{ m / s}$$

In the y direction, $\qquad 0 = m(4.33 \text{ m / s})\sin 30° + mv_{2fy}$

$$v_{2fy} = -2.17 \text{ m / s}$$

$$\mathbf{v}_{2f} = 1.25\hat{\mathbf{i}} - 2.17\hat{\mathbf{j}} \quad (\text{or } 2.50 \text{ m/s at } -60.0°) \qquad \lozenge$$

We did not have to use the fact that the collision is elastic.

35. An object of mass 3.00 kg, moving with an initial velocity of $5.00\hat{\mathbf{i}}$ m/s, collides with and sticks to an object of mass 2.00 kg with an initial velocity of $-3.00\hat{\mathbf{j}}$ m/s. Find the final velocity of the composite object.

Solution

Momentum of the two-object system is conserved, with both objects having the same final velocity:

$$m_1\mathbf{v}_{1i} + m_2\mathbf{v}_{2i} = m_1\mathbf{v}_{1f} + m_2\mathbf{v}_{2f}$$

$$(3.00 \text{ kg})(5.00\hat{\mathbf{i}} \text{ m / s}) + (2.00 \text{ kg})(-3.00\hat{\mathbf{j}} \text{ m / s}) = (3.00 \text{ kg} + 2.00 \text{ kg})\mathbf{v}_f$$

$$\mathbf{v}_f = \frac{15.0\hat{\mathbf{i}} - 6.00\hat{\mathbf{j}}}{5.00} \text{ m / s} = (3.00\hat{\mathbf{i}} - 1.20\hat{\mathbf{j}}) \text{ m / s} \qquad \lozenge$$

Related Calculation: Compute the kinetic energy of the system both before and after the collision; show that kinetic energy is not conserved.

$$K_{1i} + K_{2i} = \tfrac{1}{2}(3.00 \text{ kg})(5.00 \text{ m / s})^2 + \tfrac{1}{2}(2.00 \text{ kg})(3.00 \text{ m / s})^2$$

$$K_{1i} + K_{2i} = 46.5 \text{ J}$$

$$K_{1f} + K_{2f} = \tfrac{1}{2}(5.00 \text{ kg})\left[(3.00 \text{ m / s})^2 + (1.20 \text{ m / s})^2\right] = 26.1 \text{ J} \qquad \lozenge$$

37. An unstable atomic nucleus of mass 17.0×10^{-27} kg initially at rest disintegrates into three particles. One of the particles, of mass 5.00×10^{-27} kg, moves along the y axis with a speed of 6.00×10^6 m/s. Another particle, of mass 8.40×10^{-27} kg, moves along the x axis with a speed of 4.00×10^6 m/s. Find (a) the velocity of the third particle and (b) the total kinetic energy increase in the process.

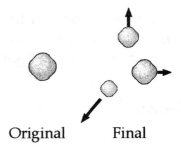

Original   Final

Solution

(a) With three particles, the total final momentum of the system is

$$m_1 \mathbf{v}_{1f} + m_2 \mathbf{v}_{2f} + m_3 \mathbf{v}_{3f}$$

and it must be zero to equal the original momentum.

The mass of the third particle is $m_3 = (17.0 - 5.00 - 8.40) \times 10^{-27}$ kg

or $\qquad\qquad m_3 = 3.60 \times 10^{-27}$ kg

The total momentum is zero: $\qquad m_1 \mathbf{v}_{1f} + m_2 \mathbf{v}_{2f} + m_3 \mathbf{v}_{3f} = 0$

Solving for $\mathbf{v}_{3f}$: $\qquad\qquad \mathbf{v}_{3f} = -\dfrac{m_1 \mathbf{v}_{1f} + m_2 \mathbf{v}_{2f}}{m_3}$

$$\mathbf{v}_{3f} = -\frac{\left(3.00\hat{\jmath} + 3.36\hat{\imath}\right) \times 10^{-20} \text{ kg} \cdot \text{m/s}}{3.60 \times 10^{-27} \text{ kg}} = \left(-9.33\hat{\imath} - 8.33\hat{\jmath}\right) \text{ Mm/s} \qquad \Diamond$$

(b) The original kinetic energy of the system is zero.

The final kinetic energy is $\qquad K = K_{1f} + K_{2f} + K_{3f}$:

$$K_{1f} = \tfrac{1}{2}\left(5.00 \times 10^{-27} \text{ kg}\right)\left(6.00 \times 10^6 \text{ m/s}\right)^2 = 9.00 \times 10^{-14} \text{ J}$$

$$K_{2f} = \tfrac{1}{2}\left(8.40 \times 10^{-27} \text{ kg}\right)\left(4.00 \times 10^6 \text{ m/s}\right)^2 = 6.72 \times 10^{-14} \text{ J}$$

$$K_{3f} = \tfrac{1}{2}\left(3.60 \times 10^{-27} \text{ kg}\right)\left(\left(-9.33 \times 10^6 \text{ m/s}\right)^2 + \left(-8.33 \times 10^6 \text{ m/s}\right)^2\right) = 28.2 \times 10^{-14} \text{ J}$$

and $\qquad\qquad\qquad\qquad\qquad K = 9.00 \times 10^{-14} \text{ J} + 6.72 \times 10^{-14} \text{ J} + 28.2 \times 10^{-14} \text{ J}$

$$K = 4.39 \times 10^{-13} \text{ J} = 439 \text{ fJ} \qquad \Diamond$$

41. A uniform piece of sheet steel is shaped as in Figure P9.41. Compute the x and y coordinates of the center of mass of the piece.

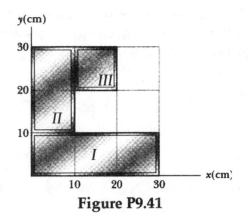

y(cm)

x(cm)

Figure P9.41

Solution

Conceptualize: By inspection, it appears that the center of mass is located at about

$$\left(12\hat{i} + 13\hat{j}\right) \text{ cm}$$

Categorize: Think of the sheet as composed of three sections, and consider the mass of each section to be at the geometric center of that section. Define the mass per unit area to be σ, and number the rectangles as shown. We can then calculate the mass and identify the center of mass of each section.

Analyze: $m_I = (30.0 \text{ cm})(10.0 \text{ cm})\sigma$ $CM_I = (15.0 \text{ cm, } 5.00 \text{ cm})$

$m_{II} = (10.0 \text{ cm})(20.0 \text{ cm})\sigma$ $CM_{II} = (5.00 \text{ cm, } 20.0 \text{ cm})$

$m_{III} = (10.0 \text{ cm})(10.0 \text{ cm})\sigma$ $CM_{III} = (15.0 \text{ cm, } 25.0 \text{ cm})$

The overall center of mass is at a point defined by the vector equation

$$\mathbf{r}_{CM} \equiv \left(\sum m_i \mathbf{r}_i\right)\Big/\sum m_i$$

Substituting the appropriate values, $\mathbf{r}_{CM}$ is calculated to be:

$$\mathbf{r}_{CM} = \frac{\sigma\left[(300)(15.0\hat{i} + 5.00\hat{j}) + (200)(5.00\hat{i} + 20.0\hat{j}) + (100)(15.0\hat{i} + 25.0\hat{j}) \text{ cm}^3\right]}{\sigma\left(300 \text{ cm}^2 + 200 \text{ cm}^2 + 100 \text{ cm}^2\right)}$$

$$\mathbf{r}_{CM} = \frac{\left(45.0\hat{i} + 15.0\hat{j} + 10.0\hat{i} + 40.0\hat{j} + 15.0\hat{i} + 25.0\hat{j}\right)}{6.00} \text{ cm}$$

$$\mathbf{r}_{CM} = \left(11.7\hat{i} + 13.3\hat{j}\right) \text{ cm} \qquad \Diamond$$

Finalize: The coordinates are close to our eyeball estimate. In solving this problem, we could have chosen to divide the original shape some other way, but the answer would be the same. This problem also shows that the center of mass can lie outside the boundary of the object.

45. A 2.00-kg particle has a velocity $(2.00\hat{\imath} - 3.00\hat{\jmath})$ m/s, and a 3.00-kg particle has a velocity $(1.00\hat{\imath} + 6.00\hat{\jmath})$ m/s. Find (a) the velocity of the center of mass and (b) the total momentum of the system.

Solution Use $\mathbf{v}_{CM} = \dfrac{m_1\mathbf{v}_1 + m_2\mathbf{v}_2}{m_1 + m_2}$ and $\mathbf{p}_{CM} = (m_1 + m_2)\mathbf{v}_{CM}$

(a) $\mathbf{v}_{CM} = \dfrac{(2.00\text{ kg})\left[(2.00\hat{\imath} - 3.00\hat{\jmath})\text{ m / s}\right] + (3.00\text{ kg})\left[(1.00\hat{\imath} + 6.00\hat{\jmath})\text{ m / s}\right]}{(2.00\text{ kg} + 3.00\text{ kg})}$

$\mathbf{v}_{CM} = (1.40\hat{\imath} + 2.40\hat{\jmath})\text{ m / s}$ ◊

(b) $\mathbf{p}_{CM} = (2.00\text{ kg} + 3.00\text{ kg})\left[(1.40\hat{\imath} + 2.40\hat{\jmath})\text{ m / s}\right] = (7.00\hat{\imath} + 12.0\hat{\jmath})\text{ kg}\cdot\text{m / s}$ ◊

---

47. Romeo (77.0 kg) entertains Juliet (55.0 kg) by playing his guitar from the rear of their boat at rest in still water, 2.70 m away from Juliet, who is in the front of the boat. After the serenade, Juliet carefully moves to the rear of the boat (away from shore) to plant a kiss on Romeo's cheek. How far does the 80.0-kg boat move toward the shore it is facing?

Solution No outside forces act on the boat-plus-lovers system, so its momentum is conserved at zero and its center of mass stays fixed: $x_{CM,i} = x_{CM,f}$.

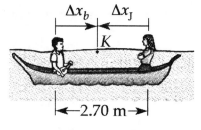

Define K to be the point where they kiss, and Δx_J and Δx_b as shown in the figure. Since Romeo moves with the boat (and thus $\Delta x_{Romeo} = \Delta x_b$), let m_b be the combined mass of Romeo and the boat.

Then, $m_J\Delta x_J + m_b\Delta x_b = 0$

Choosing the x axis to point away from the shore,

$$(55.0\text{ kg})\Delta x_J + (77.0\text{ kg} + 80.0\text{ kg})\Delta x_b = 0$$

and $\Delta x_J = -2.85\Delta x_b$

As Juliet moves away from shore, the boat and Romeo glide toward the shore until the original 2.70 m gap between them is closed:

$$\Delta x_J - \Delta x_b = 2.70\text{ m}$$

Substituting, we find $\Delta x_b = -0.700\text{ m}$ or 0.700 m towards the shore ◊

---

49. The first stage of a Saturn V space vehicle consumed fuel and oxidizer at the rate of 1.50×10^4 kg/s, with an exhaust speed of 2.60×10^3 m/s. (a) Calculate the thrust produced by these engines. (b) Find the acceleration of the vehicle just as it lifted off the launch pad, if the vehicle's initial mass was 3.00×10^6 kg. **Note:** You must include the gravitational force to solve part (b)]

Solution

Conceptualize: The thrust must be at least equal to the weight of the rocket (≈ 30 MN); otherwise the launch would not have been successful! However, since Saturn V rockets accelerate rather slowly compared to the acceleration of falling objects, the thrust should be less than about twice the rocket's weight, so that $0 < a < g$.

Categorize: Use Newton's second law to find the force and acceleration from the changing momentum.

Analyze:

(a) The thrust, F, is equal to the time rate of change of momentum as fuel is exhausted from the rocket.

$$F = \frac{dp}{dt} = \frac{d}{dt}(mv_e)$$

Since v_e is a constant exhaust velocity,

$$F = v_e(dm/dt) \quad \text{where} \quad dm/dt = 1.50 \times 10^4 \text{ kg / s}$$

$$\text{and} \quad v_e = 2.60 \times 10^3 \text{ m / s}$$

$$F = (2.60 \times 10^3 \text{ m / s})(1.50 \times 10^4 \text{ kg / s}) = 39.0 \text{ MN} \qquad \lozenge$$

(b) Applying $\Sigma F = ma$: $(3.90 \times 10^7 \text{ N}) - (3.00 \times 10^6 \text{ kg})(9.80 \text{ m / s}^2) = (3.00 \times 10^6 \text{ kg})a$

$$a = \frac{(3.90 \times 10^7 \text{ N}) - (29.4 \times 10^6 \text{ N})}{3.00 \times 10^6 \text{ kg}} = 3.20 \text{ m / s}^2 \text{ up} \qquad \lozenge$$

Finalize: As expected, the thrust is slightly greater than the weight of the rocket, and the acceleration is about $0.3g$, so the answers appear to be reasonable. This kind of rocket science is not so complicated after all!

57. An 80.0-kg astronaut is working on the engines of his ship, which is drifting through space with a constant velocity. The astronaut, wishing to get a better view of the Universe, pushes against the ship and much later finds himself 30.0 m behind the ship. Without a thruster, the only way to return to the ship is to throw his 0.500-kg wrench directly away from the ship. If he throws the wrench with a speed of 20.0 m/s relative to the ship, how long does it take the astronaut to reach the ship?

Solution

Conceptualize: We hope the momentum of the wrench provides enough recoil so that the astronaut can reach the ship before he loses life support! We might expect the elapsed time to be on the order of several minutes based on the description of the situation.

Categorize: No external force acts on the system (astronaut plus wrench), so the total momentum is constant. Since the final momentum (astronaut plus wrench) is zero,

we have Final momentum = Initial momentum = 0

Analyze: In equation form, $m_w v_w + m_a v_a = 0$

Thus, $v_a = -m_w v_w / m_a$

$$v_a = -\frac{(0.500 \text{ kg})(20.0 \text{ m / s})}{80.0 \text{ kg}} = -0.125 \text{ m / s}$$

At this speed, the time he takes to travel to the ship is

$$t = \frac{30.0 \text{ m}}{0.125 \text{ m / s}} = 240 \text{ s} = 4.00 \text{ minutes} \qquad \Diamond$$

Finalize: The astronaut is fortunate that the wrench gave him sufficient momentum to return to the ship in a reasonable amount of time! In this problem, we were told that the astronaut was not drifting away from the ship when he threw the wrench. However, this is not possible since he did not encounter an external force that would reduce his velocity away from the ship (there is no air friction beyond earth's atmosphere). If this were a real-life situation, the astronaut would have to throw the wrench hard enough to overcome his momentum caused by his original push away from the ship, and then some.

67. A 5.00-g bullet moving with an initial speed of 400 m/s is fired into and passes through a 1.00-kg block, as in Figure P9.67. The block, initially at rest on a frictionless, horizontal surface, is connected to a spring of force constant 900 N/m. If the block moves 5.00 cm to the right after impact, find (a) the speed at which the bullet emerges from the block and (b) the mechanical energy converted into internal energy in the collision.

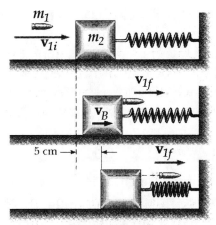

**Figure P9.67
(modified)**

Solution   First find the initial velocity of the block, using conservation of energy during compression of the spring.   Assume that the bullet has passed completely through the block before the spring has started to compress. Note that conservation of momentum does not apply here.

$$\tfrac{1}{2}m_2 v_B{}^2 = \tfrac{1}{2}kx^2 \qquad \text{becomes} \qquad \tfrac{1}{2}(1.00\ \text{kg})v_B{}^2 = \tfrac{1}{2}(900\ \text{N}/\text{m})(0.0500\ \text{m})^2$$

$$\text{yielding} \qquad v_B = \sqrt{2.25}\ \text{m}/\text{s} = 1.50\ \text{m}/\text{s}$$

(a)   When the bullet collides with the block, it is the momentum that is conserved:

$$m_1 v_{1i} + m_2 v_{2i} = m_1 v_{1f} + m_2 v_B$$

so $$v_{1f} = (m_1 v_{1i} - m_2 v_B)/m_1$$

$$v_{1f} = \frac{(5.00 \times 10^{-3}\ \text{kg})(400\ \text{m}/\text{s}) - (1.00\ \text{kg})(1.50\ \text{m}/\text{s})}{5.00 \times 10^{-3}\ \text{kg}} = 100\ \text{m}/\text{s}$$

(b)   We use the work-energy theorem to find the energy lost in the collision. Before the collision, the block is motionless, and the bullet's energy is:

$$K_1 = \tfrac{1}{2}m v_{1i}{}^2 = \tfrac{1}{2}(0.00500\ \text{kg})(400\ \text{m}/\text{s})^2 = 400\ \text{J}$$

After the collision, the energy is:

$$K_2 = \tfrac{1}{2}m_1 v_{1f}{}^2 + \tfrac{1}{2}m_2 v_B{}^2$$

$$K_2 = \tfrac{1}{2}(0.00500\ \text{kg})(100\ \text{m}/\text{s})^2 + \tfrac{1}{2}(1.00\ \text{kg})(1.50\ \text{m}/\text{s})^2 = 26.1\ \text{J}$$

Therefore, $$|\Delta K| = |K_2 - K_1| = |26.1\ \text{J} - 400\ \text{J}| = 374\ \text{J} \qquad \lozenge$$

Chapter 10
ROTATION OF A RIGID OBJECT ABOUT A FIXED AXIS

EQUATIONS AND CONCEPTS

The **arc length**, s, is the distance traveled by a particle which moves along a circular path of radius r. *The radial line from the center of the circle defined by the path to the particle sweeps out an angle θ.*

$$\theta = \frac{s}{r} \tag{10.1b}$$

The **angular displacement** θ is the ratio of two lengths (arc length to radius) and hence is a dimensionless quantity. However, it is common practice to refer to the angle as being in units of radians. In calculations, the relationship between radians and degrees is shown below the figure.

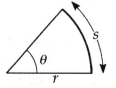

$$\theta \,(\text{rad}) = \frac{\pi}{180°}\, \theta \,(\text{deg})$$

The **average angular speed** $\bar{\omega}$ of a particle or body rotating about a fixed axis equals the ratio of the angular displacement $\Delta\theta$ to the time interval Δt, where θ is measured in radians. *In the case of a rigid body, $\Delta\theta$ is associated with the entire body as well as with each individual particle within the body.*

$$\bar{\omega} \equiv \frac{\theta_f - \theta_i}{t_f - t_i} = \frac{\Delta\theta}{\Delta t} \tag{10.2}$$

The **instantaneous angular speed** ω is defined as the limit of the average angular velocity as Δt approaches zero. ω *is considered positive when θ is increasing counterclockwise.*

$$\omega \equiv \lim_{\Delta t \to 0} \frac{\Delta\theta}{\Delta t} = \frac{d\theta}{dt} \tag{10.3}$$

The **average angular acceleration** $\bar{\alpha}$ of a rotating body is defined as the ratio of the change in angular velocity to the time interval Δt.

$$\bar{\alpha} \equiv \frac{\omega_f - \omega_i}{t_f - t_i} = \frac{\Delta \omega}{\Delta t} \qquad (10.4)$$

The **instantaneous angular acceleration** equals the limit of the average angular acceleration as Δt approaches zero. *When a rigid object is rotating about a fixed axis, every particle on the object rotates through the same angle in a given time interval and has the same angular speed and the same angular acceleration.*

$$\alpha \equiv \lim_{\Delta t \to 0} \frac{\Delta \omega}{\Delta t} = \frac{d\omega}{dt} \qquad (10.5)$$

The **equations of rotational kinematics** apply when a particle or body rotates about a fixed axis with *constant angular acceleration.*

$$\omega_f = \omega_i + \alpha t \qquad (10.6)$$

$$\theta_f = \theta_i + \omega_i t + \tfrac{1}{2}\alpha t^2 \qquad (10.7)$$

$$\omega_f^2 = \omega_i^2 + 2\alpha\left(\theta_f - \theta_i\right) \qquad (10.8)$$

$$\theta_f = \theta_i + \tfrac{1}{2}\left(\omega_i + \omega_f\right)t \qquad (10.9)$$

The **tangential speed** of any point on a rigid body a distance r from a fixed axis of rotation is related to the angular speed through the relation $v = r\omega$.

$$v = r\omega \qquad (10.10)$$

The **tangential acceleration** of any point on a rotating rigid object is related to the angular acceleration through the relation $a_t = r\alpha$. *Note that every point on the object has the same ω and α, but the values of v and a_t depend on the radial distance from the axis.*

$$a_t = r\alpha \qquad (10.11)$$

The **centripetal acceleration** is directed toward the center of rotation. *The magnitude of a_r equals a_c.*

$$a_c = \frac{v^2}{r} = r\omega^2 \tag{10.12}$$

The **total linear acceleration** has both radial and tangential components and a magnitude given by Equation 10.13.

$$a = \sqrt{a_t^2 + a_r^2} = r\sqrt{\alpha^2 + \omega^4} \tag{10.13}$$

The **moment of inertia** of a system of particles is defined by Equation 10.15, where m_i is the mass of the ith particle and r_i is its distance from a specified axis. Note that I has SI units of $\text{kg} \cdot \text{m}^2$. *Moment of inertia is a measure of the resistance of an object to changes in its rotational motion.*

$$I = \sum_i m_i r_i^2 \tag{10.15}$$

The **moment of inertia of an extended rigid object** with a high degree of symmetry can be determined by integrating over each mass element of the object. The integration can usually be simplified by introducing volumetric mass density ρ ($dm = \rho\, dV$), surface mass density σ ($dm = \sigma\, dA$), and linear mass density λ ($dm = \lambda\, dL$).

$$I = \int r^2 dm \tag{10.17}$$

The **parallel axis theorem** can be used to calculate the moment of inertia of an object of mass M about an axis which is parallel to the axis of symmetry of the object, at a distance D.

$$I = I_{\text{CM}} + MD^2 \tag{10.18}$$

The **rotational kinetic energy** of a rigid body rotating with an angular speed ω about some axis is proportional to the square of the angular speed. *Note that K_R does not represent a new form of energy. It is simply a convenient form for representing rotational kinetic energy.*

$$K_R = \frac{1}{2}I\omega^2 \qquad (10.16)$$

The **torque τ** due to an applied force has a magnitude given by the product of the force and its moment arm d, *where d equals the perpendicular distance from the rotation axis to the line of action of **F**. The line of action of a force is an imaginary line extending out of each end of the vector representing the force which gives rise to the torque. Torque is a measure of the ability of a force to rotate a body about a specified axis and should not be confused with force.*

$$\tau \equiv rF\sin\phi = Fd \qquad (10.19)$$

$$\text{where } d = r\sin\phi$$

The **rotational analog of Newton's second law** of motion is valid for a rigid body of arbitrary shape rotating about a fixed axis: the net torque is proportional to the angular acceleration. *Equation 10.21 applies when forces acting on a rigid body have radial as well as tangential components.*

$$\sum\tau = I\alpha \qquad (10.21)$$

The **instantaneous power** delivered by a force in rotating a rigid body about a fixed axis is proportional to the angular velocity. *Equation 10.23 is analogous to $\mathcal{P} = Fv$ in the case of linear motion.*

$$\mathcal{P} = \frac{dW}{dt} = \tau\omega \qquad (10.23)$$

The **work-energy theorem** says that the net work done by external forces in rotating a symmetric rigid object about a fixed axis equals the change in the object's rotational energy.

$$\sum W = \frac{1}{2}I\omega_f^2 - \frac{1}{2}I\omega_i^2 \qquad (10.24)$$

For **pure rolling motion**, the linear speed of the center of mass and the magnitude of the linear acceleration of the center of mass are given by Equations 10.25 and 10.26. *Pure rolling motion occurs when a rigid object rolls without slipping.*

$$v_{CM} = R\omega \tag{10.25}$$

$$a_{CM} = R\alpha \tag{10.26}$$

The **total kinetic energy of a rolling object** is the sum of the rotational kinetic energy about the center of mass and the translational kinetic energy of the center of mass.

$$K = \frac{1}{2}I_{CM}\omega^2 + \frac{1}{2}Mv_{CM}{}^2 \tag{10.28}$$

$$K = \frac{1}{2}\left(\frac{I_{CM}}{R^2} + M\right)v_{CM}{}^2 \tag{10.29}$$

or

$$K = \frac{1}{2}I_{CM}\omega^2 + \frac{1}{2}MR^2\omega^2$$

SUGGESTIONS, SKILLS, AND STRATEGIES

CALCULATING MOMENTS OF INERTIA

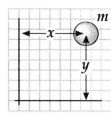

The method for calculating the moment of inertia of a system of particles about a specified axis is straightforward; apply Equation 10.15, $I = \Sigma m_i r_i^2$, where m_i is the mass of the ith particle and r_i is the distance from the axis of rotation to the particle. For mass m in the figure, the moment of inertia about the x axis is $I_x = my^2$ and about the y axis the moment of inertia is $I_y = mx^2$. A particle on the x axis would have a zero moment of inertia about that axis.

In using Equation 10.17, $I = \int r^2 dm$, to calculate the moment of inertia of a rigid body about an axis of symmetry, it is usually easier to change the integrand to sum over all volume elements. To do this let $dm = \rho dV$, where ρ is the volumetric mass density and dV is an element of volume. In the case of a plane sheet, $dm = \sigma dA$ where σ is the surface mass density and dA is an element of area; and for a long thin rod, $dm = \lambda dx$ where λ is the linear mass density and dx is an element of length. In the case of a uniform thin hoop or ring (see Example 10.5 in your textbook) it is easy to integrate directly over the mass elements since all mass elements are the same distance from the axis of symmetry. **Note:** If the density of the object is not constant over the volume, area, or length, then ρ, σ, or λ must be replaced with the expression showing the manner in which density varies.

If the moment of inertia of a rigid body about an axis through the center of mass I_{CM} is known, you can easily evaluate the moment of inertia about any axis parallel to the axis through the center of mass using the **parallel axis theorem**:

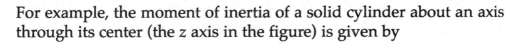

$$I = I_{CM} + MD^2$$

where M is the mass of the body and D is the distance between the two axes.

For example, the moment of inertia of a solid cylinder about an axis through its center (the z axis in the figure) is given by

$$I_z = \frac{1}{2}MR^2$$

Hence, the moment of inertia about the z' axis located a distance $D = R$ from the z axis is:

$$I_{z'} = I_z + MR^2 = \frac{1}{2}MR^2 + MR^2 = \frac{3}{2}MR^2$$

CALCULATING TORQUE

Do not confuse force and torque. Both are vector quantities. Torque is the effect of a force in causing or changing the rotation of a rigid body. The magnitude of a torque is the product of the force and the moment arm of the force (the perpendicular distance from the line of action of the force to the pivot point).

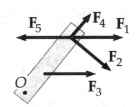

For rotation about a fixed axis, the vector nature of the torque can be represented by using signed scalar quantities. Torques which produce counterclockwise rotations are positive and those which produce clockwise rotations are negative. The correct algebraic signs must be taken into account when using Equation 10.21, $\Sigma\tau = I\alpha$. In the figure, F_5 gives rise to a positive torque about an axis through O, force F_4 produces zero torque (the moment arm is zero) and the other forces produce negative torques. Consider the moment arms and note that $|\tau_2| > |\tau_1| > |\tau_3|$.

REVIEW CHECKLIST

You should be able to:

▷ Apply the equations of rotational kinematics to make calculations of rotational displacement, angular velocity, and angular acceleration. (Section 10.2)

▷ Make calculations relating corresponding angular and linear quantities. (Section 10.3)

▷   Calculate the moment of inertia of a system of discrete point masses. (Section 10.4)

▷   Calculate the moment of inertia of solids which have a high degree of symmetry by integrating over the volume, area, or length. Use the parallel axis theorem. (Section 10.5)

▷   Calculate the torque due to a force acting on a rigid body and determine the angular acceleration due to the net force. (Sections 10.6 and 10.7)

▷   Make calculations of rotational kinetic energy and apply the work-kinetic energy theorem to rotating bodies. Calculate the power delivered by a tangential force. (Section 10.8)

▷   Calculate the kinetic energy of a rigid body in pure rolling motion. (Section 10.9)

Some important points to remember:

▷   Quantitatively, the angular displacement, speed, and acceleration for a rigid body system in rotational motion are related to the distance traveled, tangential speed, and tangential acceleration. The linear quantity is calculated by multiplying the angular quantity by the radius arm for an object or point in that system.

▷   If a body rotates about a fixed axis, every particle on the body has the same angular speed and angular acceleration. For this reason, rotational motion can be simply described using these quantities. The kinematic equations which describe angular motion are analogous to the corresponding set of equations pertaining to linear motion. The expressions for Newton's second law, kinetic energy, and power also have analogous forms for linear and rotational motion.

▷   The value of the moment of inertia I of a system of particles or a rigid body depends on the manner in which the mass is distributed relative to the axis of rotation. For example, the moment of inertia of a disk is less than that of a ring of equal mass and radius when calculated about an axis perpendicular to their plane and through the center of mass. The parallel-axis theorem is useful for calculating I about an axis parallel to one that goes through the center of mass.

▷   Torque should not be confused with force. The magnitude of a torque associated with a force depends on the magnitude of the force, the point of application of the force, and the angle between the force vector and the line joining the pivot point and the point where the force is applied.

▷   The description of the motion of an object in pure rolling motion can be simplified by considering the motion (displacement, velocity, and acceleration) of the center of mass.

Chapter 10

ANSWERS TO SELECTED CONCEPTUAL QUESTIONS

1.   What is the angular speed of the second hand of a clock? What is the direction of ω as you view a clock hanging on a vertical wall? What is the magnitude of the angular acceleration vector α of the second hand?

Answer   The second hand of a clock turns at one revolution per minute, so

$$\omega = \frac{2\pi \text{ rad}}{60 \text{ s}} = 0.105 \text{ rad / s}$$

The motion is clockwise, so the direction of the vector angular velocity is away from you. It turns steadily, so ω is constant, and α is zero.

18.   If you see an object rotating, is there necessarily a net torque acting on it?

Answer   An object rotates with constant angular momentum when zero total torque acts on it. For example, consider the Earth; it rotates at a constant rate of once per day, but there is no net torque acting on it.

20.   In a tape recorder, the tape is pulled past the read-and-write heads at a constant speed by the drive mechanism. Consider the reel from which the tape is pulled. As the tape is pulled from it, the radius of the roll of remaining tape decreases. How does the torque on the reel change with time? How does the angular speed of the reel change in time? If the drive mechanism is switched on so that the tape is suddenly jerked with a large force, is the tape more likely to break when it is being pulled from a nearly full reel or from a nearly empty reel?

Answer   Since the source reel stops almost instantly when the tape stops playing, the friction on the source reel axle must be fairly large. Since the source reel appears to us to rotate at almost constant angular velocity, the angular acceleration must be very small. Therefore, the torque on the source reel due to the tension in the tape must almost exactly balance the frictional torque. In turn, the frictional torque is nearly constant because kinetic friction forces don't depend on velocity, and the radius of the axle where the friction is applied is constant. Thus we conclude that the torque exerted by the tape on the source reel is essentially constant in time as the tape plays.

As the source reel radius R shrinks, the reel's angular speed $\omega = v/R$ must increase to keep the tape speed v constant. But the biggest change is to the reel's moment of inertia. We model the reel as a roll of tape, ignoring any spool or platter carrying the tape. If we

155

think of the roll of tape as a uniform disk, then its moment of inertia is $I = \frac{1}{2}MR^2$. But the roll's mass is proportional to its base area πR^2. Thus, on the whole the moment of inertia is proportional to R^4. The moment of inertia decreases very rapidly as the reel shrinks!

The tension in the tape coming into the read-and-write heads is normally dominated by balancing the constant frictional torque on the source reel, according to $TR \approx \tau_{\text{friction}}$. Therefore, as the tape plays the tension is largest when the reel is smallest. However, in the case of a sudden jerk on the tape, the rotational dynamics of the source reel becomes important. If the source reel is full, then the moment of inertia, proportional to R^4, will be so large that higher tension in the tape will be required to give the source reel its angular acceleration. If the reel is nearly empty, then the same tape acceleration will require a smaller tension. Thus, the tape will be more likely to break when the source reel is nearly **full**. One sees the same effect in the case of paper towels; it is easier to snap a towel free when the roll is new than when it is nearly empty.

□ □ □ □

SOLUTIONS TO SELECTED END-OF-CHAPTER PROBLEMS

3. A wheel starts from rest and rotates with constant angular acceleration and to reach an angular speed of 12.0 rad/s in 3.00 s. Find (a) the magnitude of the angular acceleration of the wheel and (b) the angle in radians through which it rotates in this time.

Solution

(a) $\quad \alpha = \dfrac{\omega_f - \omega_i}{t} = \dfrac{(12.0 - 0)\ \text{rad}/\text{s}}{3.00\ \text{s}} = 4.00\ \text{rad}/\text{s}^2$ ◊

(b) $\quad \theta_f = \omega_i t + \frac{1}{2}\alpha t^2 = \frac{1}{2}\left(4.00\ \text{rad}/\text{s}^2\right)(3.00\ \text{s})^2 = 18.0\ \text{rad}$ ◊

5. An electric motor rotating a grinding wheel at 100 rev/min is switched off. With constant negative angular acceleration of magnitude 2.00 rad/s², (a) how long does it take the wheel to stop? (b) Through how many radians does it turn while it is slowing down?

Solution We use the rigid body under constant angular acceleration model.

We are given $\qquad\qquad \alpha = -2.00 \text{ rad} / \text{s}^2$

$$\omega_f = 0 \qquad \text{and} \qquad \omega_i = 100 \, \frac{\text{rev}}{\text{min}} \left(2\pi \, \frac{\text{rad}}{\text{rev}} \right)\left(\frac{1 \text{ min}}{60.0 \text{ s}} \right) = 10.47 \text{ rad} / \text{s}$$

(a) $\omega_f = \omega_i + \alpha t$: $\qquad\qquad t = \dfrac{\omega_f - \omega_i}{\alpha} = \dfrac{0 - (10.5 \text{ rad} / \text{s})}{-2.00 \text{ rad} / \text{s}^2} = 5.24 \text{ s}$ ◊

(b) $\omega_f^2 - \omega_i^2 = 2\alpha\!\left(\theta_f - \theta_i\right)$: $\qquad \theta_f - \theta_i = \dfrac{\omega_f^2 - \omega_i^2}{2\alpha} = \dfrac{0 - (10.5 \text{ rad} / \text{s})^2}{2\!\left(-2.00 \text{ rad} / \text{s}^2\right)} = 27.4 \text{ rad}$ ◊

Note also in part (b) that since a constant acceleration is acting for time t,

$$\theta_f - \theta_i = \overline{\omega} t = \left(\frac{10.5 + 0 \text{ rad} / \text{s}}{2} \right)(5.24 \text{ s}) = 27.4 \text{ rad} \qquad ◊$$

---

13. A wheel 2.00 m in diameter lies in a vertical plane and rotates with a constant angular acceleration of 4.00 rad/s². The wheel starts at rest at $t=0$, and the radius vector of a certain point P on the rim makes an angle of 57.3° with the horizontal at this time. At $t=2.00$ s, find (a) the angular speed of the wheel, (b) the tangential speed and total acceleration of the point P, and (c) the angular position of the point P.

Solution

$$r = 1.00 \text{ m}, \ \alpha = 4.00 \text{ rad} / \text{s}^2, \ \omega_i = 0, \text{ and } \ \theta_i = 57.3° = 1 \text{ rad}$$

(a) $\qquad\qquad\qquad \omega_f = \omega_i + \alpha t = 0 + \alpha t$ ◊

At $t = 2.00$ s, $\qquad \omega_f = \left(4.00 \text{ rad} / \text{s}^2\right)(2.00 \text{ s}) = 8.00 \text{ rad} / \text{s}$ ◊

(b) $v = r\omega = (1.00 \text{ m})(8.00 \text{ rad} / \text{s}) = 8.00 \text{ m} / \text{s}$ ◊

$a_c = r\omega^2 = (1.00 \text{ m})(8.00 \text{ rad} / \text{s})^2 = 64.0 \text{ m} / \text{s}^2$ ◊

$a_t = r\alpha = (1.00 \text{ m})\!\left(4.00 \text{ rad} / \text{s}^2\right) = 4.00 \text{ m} / \text{s}^2$ ◊

The magnitude of the total acceleration is:

$$a = \sqrt{a_c^2 + a_t^2} = \sqrt{\left(64.0 \text{ m} / \text{s}^2\right)^2 + \left(4.00 \text{ m} / \text{s}^2\right)^2} = 64.1 \text{ m} / \text{s}^2 \qquad ◊$$

The direction of the total acceleration vector makes an angle ϕ with respect to the radius to point P:

$$\phi = \tan^{-1}\left(\frac{a_t}{a_c}\right) = \tan^{-1}\left(\frac{4.00 \text{ m / s}^2}{64.0 \text{ m / s}^2}\right) = 3.58° \qquad \diamond$$

(c) $\quad \theta_f = \theta_i + \omega_i t + \frac{1}{2}\alpha t^2 = (1.00 \text{ rad}) + \frac{1}{2}(4.00 \text{ rad / s}^2)(2.00 \text{ s})^2 = 9.00 \text{ rad} \qquad \diamond$

$\theta_f - \theta_i$ is the total angle through which point P has passed, and is greater than one revolution. The position we attribute to point P is found by subtracting one revolution from θ_f. Therefore P is at

$9.00 \text{ rad} - 2\pi \text{ rad} = 2.72 \text{ rad}$. $\qquad \diamond$

---

17. A disk 8.00 cm in radius rotates at a constant rate of 1 200 rev/min about its central axis. Determine (a) its angular speed, (b) the tangential speed at a point 3.00 cm from its center, (c) the radial acceleration of a point on the rim, and (d) the total distance a point on the rim moves in 2.00 s.

Solution

(a) $\quad \omega = 2\pi f = (2\pi \text{ rad / rev})\left(\dfrac{1\,200 \text{ rev / min}}{60 \text{ s / min}}\right) = 125.7 \text{ rad / s} = 126 \text{ rad / s} \qquad \diamond$

(b) $\quad v = \omega R = (125.7 \text{ rad / s})(0.030\,0 \text{ m}) = 3.77 \text{ m / s} \qquad \diamond$

(c) $\quad a_c = \omega^2 R = (125.7 \text{ rad / s})^2(0.080\,0 \text{ m}) = 1.26 \times 10^3 \text{ m / s}^2 \qquad \diamond$

(d) $\quad s = R\theta = R\omega t = (8.00 \times 10^{-2} \text{ m})(125.7 \text{ rad / s})(2.00 \text{ s}) = 20.1 \text{ m} \qquad \diamond$

---

21. The four particles in Figure P10.21 are connected by rigid rods of negligible mass. The origin is at the center of the rectangle. If the system rotates in the xy plane about the z axis with an angular speed of 6.00 rad/s, calculate (a) the moment of inertia of the system about the z axis and (b) the rotational kinetic energy of the system.

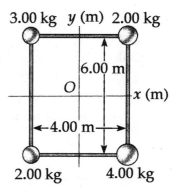

Figure P10.21

Solution

(a) All four particles are at a distance r from the z axis,

so $r^2 = (3.00 \text{ m})^2 + (2.00 \text{ m})^2 = 13.00 \text{ m}^2$

Thus, $I_z = \sum m_i r_i^2 = (3.00 \text{ kg})(13.00 \text{ m}^2) + (2.00 \text{ kg})(13.00 \text{ m}^2) +$

$$+ (4.00 \text{ kg})(13.00 \text{ m}^2) + (2.00 \text{ kg})(13.00 \text{ m}^2)$$

$I_z = 143 \text{ kg} \cdot \text{m}^2$ ◊

(b) $K_R = \frac{1}{2} I_z \omega^2 = \frac{1}{2}(143 \text{ kg} \cdot \text{m}^2)(6.00 \text{ rad} / \text{s})^2 = 2.57 \text{ kJ}$ ◊

---

31. Find the net torque on the wheel in Figure P10.31 about the axle through O if $a = 10.0$ cm and $b = 25.0$ cm.

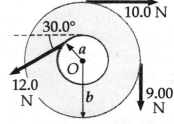

Figure P10.31

Solution

Conceptualize: By examining the magnitudes of the forces and their respective lever arms, it appears that the wheel will rotate clockwise, and the net torque appears to be about $5 \text{ N} \cdot \text{m}$.

Categorize: To find the net torque, we add the individual torques, remembering to apply the convention that a torque producing clockwise rotation is negative and a counterclockwise rotation is positive.

Analyze: $\sum \tau = \sum Fd = +(12.0 \text{ N})(0.100 \text{ m}) - (10.0 \text{ N})(0.250 \text{ m}) - (9.00 \text{ N})(0.250 \text{ m})$

$\sum \tau = -3.55 \text{ N} \cdot \text{m}$ ◊

(This is $3.55 \text{ N} \cdot \text{m}$ into the plane of the page, a clockwise torque.)

Finalize: The resulting torque has a reasonable magnitude and produces clockwise rotation as expected. Note that the 30° angle was not required for the solution since each force acted perpendicular to its lever arm. The 10-N force is to the right, but its torque is negative — that is, clockwise, just like the torque of the downward 9-N force.

---

35. A model airplane with mass 0.750 kg is tethered by a wire so that it flies in a circle 30.0 m in radius. The airplane engine provides a net thrust of 0.800 N perpendicular to the tethering wire. (a) Find the torque the net thrust produces about the center of the circle. (b) Find the angular acceleration of the airplane when it is in level flight. (c) Find the linear acceleration of the airplane tangent to its flight path.

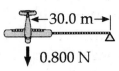

Solution

(a) $\tau = Fd$: $\qquad \tau = (0.800 \text{ N})(30.0 \text{ m}) = 24.0 \text{ N} \cdot \text{m}$ ◊

(b) $I = mr^2$: $\qquad I = (0.750 \text{ kg})(30.0 \text{ m})^2 = 675 \text{ kg} \cdot \text{m}^2$

The angular acceleration is found from

$$\sum \tau = I\alpha: \quad \alpha = \frac{\sum \tau}{I} = \frac{24.0 \text{ N} \cdot \text{m}}{675 \text{ kg} \cdot \text{m}^2} = 0.035\,6 \text{ rad} / \text{s}^2 \qquad ◊$$

(c) $a = r\alpha$: $\qquad a = (30.0 \text{ m})(0.035\,6 / \text{s}^2) = 1.07 \text{ m} / \text{s}^2$ ◊

We could also find this linear acceleration from $\Sigma F = ma$:

$$a = \frac{\sum F}{m} = \frac{0.800 \text{ N}}{0.750 \text{ kg}} = 1.07 \text{ m} / \text{s}^2$$

---

45. An object with a weight of 50.0 N is attached to the free end of a light string wrapped around a reel of 0.250 m and mass 3.00 kg. The reel is a solid disk, free to rotate in a vertical plane about the horizontal axis passing through its center. The suspended object is released 6.00 m above the floor. (a) Determine the tension in the string, the acceleration of the object, and the speed with which the object hits the floor. (b) Verify your answers by using the principle of conservation of energy to find the speed with which the object hits the floor.

Solution

Conceptualize: Since the rotational inertia of the reel will slow the fall of the weight, we should expect the downward acceleration to be less than g. If the reel did not rotate, the tension in the string would be equal to the weight of the object; and if the reel disappeared, the tension would be zero. Therefore, $T < mg$ for the given problem. With similar reasoning, the final speed must be less than if the weight were to fall freely:

$$v_f < \sqrt{2g\Delta y} \approx 11 \text{ m} / \text{s}$$

Categorize: We can find the acceleration and tension using the rotational form of Newton's second law. The final speed can be found from the kinematics equation stated above and from conservation of energy. Free-body diagrams will greatly assist in analyzing the forces.

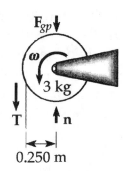

Analyze:

(a) Use $\Sigma \tau = I\alpha$ to find T and a.

First find I for the reel, which we assume to be a uniform disk.

$$I = \tfrac{1}{2}MR^2 = \tfrac{1}{2}3.00 \text{ kg}(0.250 \text{ m})^2 = 0.093\,8 \text{ kg} \cdot \text{m}^2$$

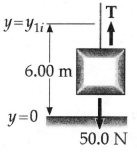

The forces on it are shown, including a normal force exerted by its axle. From the diagram, we can see that the tension is the only unbalanced force causing the wheel to rotate.

$\Sigma \tau = I\alpha$ becomes

$$n(0) + F_g(0) + T(0.250 \text{ m}) = \left(0.093\,8 \text{ kg} \cdot \text{m}^2\right)(a/0.250 \text{ m}) \qquad \textbf{[1]}$$

where we have applied $a_t = r\alpha$ to the point of contact between string and pulley.

The falling weight has mass $\qquad m = \dfrac{F_g}{g} = \dfrac{50.0 \text{ N}}{9.80 \text{ m}/\text{s}^2} = 5.10 \text{ kg} \qquad \lozenge$

For this mass, $\Sigma F_y = ma_y$ becomes $\quad 50.0 \text{ N} - T = (5.10 \text{ kg})a \qquad \textbf{[2]}$

Note that we have defined downwards to be positive, so that positive linear acceleration of the object corresponds to positive angular acceleration of the reel. We now have our two equations in the unknowns T and a for the two connected objects. Substituting T from Equation [2] into Equation [1], we have

$$\left[50.0 \text{ N} - (5.10 \text{ kg})a\right](0.250 \text{ m}) = \left(0.093\,8 \text{ kg} \cdot \text{m}^2\right)(a/0.250 \text{ m})$$

$$12.5 \text{ N} \cdot \text{m} - (1.28 \text{ kg} \cdot \text{m})a = (0.375 \text{ kg} \cdot \text{m})a$$

$12.5 \text{ N} \cdot \text{m} = a(1.65 \text{ kg} \cdot \text{m})$ or $a = 7.57 \text{ m}/\text{s}^2 \qquad \lozenge$

and $\qquad\qquad\qquad\qquad T = 50.0 \text{ N} - 5.10 \text{ kg}\left(7.57 \text{ m}/\text{s}^2\right) = 11.4 \text{ N} \qquad \lozenge$

For the motion of the weight, $\quad v_f^2 = v_i^2 + 2a(x_f - x_i) = 0^2 + 2\left(7.57 \text{ m}/\text{s}^2\right)(6.00 \text{ m})$

$$v_f = 9.53 \text{ m}/\text{s} \text{ (down)} \qquad \lozenge$$

(b) The work-kinetic energy theorem can take account of multiple objects more easily than Newton's second law. Like your bratty cousins, the work-kinetic energy theorem grows between visits; now it reads:

$$\left(K_1 + K_2 + U_g\right)_i = \left(K_1 + K_2 + U_g\right)_f$$

$$0 + 0 + m_1 g y_{1i} = \frac{1}{2} m_1 v_{1f}^2 + \frac{1}{2} I_2 \omega_{2f}^2 + 0$$

Now note that $\omega = v / r$ as the string unwinds from the reel.

Substituting, $50.0 \text{ N}(6.00 \text{ m}) = \frac{1}{2}(5.10 \text{ kg})v_f^2 + \frac{1}{2}(0.0938 \text{ kg} \cdot \text{m}^2)\left(\dfrac{v_f}{0.250 \text{ m}}\right)^2$

$$300 \text{ N} \cdot \text{m} = \frac{1}{2}(5.10 \text{ kg})v_f^2 + \frac{1}{2}(1.50 \text{ kg})v_f^2$$

$$v_f = \sqrt{\dfrac{2(300 \text{ N} \cdot \text{m})}{6.60 \text{ kg}}} = 9.53 \text{ m / s}$$ ◊

Finalize: As we should expect, both methods give the same final speed for the falling object, but the energy method is simpler. The acceleration is less than g, and the tension is less than the object's weight as we predicted. Now that we understand the effect of the reel's moment of inertia, this problem solution could be applied to solve other real-world pulley systems with masses that should not be ignored.

---

47. This problem describes one experimental method for determining the moment of inertia of an irregularly shaped object such as the payload for a satellite. Figure P10.47 shows a counterweight of mass m suspended by a cord wound around a spool of radius r, forming part of a turntable supporting the object. The turntable can rotate without friction. When the counterweight is released from rest, it descends through a distance h, acquiring a speed v. Show that the moment of inertia I of the rotating apparatus (including the turntable) is

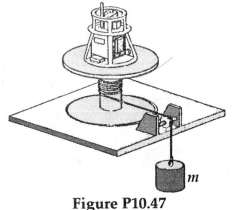

Figure P10.47

$$mr^2\left(2gh / v^2 - 1\right)$$

Solution   If the friction is negligible, then the energy of the counterweight-payload-turntable-Earth system is conserved as the counterweight unwinds. Each point on the cord moves at a linear speed of $v = \omega r$, where r is the radius of the spool. The energy conservation equation gives us:

$$\left(K_1 + K_2 + U_g\right)_i + W_{\text{other}} = \left(K_1 + K_2 + U_g\right)_f$$

Solving, we have

$$0 + 0 + mgh + 0 + 0 = \tfrac{1}{2}mv^2 + \tfrac{1}{2}I\omega^2 + 0 + 0$$

$$mgh = \tfrac{1}{2}mv^2 + \tfrac{1}{2}\frac{Iv^2}{r^2}$$

$$2mgh - mv^2 = I\frac{v^2}{r^2}$$

and finally,

$$I = mr^2\left(\frac{2gh}{v^2} - 1\right) \qquad \Diamond$$

---

49. (a) A uniform, solid disk of radius R and mass M is free to rotate on a frictionless pivot through a point on its rim (Fig. P10.49). If the disk is released from rest in the position shown by the blue circle, what is the speed of its center of mass when the disk reaches the position indicated by the dashed circle? (b) What is the speed of the lowest point on the disk in the dashed position? (c) **What if?** Repeat part (a) using a uniform hoop.

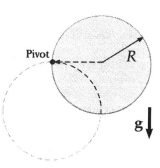

Pivot R

g

Figure P10.49

Solution

We cannot use the equation $\omega_f^2 - \omega_i^2 = 2\alpha(\pi/2)$ to find ω_f, because α is not constant. Instead, we use conservation of energy. To identify the change in gravitational energy, think of the height through which the center of mass falls. Using the parallel axis theorem, the moment of inertia of the disk about the pivot point on the disk is:

$$I = I_c + MD^2 = \tfrac{1}{2}MR^2 + MR^2 = \tfrac{3}{2}MR^2$$

The pivot point is fixed, so the kinetic energy is entirely rotational around the pivot:

$$(K + U)_i = (K + U)_f: \qquad MgR = \tfrac{1}{2}\left(\tfrac{3}{2}MR^2\right)\omega^2 + 0$$

Solving for ω,

$$\omega = \sqrt{\frac{4g}{3R}}$$

(a)   At the center of mass,

$$v = R\omega = 2\sqrt{\frac{Rg}{3}} \qquad \Diamond$$

(b)   At the lowest point on the rim,

$$v = 2R\omega = 4\sqrt{\frac{Rg}{3}} \qquad \Diamond$$

Chapter 10

(c) For a hoop, $$I_c = MR^2 \text{ and } I_{rim} = 2MR^2$$

By conservation of energy then, $$MgR = \tfrac{1}{2}\left(2MR^2\right)\omega^2 + 0 \quad \text{so} \quad \omega = \sqrt{\frac{g}{R}}$$

and the center of mass moves at $$v_{CM} = R\omega = \sqrt{gR}, \text{ slower than the disk.} \qquad \Diamond$$

---

51. A cylinder of mass 10.0 kg rolls without slipping on a horizontal surface. At the instant its center of mass has a speed of 10.0 m/s, determine (a) the translational kinetic energy of its center of mass, (b) the rotational kinetic energy about its center of mass, and (c) its total energy.

Solution

(a) $$K_{trans} = \tfrac{1}{2}mv_{CM}^2 = \tfrac{1}{2}(10.0 \text{ kg})(10.0 \text{ m / s})^2 = 500 \text{ J} \qquad \Diamond$$

(b) Call the radius of the cylinder R. An observer at the center sees the rough surface and the circumference of the cylinder moving at 10.0 m/s, so the angular speed of the cylinder is:

$$\omega = \frac{v_{CM}}{R} = \frac{10.0 \text{ m / s}}{R}$$

The moment of inertia about an axis through the center of mass is:

$$I_{CM} = \tfrac{1}{2}mR^2 \quad \text{so} \quad K_{rot} = \tfrac{1}{2}I_{CM}\omega^2 = \left(\tfrac{1}{2}\right)\left[\tfrac{1}{2}(10.0 \text{ kg})R^2\right]\left(\frac{10.0 \text{ m / s}}{R}\right)^2 = 250 \text{ J} \qquad \Diamond$$

(c) We can now add up the total energy: $$K_{tot} = 500 \text{ J} + 250 \text{ J} = 750 \text{ J} \qquad \Diamond$$

---

53. (a) Determine the acceleration of the center of mass of a uniform solid disk rolling down an incline making angle θ with the horizontal. Compare this acceleration with that of a uniform hoop. (b) What is the minimum coefficient of friction required to maintain pure rolling motion for the disk?

Solution

Conceptualize: The acceleration of the disk will depend on the angle of the incline. In fact, it should be proportional to $g\sin\theta$ since the disk should not accelerate when the incline angle is zero. The acceleration of the disk should also be greater than a hoop since the mass of the disk is closer to its center, giving it less rotational inertia so that it can roll faster than the hoop. The required coefficient of friction is difficult to predict, but is probably between 0 and 1 since this is a typical range for μ.

Categorize: We can find the acceleration by applying Newton's second law and considering both the linear and rotational motion. A free-body diagram (left-hand disk) and a motion diagram (right-hand disk) will greatly assist us in defining our variables and seeing how the forces are related.

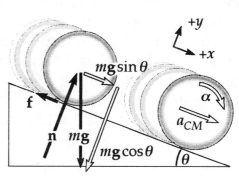

Analyze:

$\sum F_x = ma_x$ becomes $mg\sin\theta - f = ma_{CM}$ [1]

$\sum F_y = ma_y$ yields $n - mg\cos\theta = 0$ [2]

$\sum \tau = I_{CM}\alpha$ gives $fr = I_{CM}a_{CM}/r$ [3]

(a)   For a disk, $(I_{CM})_{disk} = \frac{1}{2}mr^2$

From [3] we find $f = \left(\frac{1}{2}mr^2\right)(a_{CM})/r^2 = \frac{1}{2}ma_{CM}$

Substituting this into [1], $mg\sin\theta - \frac{1}{2}ma_{CM} = ma_{CM}$

so that $(a_{CM})_{disk} = \frac{2}{3}g\sin\theta$ ◊

For a hoop, $(I_{CM})_{hoop} = mr^2$

From [3], $f = mr^2 a_{CM}/r^2 = ma_{CM}$

Substituting this into [1], $mg\sin\theta - ma_{CM} = ma_{CM}$

so $(a_{CM})_{hoop} = \frac{1}{2}g\sin\theta$

Therefore, $\dfrac{(a_{CM})_{disk}}{(a_{CM})_{hoop}} = \dfrac{\frac{2}{3}g\sin\theta}{\frac{1}{2}g\sin\theta} = \dfrac{4}{3}$ ◊

(b)   From [2] we find $n = mg\cos\theta$: $f \le \mu_s n = \mu mg\cos\theta$

Likewise, from Equation [1], $f = (mg\sin\theta) - ma_{CM}$

Setting these equations equal, $\mu_s mg\cos\theta \ge mg\sin\theta - \frac{2}{3}mg\sin\theta$

so $\mu_s \ge \frac{1}{3}\left(\dfrac{\sin\theta}{\cos\theta}\right) = \frac{1}{3}\tan\theta$ ◊

165

Finalize: As expected, the acceleration of the disk is proportional to $g\sin\theta$ and is slightly greater than the acceleration of the hoop. The minimum coefficient of friction result is similar to the result found for a block on an incline plane, where $\mu = \tan\theta$. However, μ is not always between 0 and 1 as predicted. For angles greater than 72° the coefficient of friction must be larger than 1. For angles greater than 80°, μ must be extremely large to make the disk roll without slipping.

---

59. A 4.00-m length of light nylon cord is wound around a uniform cylindrical spool of radius 0.500 m and mass 1.00 kg. The spool is mounted on a frictionless axle and is initially at rest. The cord is pulled from the spool with a constant acceleration of magnitude 2.50 m/s². (a) How much work has been done on the spool when it reaches an angular speed of 8.00 rad/s? (b) Assuming that there is enough cord on the spool, how long does it take the spool to reach this angular speed? (c) Is there enough cord on the spool?

Solution

(a) $\quad W = \Delta K_R = \frac{1}{2}I\omega_f^2 - \frac{1}{2}I\omega_i^2 = \frac{1}{2}I\left(\omega_f^2 - \omega_i^2\right)$ $\qquad$ where $\qquad I = \frac{1}{2}mR^2$

$\quad W = \left(\frac{1}{2}\right)\left(\frac{1}{2}\right)(1.00\text{ kg})(0.500\text{ m})^2\left[(8.00\text{ rad / s})^2 - 0\right] = 4.00\text{ J}$ $\qquad\qquad$ ◊

(b) $\quad \omega_f = \omega_i + \alpha t \qquad$ where $\qquad\qquad \alpha = \dfrac{a}{r} = \dfrac{2.50\text{ m / s}^2}{0.500\text{ m}} = 5.00\text{ rad / s}^2$

$\qquad\qquad\qquad\qquad\qquad\qquad\qquad t = \dfrac{\omega_f - \omega_i}{\alpha} = \dfrac{8.00\text{ rad / s} - 0}{5.00\text{ rad / s}^2} = 1.60\text{ s}$ $\qquad$ ◊

(c) $\quad \theta_f = \theta_i + \omega_i t + \frac{1}{2}\alpha t^2 \qquad\qquad \theta_f = 0 + 0 + \frac{1}{2}\left(5.00\text{ rad / s}^2\right)(1.60\text{ s})^2 = 6.40\text{ rad}$

The length pulled from the spool is

$$s = r\theta = (0.500\text{ m})(6.40\text{ rad}) = 3.20\text{ m}$$

When the spool reaches an angular velocity of 8.00 rad/s, 1.60 s will have elapsed and 3.20 m of cord will have been removed from the spool. Our answer is **yes**. ◊

---

61. A long uniform rod of length L and mass M is pivoted about a horizontal, frictionless pin passing through one end. The rod is released from rest in a vertical position, as in Figure P10.61. At the instant the rod is horizontal, find (a) its angular speed, (b) the magnitude of its angular acceleration, (c) the x and y components of the acceleration of its center of mass, and (d) the components of the reaction force at the pivot.

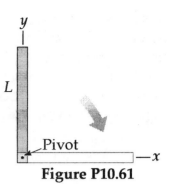

Figure P10.61

Solution

(a) Since only conservative forces are acting on the bar, use conservation of energy of the bar-Earth system:

$$\Delta K + \Delta U = 0 \qquad \text{or} \qquad K_f - K_i + U_f - U_i = 0$$

For evaluation of its gravitational energy, a rigid body can be modeled as a particle at its center of mass. Take the zero configuration for potential energy with the bar horizontal.

Under these conditions $\qquad U_f = 0 \quad$ and $\quad U_i = MgL / 2$

Using the equation above, $\quad \left(\frac{1}{2} I \omega_f^2 - 0\right) + \left(0 - \frac{1}{2} MgL\right) = 0 \quad$ and $\quad \omega_f = \sqrt{MgL / I}$

For a bar rotating about an axis through one end, $I = ML^2 / 3$

Therefore, $\qquad\qquad\qquad \omega_f = \sqrt{(MgL)/\frac{1}{3}ML^2} = \sqrt{3g / L}$ ◊

Note that we have chosen clockwise rotation as positive.

(b) $\sum \tau = I\alpha :$ $\qquad\qquad Mg(L / 2) = \left(\frac{1}{3} ML^2\right)\alpha \quad$ and $\quad \alpha = 3g/2L$ ◊

(c) $\qquad\qquad\qquad a_x = -a_c = -r\omega_f^2 = -\left(\frac{L}{2}\right)\left(\frac{3g}{L}\right) = -\frac{3g}{2}$ ◊

Since this is **centripetal** acceleration, it is directed along the **negative** horizontal.

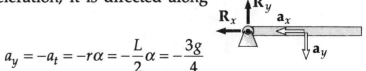

$$a_y = -a_t = -r\alpha = -\frac{L}{2}\alpha = -\frac{3g}{4}$$

(d) Using $\sum \mathbf{F} = m\mathbf{a}$, we have $\quad R_x = Ma_x = -3Mg / 2$ in the **negative** direction ◊

$R_y - Mg = Ma_y \quad$ so $\qquad R_y = M\left(g + a_y\right) = M(g - 3g / 4) = Mg / 4$ ◊

71. Two blocks, as shown in Figure P10.71, are connected by a string of negligible mass passing over a pulley of radius 0.250 m and moment of inertia I. The block on the frictionless incline is moving up with a constant acceleration of 2.00 m/s^2. (a) Determine T_1 and T_2, the tensions in the two parts of the string. (b) Find the moment of inertia of the pulley.

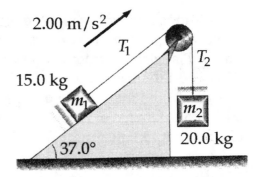

Figure P10.71

Solution

Conceptualize: In earlier problems, we assumed that the tension in a string was the same on either side of a pulley. Here we see that the moment of inertia changes that assumption, but we should still expect the tensions to be similar in magnitude (about the weight of each mass ~150 N), and $T_2 > T_1$ for the pulley to rotate clockwise as shown.

If we knew the mass of the pulley, we could calculate its moment of inertia, but since we only know the acceleration, it is difficult to estimate I. We at least know that I must have units of kg·m^2, and a 50-cm disk probably has a mass less than 10 kg, so I is probably less than 0.3 kg·m^2.

Categorize: For each block, we know its mass and acceleration, so we can use Newton's second law to find the net force, and from it the tension. The difference in the two tensions causes the pulley to rotate, so this net torque and the resulting angular acceleration can be used to find the pulley's moment of inertia.

Analyze:

(a) Apply $\Sigma F = ma$ to each block to find each string tension. The forces acting on the 15-kg block are its weight, the normal support from the incline, and T_1. Taking the positive x axis as directed up the incline ,

$$\Sigma F_x = ma_x \text{ yields:} \quad -(m_1 g)_x + T_1 = m_1(+a)$$

Substituting known values, and solving for T_1,

$$-(15.0 \text{ kg})(9.80 \text{ m / s}^2)\sin 37° + T_1 = (15.0 \text{ kg})(2.00 \text{ m / s}^2)$$

$$T_1 = 118 \text{ N}$$

Similarly, for the counterweight, we have

$$\sum F_y = ma_y \qquad \text{or} \qquad T_2 - m_2 g = m_2(-a)$$

$$T_2 - (20.0 \text{ kg})(9.80 \text{ m} / \text{s}^2) = (20.0 \text{ kg})(-2.00 \text{ m} / \text{s}^2)$$

So, $\qquad T_2 = 156$ N $\qquad\qquad\qquad\qquad\qquad\qquad\qquad\qquad$ ◊

(b)   Now for the pulley, $\sum \tau = r(T_2 - T_1) = I\alpha$

We may choose to call clockwise positive.   The angular acceleration is:

$$\alpha = \frac{a}{r} = \frac{2.00 \text{ m} / \text{s}^2}{0.250 \text{ m}} = 8.00 \text{ rad} / \text{s}^2$$

$\sum \tau = I\alpha:$ $\qquad\qquad (-118 \text{ N})(0.250 \text{ m}) + (156 \text{ N})(0.250 \text{ m}) = I(8.00 \text{ rad} / \text{s}^2)$

$$I = \frac{9.38 \text{ N} \cdot \text{m}}{8.00 \text{ rad} / \text{s}^2} = 1.17 \text{ kg} \cdot \text{m}^2 \qquad\qquad\qquad ◊$$

Finalize: The tensions are close to the weight of each mass and $T_2 > T_1$ as expected. However, the moment of inertia for the pulley is about 4 times greater than expected. Unless we made a mistake in solving this problem, our result means that the pulley has a mass of 37.4 kg (about 80 lb), which means that the pulley is probably made of a dense material, like steel. This is certainly not a problem where the mass of the pulley can be ignored since the pulley has more mass than the combination of the two blocks!

---

73.   As a result of friction, the angular speed of a wheel changes with time according to

$$\omega = \frac{d\theta}{dt} = \omega_0 e^{-\sigma t} \qquad\qquad \text{where } \omega_0 \text{ and } \sigma \text{ are constants.}$$

The angular speed changes from 3.50 rad/s at $t=0$ to 2.00 rad/s at $t=9.30$ s. Use this information to determine σ and ω_0. Then determine (a) the magnitude of the angular acceleration at $t=3.00$ s,   (b) the number of revolutions the wheel makes in the first 2.50 s, and (c) the number of revolutions it makes before coming to rest.

Solution

When $t=0$, $\omega = 3.50 \text{ rad/s}$ and $e^0 = 1$

so $\omega = \omega_0 e^{-\sigma t}$ becomes $\omega_{t=0} = \omega_0 e^{-\sigma(t=0)} = 3.50 \text{ rad/s}$

which gives $\omega_0 = 3.50 \text{ rad/s}$ ◊

We now calculate σ: $2.00 \text{ rad/s} = (3.50 \text{ rad/s})e^{-\sigma(9.30 \text{ s})}$

$$0.571 = e^{-\sigma(9.30 \text{ s})}$$

$$\ln(0.571) = -0.560 = \ln\left(e^{-9.30\sigma}\right) = -9.30\sigma$$

and $\sigma = 0.0602 \text{ s}^{-1}$ ◊

(a)   At all times $\alpha = \dfrac{d\omega}{dt} = \dfrac{d}{dt}\left[\omega_0 e^{-\sigma t}\right] = -\sigma\omega_0 e^{-\sigma t}$

At $t = 3.00 \text{ s}$, $\alpha = -\left(0.060\ 2 \text{ s}^{-1}\right)(3.50 \text{ rad/s})e^{-0.181} = -0.176 \text{ rad/s}^2$ ◊

(b)   From the given equation, $d\theta = \omega_0 e^{-\sigma t}dt$

and $\theta = \displaystyle\int_{0\text{ s}}^{2.50\text{ s}} \omega_0 e^{-\sigma t}dt = \dfrac{\omega_0}{-\sigma}e^{-\sigma t}\Big|_{0\text{ s}}^{2.50\text{ s}} = \dfrac{\omega_0}{-\sigma}\left(e^{-2.50\sigma} - 1\right)$

Substituting and solving, $\theta = -58.2(0.860 - 1) \text{ rad} = 8.12 \text{ rad}$

or $\theta = (8.12 \text{ rad})\left(\dfrac{1 \text{ rev}}{2\pi \text{ rad}}\right) = 1.29 \text{ rev}$ ◊

(c)   The motion continues to a finite limit, as ω approaches zero and t goes to infinity. From part (b), the total angular displacement is

$$\theta = \dfrac{\omega_0}{-\sigma}e^{-\sigma t}\Big|_0^{\infty} = \dfrac{\omega_0}{-\sigma}\left(e^{-\infty} - e^0\right) = \dfrac{\omega_0}{-\sigma}(0-1) = \dfrac{\omega_0}{\sigma}$$

Substituting, $\theta = 58.2 \text{ rad}$ or $\theta = \left(\dfrac{1 \text{ rev}}{2\pi \text{ rad}}\right)(58.2 \text{ rad}) = 9.26 \text{ rev}$ ◊

77. A string is wound around a uniform disk of radius R and mass M. The disk is released from rest with the string vertical and its top end tied to a fixed bar (Fig. P10.77). Show that (a) the tension in the string is one third of the weight of the disk, (b) the magnitude of the acceleration of the center of mass is $2g/3$, and (c) the speed of the center of mass is $(4gh/3)^{1/2}$ after the disk has descended through a distance h. Verify your answer to (c) using the energy approach.

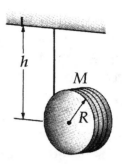

Figure P10.77

Solution Choosing positive linear quantities to be downwards and positive angular quantities to be clockwise,

$\sum \mathbf{F} = m\mathbf{a}$ yields

$$\sum F = Mg - T = Ma$$

or

$$a = \frac{Mg - T}{M}$$

$\Sigma\tau = I\alpha$ becomes

$$\sum \tau = TR = I\alpha = \frac{1}{2}MR^2\left(\frac{a}{R}\right)$$

so

$$a = \frac{2T}{M}$$

(a) Setting these two equations equal, $T = Mg/3$ ◊

(b)

$$a = \frac{2T}{M} = \frac{2Mg}{3M}$$

or

$$a = \frac{2}{3}g$$ ◊

(c) Since $v_i = 0$ and $a = \frac{2}{3}g$,

$$v_f^2 = v_i^2 + 2ah$$

gives us

$$v_f^2 = 0 + 2\left(\frac{2}{3}g\right)h$$

or

$$v_f = \sqrt{4gh/3}$$ ◊

Now we verify this answer. Requiring conservation of mechanical energy, we have

$$\Delta U + \Delta K_{\text{rot}} + \Delta K_{\text{trans}} = 0$$

$$mg\Delta h + \frac{1}{2}I\omega^2 + \frac{1}{2}mv^2 = 0$$

$$(0 - mgh) + \frac{1}{2}\left(\frac{1}{2}MR^2\right)\omega^2 - 0 + \left(\frac{1}{2}Mv^2 - 0\right) = 0$$

When there is no slipping, $\omega = \frac{v}{R}$ and $v = \sqrt{\frac{4gh}{3}}$ ◊

85. A spool of wire of mass M and radius R is unwound under a constant force F (Fig. P10.85). Assuming the spool is a uniform solid cylinder that doesn't slip, show that (a) the acceleration of the center of mass is $4F/3M$ and (b) the force of friction is to the **right** and is equal in magnitude to $F/3$. (c) If the cylinder starts from rest and rolls without slipping, what is the speed of its center of mass after it has rolled through a distance d?

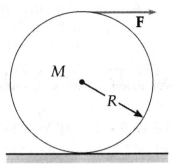

Figure P10.85

Solution

To keep the spool from slipping, there must be static friction acting at its contact point with the floor. Assume friction is directed to the left. Then the full set of Newton's second-law equations are:

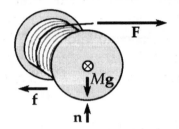

$$\sum F_x = ma_x: \qquad\qquad F - f = Ma$$

$$\sum F_y = ma_y: \qquad\qquad -Mg + n = 0$$

$$\sum \tau_{CM} = I_{CM}\alpha_{CM}: \qquad\qquad -FR + Mg(0) + n(0) - fR = \tfrac{1}{2}MR^2\left(-\frac{a}{R}\right)$$

We have written $\alpha_{CM} = -a/R$ to describe the clockwise rotation. Regarding F, M, and R as known, the first and third of these equations allow us to solve for f and a:

$$f = F - Ma \qquad\qquad \text{and} \qquad -F - f = -\tfrac{1}{2}Ma$$

(a) Substituting for f, $\qquad\qquad F + F - Ma = \tfrac{1}{2}Ma$

Thus, $2F = \dfrac{3}{2}Ma$ $\qquad$ and $\qquad a = \dfrac{4F}{3M}$ $\qquad\qquad\qquad$ ◊

(b) Then solving for f, $\qquad f = F - M\left(\dfrac{4F}{3M}\right) = F - \dfrac{4F}{3} = -\dfrac{F}{3}$ $\qquad$ ◊

The negative sign means that the force of friction is opposite the direction we assumed. That is, **f** is to the right, so the spool does not spin like the wheel of a car stuck in the snow.

(c) Since a is constant, we can use $\qquad v_f{}^2 = v_i{}^2 + 2a(x_f - x_i) = 0 + 2\left(\dfrac{4F}{3M}\right)d$

$$v_f = \sqrt{\frac{8Fd}{3M}} \qquad\qquad\qquad\qquad ◊$$

Chapter 11
ANGULAR MOMENTUM

EQUATIONS AND CONCEPTS

The **vector torque** due to an applied force can be expressed as a vector product. The position vector **r** locates the point of application of the force. The force **F** will tend to produce a rotation about an axis perpendicular to the plane formed by **r** and **F**. *Torque depends on the choice of the origin and has SI units of N · m.*

$$\boldsymbol{\tau} \equiv \mathbf{r} \times \mathbf{F} \tag{11.1}$$

The **vector product or cross product** of any two vectors **A** and **B** is a vector **C** whose magnitude is given by $C = AB\sin\theta$. The direction of **C** is perpendicular to the plane formed by **A** and **B** and θ is the angle between **A** and **B**. The cross product of **A** and **B** is written $\mathbf{A} \times \mathbf{B}$ and is read "A cross B". *The sense of C can be determined from the right-hand rule.* See the figure in Suggestions, Skills, and Strategies.

$$\mathbf{C} = \mathbf{A} \times \mathbf{B} \tag{11.2}$$

$$C \equiv AB\sin\theta \tag{11.3}$$

The vector cross product of two vectors can be expressed in terms of the components of the vectors and unit vectors.

$$\mathbf{A} \times \mathbf{B} =$$
$$\left(A_y B_z - A_z B_y\right)\hat{\mathbf{i}} - \left(A_x B_z - A_z B_x\right)\hat{\mathbf{j}} + \left(A_x B_y - A_y B_x\right)\hat{\mathbf{k}} \tag{11.8}$$

The **angular momentum of a particle** relative to the origin is defined as $\mathbf{L} \equiv \mathbf{r} \times \mathbf{p}$; the cross product of the instantaneous position vector and the instantaneous momentum. The SI unit of angular momentum is $\text{kg} \cdot \text{m}^2 / \text{s}$. *Note that both the magnitude and direction of L depend on the choice of origin.*

$$\mathbf{L} \equiv \mathbf{r} \times \mathbf{p} \tag{11.10}$$

173

The **torque (τ) on a particle** equals the time rate of change of its angular momentum when the same origin is used to define **L** and τ. Compare Equation 11.11, $\Sigma\tau = d\mathbf{L}/dt$ to $\Sigma\mathbf{F} = d\mathbf{p}/dt$ and note that force causes a change in linear momentum and torque causes a change in angular momentum. *This expression is the basic equation for treating rotating rigid bodies and rotating particles. Equation 11.11 is valid for any origin fixed in an inertial frame.*

$$\sum \tau = \frac{d\mathbf{L}}{dt} \qquad (11.11)$$

The **magnitude of a particle's angular momentum** about a specified origin depends on the magnitude mv of its momentum, the magnitude r of the position vector, and the angle ϕ between the directions of **v** and **r**. The direction of **L** is perpendicular to both the direction of **r** and the direction of **v**. *When the linear momentum of a particle is along a line that passes through the origin, the particle has zero angular momentum. Note that in the figure, the correct angle between the vectors is larger than 90°. However, also note that since $\sin(180° - \phi) = \sin\phi$, the calculation will yield the same result if the angle less than 90° is used.*

$$L = mvr \sin\phi \qquad (11.12)$$

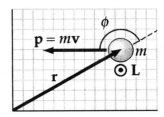

In the figure, **L** is directed out of the page.

The **net external torque acting on a system of particles** equals the time rate of change of the total angular momentum of the system. The angular momentum of a system of particles is obtained by taking the vector sum of the individual angular momenta about some point in an inertial frame. *The resultant torque acting on a system about an axis through the center of mass equals the time rate of the change of angular momentum of the system regardless of the motion of the center of mass.*

$$\sum \tau_{ext} = \frac{d\mathbf{L}_{tot}}{dt} \qquad (11.13)$$

The **magnitude of the angular momentum of a rigid body** rotating in the xy plane about a *fixed axis* (the z axis) is given by the product $I\omega$, where I is the moment of inertia about the axis of rotation and ω is the angular speed. *The direction of ω and L are along the z axis (perpendicular to the xy plane).*

$$L_z = I\omega \qquad (11.14)$$

The **rotational form of Newton's second law** states that the net torque acting on a rigid object rotating about a fixed axis equals the product of the moment of inertia about the axis of rotation and the angular acceleration relative to that axis. *This equation is also valid for a rigid object rotating about a moving axis if the axis of rotation is an axis of symmetry and also passes through the center of mass.*

$$\sum \tau_{\text{ext}} = I\alpha \qquad (11.16)$$

The **law of conservation of angular momentum** states that if the resultant external torque acting on a system is zero, the total angular momentum is constant in both magnitude and direction.

If $\quad \sum \tau_{\text{ext}} = \dfrac{d\mathbf{L}_{\text{tot}}}{dt} = 0 \qquad (11.17)$

then $\quad \mathbf{L}_{\text{tot}} = \text{constant}$

or $\quad \mathbf{L}_i = \mathbf{L}_f \qquad (11.18)$

If a **redistribution of mass in an isolated system** ($\Sigma \tau = 0$) results in a change in the moment of inertia of the system, conservation of angular momentum can be used to find the final angular speed in terms of the initial angular speed.

$$I_i \omega_i = I_f \omega_f = \text{constant} \qquad (11.19)$$

In an **isolated system**, three quantities are conserved: total mechanical energy, linear momentum, and angular momentum.

$$E_i = E_f$$
$$\mathbf{p}_i = \mathbf{p}_f$$
$$\mathbf{L}_i = \mathbf{L}_f$$

The **precessional frequency of a gyroscope** is the rate at which the axle of the gyroscope rotates about the vertical axis. The gyroscope has an angular velocity ω, and the center of mass of the gyroscope is a distance h from the pivot point. The axle of the gyroscope sweeps out an angle $d\phi$ in a time interval dt. **The direction of the angular momentum vector changes while the magnitude remains constant.** *Equation 11.20 is valid only when $\omega_p \ll \omega$. The precessional motion decreases as the spinning rate increases.*

$$\omega_p = \frac{d\phi}{dt} = \frac{Mgh}{I\omega} \qquad (11.20)$$

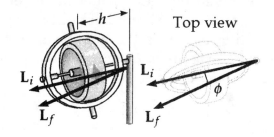

SUGGESTIONS, SKILLS, AND STRATEGIES

The operation of the vector or cross product is used for the first time in this chapter. (Recall that the angular momentum **L** of a particle is defined as $\mathbf{L} = \mathbf{r} \times \mathbf{p}$, while torque is defined by the expression $\boldsymbol{\tau} = \mathbf{r} \times \mathbf{F}$.) Let us briefly review the cross-product operation and some of its properties.

If **A** and **B** are any two vectors, their cross product, written as $\mathbf{A} \times \mathbf{B}$, is also a vector **C**.

That is, $\mathbf{C} = \mathbf{A} \times \mathbf{B}$, where the magnitude of **C** is given by $C = |\mathbf{C}| = AB\sin\theta$, and θ is the angle between **A** and **B** as in the figure below.

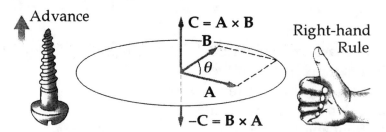

The direction of **C** is perpendicular to the plane formed by **A** and **B**, and its sense is determined by the right-hand rule. You should practice this rule for various choices of vector pairs. Note that $\mathbf{B} \times \mathbf{A}$ is directed opposite to $\mathbf{A} \times \mathbf{B}$. That is, $\mathbf{A} \times \mathbf{B} = -\mathbf{B} \times \mathbf{A}$. This follows from the right-hand rule. You should not confuse the cross product of two vectors, which is a vector quantity, with the dot product of two vectors, which is a scalar quantity. (Recall that the dot product is defined as $\mathbf{A} \cdot \mathbf{B} = AB\cos\theta$.)

Note that the cross product of any vector with itself is zero. That is, $\mathbf{A} \times \mathbf{A} = 0$ since $\theta = 0$, and $\sin(0) = 0$.

Very often, vectors will be expressed in unit vector form, and it is convenient to make use of the multiplication table for unit vectors. Note that $\hat{i}$, $\hat{j}$, and $\hat{k}$ represent a set of mutually orthogonal vectors as shown below.

$$\hat{i} \times \hat{i} = \hat{j} \times \hat{j} = \hat{k} \times \hat{k} = 0$$

$$\hat{i} \times \hat{j} = -\hat{j} \times \hat{i} = \hat{k}$$

$$\hat{j} \times \hat{k} = -\hat{k} \times \hat{j} = \hat{i}$$

$$\hat{k} \times \hat{i} = -\hat{i} \times \hat{k} = \hat{j}$$

The cross product of two vectors can be expressed in determinant form. When the determinant is expanded, the cross product can be calculated in terms of the vector components:

$$\mathbf{A} \times \mathbf{B} = \left(A_y B_z - A_z B_y\right)\hat{i} - \left(A_x B_z - A_z B_x\right)\hat{j} + \left(A_x B_y - A_y B_x\right)\hat{k}$$

For example, consider two vectors in the xy plane:

$$\mathbf{A} = 3.00\hat{i} + 5.00\hat{j} \quad \text{and} \quad \mathbf{B} = 4.00\hat{j}$$

The cross-product can be found using the expression above:

Let $\mathbf{C} = \mathbf{A} \times \mathbf{B}$

$$\mathbf{C} = \mathbf{A} \times \mathbf{B} = \left[(5.00)(0) - (0)(4.00)\right]\hat{i} - \left[(3.00)(0) - (0)(0)\right]\hat{j} + \left[(3.00)(4.00) - (5.00)(0)\right]\hat{k}$$

$$\mathbf{C} = \mathbf{A} \times \mathbf{B} = 12\hat{k}$$

As expected from the right hand rule, the direction of $\mathbf{C}$ is along $\hat{k}$, the z axis.

REVIEW CHECKLIST

You should be able to:

▷ Determine the magnitude and direction of the cross product of two vectors and find the angle between two vectors. (Section 11.1)

▷ Given the position vector of a particle, calculate the net torque on the particle due to several specified forces. (Section 11.1)

▷ Determine the angular momentum of a system of particles rotating about a specified axis. (Section 11.2)

▷ Calculate the angular momentum **L** of a particle moving with a velocity **v** relative to a specified origin. Note that both **L** and $\boldsymbol{\tau}$ are quantities which depend on the choice of the origin since they involve the vector position **r** of the particle: $\mathbf{L} = \mathbf{r} \times \mathbf{p}$ and $\boldsymbol{\tau} = \mathbf{r} \times \mathbf{F}$. (Section 11.2)

▷ Apply equation 11.13 to calculate the change in angular momentum of a system due to the resultant torque acting on the system. (Section 11.2)

▷ Calculate the angular acceleration of a rigid body about an axis due to external torque. (Section 11.3)

▷ Apply the conservation of angular momentum principle to a body rotating about a fixed axis when the moment of inertia changes due to a change in the mass distribution. (Section 11.4)

▷ Calculate the precessional frequency of a gyroscope. (Section 11.5)

ANSWERS TO SELECTED CONCEPTUAL QUESTIONS

11. Why does a long pole help a tightrope walker stay balanced?

Answer The long pole increases the tightrope walker's moment of inertia, and therefore decreases his angular acceleration, under any given torque. That gives him more time to respond, and in essence, aids his reflexes.

□ □ □ □

15. If global warming occurs over the next century, it is likely that some polar ice will melt and the water will be distributed closer to the Equator. How would this change the moment of inertia of the Earth? Would the length of the day (one revolution) increase or decrease?

Answer The Earth already bulges slightly at the Equator, and is slightly flat at the poles. If more mass moved towards the Equator, it would essentially move the mass to a greater distance from the axis of rotation, and increase the moment of inertia. Since conservation of angular momentum requires that $\omega_z I_z = $ const, an increase in the moment of inertia would decrease the angular velocity, and slow down the spinning of the Earth. Thus, the length of each day would increase.

□ □ □ □

SOLUTIONS TO SELECTED END-OF-CHAPTER PROBLEMS

3. Two vectors are given by $\mathbf{A} = -3\hat{\mathbf{i}} + 4\hat{\mathbf{j}}$ and $\mathbf{B} = 2\hat{\mathbf{i}} + 3\hat{\mathbf{j}}$. Find (a) $\mathbf{A} \times \mathbf{B}$ and (b) the angle between $\mathbf{A}$ and $\mathbf{B}$.

Solution We assume the data are known to three significant digits.

(a) $\mathbf{A} \times \mathbf{B} = \left(-3\hat{\mathbf{i}} + 4\hat{\mathbf{j}}\right) \times \left(2\hat{\mathbf{i}} + 3\hat{\mathbf{j}}\right)$

  $\mathbf{A} \times \mathbf{B} = \left(-6\hat{\mathbf{i}} \times \hat{\mathbf{i}}\right) - \left(9\hat{\mathbf{i}} \times \hat{\mathbf{j}}\right) + \left(8\hat{\mathbf{j}} \times \hat{\mathbf{i}}\right) + \left(12\hat{\mathbf{j}} \times \hat{\mathbf{j}}\right) = 0 - 9\hat{\mathbf{k}} + 8\left(-\hat{\mathbf{k}}\right) + 0 = -17\hat{\mathbf{k}}$ ◊

(b) Since $\left|\mathbf{A} \times \mathbf{B}\right| = \left|AB\sin\theta\right|$

  $\theta = \sin^{-1}\dfrac{\left|\mathbf{A} \times \mathbf{B}\right|}{AB} = \sin^{-1}\left(\dfrac{17}{\sqrt{3^2 + 4^2}\sqrt{2^2 + 3^2}}\right) = 70.6°$ ◊

Related Calculation: To solidify your understanding, review taking the dot product of these same two vectors, and again find the angle between them.

  $\mathbf{A} \cdot \mathbf{B} = \left(-3.00\hat{\mathbf{i}} + 4.00\hat{\mathbf{j}}\right) \cdot \left(2.00\hat{\mathbf{i}} + 3.00\hat{\mathbf{j}}\right) = -6.00 + 12.0 = 6.00$ ◊

  $\mathbf{A} \cdot \mathbf{B} = AB\cos\theta$

  $\theta = \cos^{-1}\dfrac{\mathbf{A} \cdot \mathbf{B}}{AB} = \cos^{-1}\left(\dfrac{6.00}{\sqrt{3^2 + 4^2}\sqrt{2^2 + 3^2}}\right) = 70.6°$ ◊

The answers agree. The method based upon the dot product is more general. It gives an unambiguous answer when the angle between the vectors is greater than 90°.

7. If $\left|\mathbf{A} \times \mathbf{B}\right| = \mathbf{A} \cdot \mathbf{B}$, what is the angle between $\mathbf{A}$ and $\mathbf{B}$?

Solution We are given the condition $\left|\mathbf{A} \times \mathbf{B}\right| = \mathbf{A} \cdot \mathbf{B}$

  This says that $AB\sin\theta = AB\cos\theta$

  and $\tan\theta = 1$

  $\theta = 45°$ satisfies this condition. ◊

11. A light rigid rod 1.00 m in length joins two particles, with masses 4.00 kg and 3.00 kg, at its ends. The combination rotates in the xy plane about a pivot through the center of the rod (Figure P11.11). Determine the angular momentum of the system about the origin when the speed of each particle is 5.00 m/s.

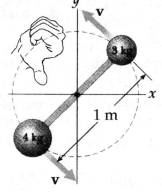

Figure P11.11

Solution   Taking the origin of the compound object to be the pivot, the angular speed and the moment of inertia are

$$\omega = v/r = (5.00 \text{ m / s})/0.500 \text{ m} = 10.0 \text{ rad / s}$$

$$I = \sum mr^2 = (4.00 \text{ kg})(0.500 \text{ m})^2 + (3.00 \text{ kg})(0.500 \text{ m})^2 = 1.75 \text{ kg} \cdot \text{m}^2$$

By the right-hand rule (shown in the modification to the figure), we find that the angular velocity is directed out of the plane. directed out of the plane. So the object's angular momentum is

$$L = I\omega = (1.75 \text{ kg} \cdot \text{m}^2)(10.0 \text{ rad / s}): \qquad \mathbf{L} = 17.5 \text{ kg} \cdot \text{m}^2/\text{s out of the plane} \lozenge$$

Note: Alternatively, we could solve this using $\mathbf{L} = \sum m\mathbf{r} \times \mathbf{v}$:

$$L = (4.00 \text{ kg})(0.500 \text{ m})(5.00 \text{ m / s}) + (3.00 \text{ kg})(0.500 \text{ m})(5.00 \text{ m / s})$$

$$L = 10.0 \text{ kg} \cdot \text{m}^2/\text{s} + 7.50 \text{ kg} \cdot \text{m}^2/\text{s}: \qquad \mathbf{L} = 17.5 \text{ kg} \cdot \text{m}^2/\text{s out of the plane} \lozenge$$

---

13. The position vector of a particle of mass 2.00 kg is given as a function of time by $\mathbf{r} = (6.00\hat{\mathbf{i}} + 5.00t\hat{\mathbf{j}})$ m.   Determine the angular momentum of the particle about the origin, as a function of time.

Solution

The velocity of the particle is $\mathbf{v} = \dfrac{d\mathbf{r}}{dt} = \dfrac{d}{dt}(6.00\hat{\mathbf{i}} \text{ m} + 5.00t\hat{\mathbf{j}} \text{ m}) = 5.00\hat{\mathbf{j}} \text{ m / s}$

The angular momentum is

$$\mathbf{L} = \mathbf{r} \times \mathbf{p} = m\mathbf{r} \times \mathbf{v} = (2.00 \text{ kg})(6.00\hat{\mathbf{i}} \text{ m} + 5.00t\hat{\mathbf{j}} \text{ m}) \times 5.00\hat{\mathbf{j}} \text{ m / s}$$

$$\mathbf{L} = (60.0 \text{ kg} \cdot \text{m}^2 / \text{s})\hat{\mathbf{i}} \times \hat{\mathbf{j}} + (50.0t \text{ kg} \cdot \text{m}^2 / \text{s})\hat{\mathbf{j}} \times \hat{\mathbf{j}}$$

$$\mathbf{L} = (60.0\hat{\mathbf{k}}) \text{ kg} \cdot \text{m}^2 / \text{s, constant in time} \qquad \lozenge$$

---

17. A particle of mass m is shot with an initial velocity $\mathbf{v}_i$ making an angle θ with the horizontal as shown in Figure P11.17. The particle moves in the gravitational field of the Earth. Find the angular momentum of the particle about the origin when the particle is (a) at the origin, (b) at the highest point of its trajectory, and (c) just before it hits the ground. (d) What torque causes its angular momentum to change?

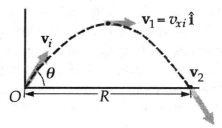

Figure P11.17

Solution   The angular momentum is $\mathbf{L} = \mathbf{r} \times m\mathbf{v}$

(a)   Since $\mathbf{r} = 0$, $\mathbf{L} = 0$ ◊

(b)   At the highest point of the trajectory,

$$\mathbf{L} = \left(\frac{v_i^2 \sin 2\theta}{2g} \hat{\mathbf{i}} + \frac{(v_i \sin \theta)^2}{2g} \hat{\mathbf{j}} \right) \times m v_{xi} \hat{\mathbf{i}} = \frac{-m(v_i \sin \theta)^2 v_i \cos \theta}{2g} \hat{\mathbf{k}}$$ ◊

(c) $\mathbf{L} = R\hat{\mathbf{i}} \times m\mathbf{v}_2$, where $R = \dfrac{v_i^2 \sin 2\theta}{g}$

$$\mathbf{L} = m\left[R\hat{\mathbf{i}} \times (v_i \cos \theta \hat{\mathbf{i}} - v_i \sin \theta \hat{\mathbf{j}}) \right] = -mRv_i \hat{\mathbf{k}} \sin \theta = \frac{-mv_i^3 \sin 2\theta \sin \theta}{g} \hat{\mathbf{k}}$$ ◊

(d)   The downward force of gravity exerts a torque in the $-z$ direction.   ◊

---

25. A particle of mass of 0.400 kg is attached to the 100-cm mark of a meter stick of mass 0.100 kg. The meter stick rotates on a horizontal, frictionless table with an angular speed of 4.00 rad/s. Calculate the angular momentum of the system when the stick is pivoted about an axis (a) perpendicular to the table through the 50.0-cm mark and (b) perpendicular to the table through the 0-cm mark.

Solution

Conceptualize: Since the angular speed is constant, the angular momentum will be greater when the center of mass of the system is farther from the axis of rotation (larger I).

Categorize: Use the equation $L = I\omega$ to find L for each I.

Analyze: Defining the distance from the pivot to the center of mass as d, we first find the rotational inertia of the system for each case, from the information

$$M = 0.100 \text{ kg}, \quad m = 0.400 \text{ kg}, \quad \text{and} \quad D = 1.00 \text{ m}$$

(a)   For the meter stick rotated about its center, $I_m = \frac{1}{12}MD^2$.

For the weight, $I_w = md^2 = m\left(\frac{1}{2}D\right)^2$

$$I = I_m + I_w = \left(\frac{1}{12}MD^2\right) + \left(\frac{1}{4}mD^2\right)$$

$$I = \frac{(0.100 \text{ kg})(1.00 \text{ m})^2}{12} + \frac{(0.400 \text{ kg})(1.00 \text{ m})^2}{4} = 0.108 \text{ kg} \cdot \text{m}^2$$

$$L = I\omega = (0.108 \text{ kg} \cdot \text{m}^2)(4.00 \text{ rad / s}) = 0.433 \text{ kg} \cdot \text{m}^2/\text{s} \qquad \lozenge$$

(b)   For a stick rotated about a point at one end,

$$I_m = \frac{1}{3}mD^2$$

$$I_m = \frac{1}{3}(0.100 \text{ kg})(1.00 \text{ m})^2 = 0.033\ 3 \text{ kg} \cdot \text{m}^2$$

For a point mass, $I_w = mD^2 = (0.400 \text{ kg})(1.00 \text{ m})^2 = 0.400 \text{ kg} \cdot \text{m}^2$

$$I = I_m + I_w = 0.433 \text{ kg} \cdot \text{m}^2$$

Thus, $L = I\omega = (0.433 \text{ kg} \cdot \text{m}^2)(4.00 \text{ rad / s}) = 1.73 \text{ kg} \cdot \text{m}^2/\text{s} \qquad \lozenge$

Finalize: As we expected, the angular momentum is larger when the center of mass is further from the rotational axis. In fact, the angular momentum for part (b) is 4 times greater than for part (a). The units of $\text{kg} \cdot \text{m}^2/\text{s}$ also make sense since $[L] = [r] \times [p]$.

---

33.   A 60.0-kg woman stands at the rim of a horizontal turntable having a moment of inertia of 500 $\text{kg} \cdot \text{m}^2$ and a radius of 2.00 m. The turntable is initially at rest and is free to rotate about a frictionless, vertical axle through its center. The woman then starts walking around the rim clockwise (as viewed from above the system) at a constant speed of 1.50 m/s relative to the Earth. (a) In what direction and with what angular speed does the turntable rotate? (b) How much work does the woman do to set herself and the turntable into motion?

Solution   The table rotates in a direction opposite to that in which the woman walks. We use the angular momentum version of the isolated system model. There are no external torques acting on the woman-turntable system; therefore, from conservation of angular momentum, we have $L_f = L_i = 0$

(a)   Therefore, $\qquad L_f = I_w\omega_w + I_t\omega_t = 0 \qquad$ and $\qquad \omega_t = -\dfrac{I_w}{I_t}\omega_w$

$\qquad$ Solving. $\qquad \omega_t = -\left(\dfrac{m_w r^2}{I_t}\right)\left(\dfrac{v_w}{r}\right) = -\dfrac{(60.0\text{ kg})(2.00\text{ m})(1.50\text{ m}/\text{s})}{500\text{ kg}\cdot\text{m}^2}$

$\qquad\qquad \omega_t = -0.360\text{ rad}/\text{s} = 0.360\text{ rad}/\text{s CCW} \qquad\qquad\qquad\qquad \Diamond$

(b)   Work done $= \Delta K$: $\quad W = K_f - 0 = \dfrac{1}{2}m_{\text{woman}}v_{\text{woman}}^2 + \dfrac{1}{2}I_{\text{table}}\omega_{\text{table}}^2$

$W = \dfrac{1}{2}(60.0\text{ kg})(1.50\text{ m}/\text{s})^2 + \dfrac{1}{2}\left(500\text{ kg}\cdot\text{m}^2\right)(0.360\text{ rad}/\text{s})^2 = 99.9\text{ J} \qquad\qquad \Diamond$

Related Questions: (a) Why is the angular momentum of the woman-turntable system conserved? (b) Why is the mechanical energy of this system not conserved? (c) Is the linear momentum of this system conserved?

(a) Because the axle exerts no torque on the woman-plus-turntable system; only torques from outside the system can change the total angular momentum.

(b) The internal forces, of the woman pushing backward on the turntable and of the turntable pushing forward on the woman, both do positive work, converting chemical into kinetic energy.

(c) No. If the woman starts walking north, she pushes south on the turntable. Its axle holds it still against linear motion by pushing north on it, and this outside force delivers northward linear momentum into the system.

---

35.   A wooden block of mass M resting on a frictionless horizontal surface is attached to a rigid rod of length ℓ and of negligible mass (Fig. P11.35). The rod is pivoted at the other end. A bullet of mass m traveling parallel to the horizontal surface and perpendicular to the rod with speed v hits the block and becomes embedded in it. (a) What is the angular momentum of the bullet-block system?   (b) What fraction of the original kinetic energy is lost in the collision?

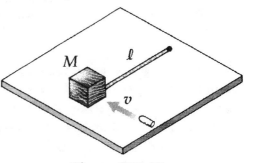

Figure P11.35

Solution

Conceptualize: Since there are no external torques acting on the bullet-block system, the angular momentum of the system will be constant and will simply be that of the bullet before it hits the block.

We should expect there to be a significant, but not a total loss of kinetic energy in this perfectly inelastic "angular collision", since the block and bullet will move together after the collision with a velocity that can be found from conservation of angular momentum.

Categorize: We have practically solved this problem already! We just need to work out the details for the loss of kinetic energy.

Analyze: Taking the origin at the pivot point,

(a) Note that $\mathbf{r}$ is perpendicular to $\mathbf{v}$, so $\sin(\theta) = 1$

   and $L = rmv\sin(\theta) = \ell mv$ ◊

(b) Taking v_f to be the speed of the bullet and the block together, we first apply conservation of angular momentum. $L_i = L_f$ becomes:

   $$\ell mv = \ell(m + M)v_f \quad \text{or} \quad v_f = \left(\frac{m}{m + M}\right)v$$

   The total kinetic energies before and after the collision are

   $$K_i = \frac{1}{2}mv^2$$

   $$K_f = \frac{1}{2}(m + M)v_f^2 = \frac{1}{2}(m + M)\left(\frac{m}{m + M}\right)^2 v^2 = \frac{1}{2}\left(\frac{m^2}{m + M}\right)v^2$$

   So the fraction of the kinetic energy that is "lost" will be:

   $$\text{Fraction} = \frac{-\Delta K}{K_i} = \frac{K_i - K_f}{K_i} = \frac{\frac{1}{2}mv^2 - \frac{1}{2}\left(\frac{m^2}{m + M}\right)v^2}{\frac{1}{2}mv^2} = \frac{M}{m + M}$$ ◊

Finalize: We could have also used conservation of linear momentum to solve part (b), instead of conservation of angular momentum; the answer would be the same. From the final equation, we can see that a larger fraction of kinetic energy is lost for a smaller bullet. This is consistent with reasoning that if the bullet were so small that the block barely moved after the collision, then nearly all the initial kinetic energy would disappear into internal energy.

49. A puck of mass m is attached to a cord passing through a small hole in a frictionless, horizontal surface (Fig. P11.49). The puck is initially orbiting with speed v_i in a circle of radius r_i. The cord is then slowly pulled from below, decreasing the radius of the circle to r. (a) What is the speed of the puck when the radius is r? (b) Find the tension in the cord as a function of r. (c) How much work W is done in moving m from r_i to r? (**Note:** The tension depends on r.) (d) Obtain numerical values for v, T, and W when $r = 0.100$ m, $m = 50.0$ g, $r_i = 0.300$ m, and $v_i = 1.50$ m/s.

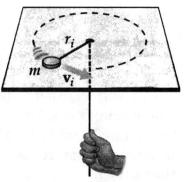

Figure P11.49

Solution

(a) Although an external force (tension of rope) acts on the mass, no external torques act.

Therefore $\qquad\qquad\qquad\qquad\qquad\qquad\qquad$ $L = \text{constant}$

and at any time $\qquad\qquad\qquad\qquad\qquad\qquad$ $mvr = mv_i r_i$

gives us $\qquad\qquad\qquad\qquad\qquad\qquad\qquad$ $v = \dfrac{v_i r_i}{r}$ $\qquad\qquad$ ◊

(b) $\qquad\qquad\qquad\qquad\qquad\qquad\qquad\qquad$ $T = \dfrac{mv^2}{r}$

Substituting for v from (a), we find that $\qquad$ $T = m\dfrac{v_i^2 r_i^2}{r^3}$ $\qquad\qquad$ ◊

(c) $\quad W = \Delta K = \dfrac{1}{2}m\left(v^2 - v_i^2\right)$ $\qquad\qquad\qquad$ $W = \dfrac{mv_i^2}{2}\left(\dfrac{r_i^2}{r^2} - 1\right)$ $\qquad$ ◊

(d) Substituting the given values into the previous equations,

we find $\qquad\qquad\qquad\qquad\qquad\qquad\qquad$ $v = 4.50$ m / s $\qquad\qquad$ ◊

$\qquad\qquad\qquad\qquad\qquad\qquad\qquad\qquad\qquad$ $T = 10.1$ N $\qquad\qquad$ ◊

and $\qquad\qquad\qquad\qquad\qquad\qquad\qquad\qquad$ $W = 0.450$ J $\qquad\qquad$ ◊

51. Two astronauts (Fig. P11.51), each having a mass of 75.0 kg, are connected by a 10.0-m rope of negligible mass. They are isolated in space, orbiting their center of mass at speeds of 5.00 m/s. Treating the astronauts as particles, (a) calculate the magnitude of the angular momentum of the system and (b) the rotational energy of the system. By pulling on the rope, one of the astronauts shortens the distance between them to 5.00 m. (c) What is the new angular momentum of the system? (d) What are the astronauts' new speeds? (e) What is the new rotational energy of the system? (f) How much work does the astronaut do in shortening the rope?

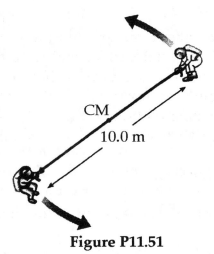

CM

10.0 m

Figure P11.51

Solution

(a) $|\mathbf{L}| = |m\mathbf{r} \times \mathbf{v}|$

In this case, r and v are perpendicular, so the magnitude of $\mathbf{L}$ about CM is

$$L = \sum mrv = 2(75.0 \text{ kg})(5.00 \text{ m})(5.00 \text{ m / s}) = 3.75 \times 10^3 \text{ kg} \cdot \text{m}^2\text{/ s} \qquad \Diamond$$

(b) $K = \frac{1}{2}mv^2 + \frac{1}{2}mv^2 = \frac{1}{2}(75.0 \text{ kg})(5.00 \text{ m / s})^2(2) = 1.88 \times 10^3 \text{ J} \qquad \Diamond$

(c)   With a lever arm of zero, the rope tension generates no torque about CM. Thus, the angular momentum is unchanged:

$$L = 3.75 \times 10^3 \text{ kg} \cdot \text{m}^2\text{/ s} \qquad \Diamond$$

(d)   Again, $L = 2mrv$: $3.75 \times 10^3 \text{ kg} \cdot \text{m}^2\text{/ s} = 2(75.0 \text{ kg})(2.50 \text{ m})(v \sin 90°)$

$$v = 10.0 \text{ m / s} \qquad \Diamond$$

(e) $K = 2\left(\frac{1}{2}mv^2\right) = 2\left(\frac{1}{2}75 \text{ kg}\right)(10 \text{ m / s})^2 = 7.50 \times 10^3 \text{ J} \qquad \Diamond$

(f) $W_{nc} = K_f - K_i = 7.50 \times 10^3 \text{ J} - 1.88 \times 10^3 \text{ J}$

$$W_{nc} = 5.62 \times 10^3 \text{ J} \qquad \Diamond$$

57. Suppose a solid disk of radius R is given an angular speed ω_i about an axis through its center and then lowered to a horizontal surface and released, as in Problem 56 (Fig. P11.56). Furthermore, assume that the coefficient of friction between disk and surface is μ. (a) Show that the time interval before pure rolling motion occurs is $R\omega_i/3\mu g$. (b) Show that the distance the disk travels before pure rolling occurs is $R^2\omega_i^2/18\mu g$.

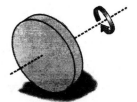

Figure P11.56

Solution

(a) If $v_i = 0$, then at any time

$$v = at$$

From Problem 56, once pure rolling occurs,

$$\omega = \frac{1}{3}\omega_i$$

so that

$$v = \frac{1}{3}R\omega_i$$

Equating these expressions for v, we find:

$$\frac{1}{3}R\omega_i = at$$

where

$$a = \frac{F}{m} = \frac{\mu m g}{m} = \mu g$$

Therefore,

$$t = \frac{R\omega_i}{3\mu g} \qquad \Diamond$$

(b) The distance of travel is

$$\Delta x = \frac{1}{2}at^2.$$

Using the result from part (a), we find

$$\Delta x = \frac{1}{2}(\mu g)\left(\frac{\frac{1}{3}R\omega_i}{\mu g}\right)^2 = \frac{R^2\omega_i^2}{18\mu g} \qquad \Diamond$$

Chapter 12

STATIC EQUILIBRIUM AND ELASTICITY

EQUATIONS AND CONCEPTS

The **first condition of equilibrium** corresponds to the condition of translational equilibrium. *When an object is at rest or is moving with constant velocity the resultant force on the object must be zero.*

$$\sum \mathbf{F} = 0 \qquad (12.1)$$

Equation 12.1 is a **vector form of three component equations**. This necessarily implies that the sum of the x, y, and z components separately must be zero.

$$\sum F_x = 0$$
$$\sum F_y = 0$$
$$\sum F_z = 0$$

The **second condition of equilibrium** of a rigid body requires that the vector sum of the torques relative to any origin must be zero. *This is a statement of rotational equilibrium and requires that the angular acceleration about any axis be zero.*

$$\sum \boldsymbol{\tau} = 0 \qquad (12.2)$$

In the **case of coplanar forces** (i.e. all forces in the xy plane) three equations specify equilibrium. Two equations correspond to the first condition of equilibrium and the third comes from the second condition (the torque equation). *In this case, the torque vector lies along a line parallel to the z axis. All problems in this chapter fall into this category.*

$$\sum F_x = 0$$
$$\sum F_y = 0 \qquad (12.3)$$
$$\sum \tau_z = 0$$

The **center of gravity** of an object is located at the center of mass when the value of g is uniform over the entire object. *In order to compute the torque due to the weight (gravitational force), all the weight can be considered to be concentrated at a single point called the center of gravity.*

$$x_{CG} = \frac{m_1 x_1 + m_2 x_2 + m_3 x_3 + \cdots}{m_1 + m_2 + m_3 + \cdots} \quad (12.4)$$

$$x_{CG} = \frac{\sum m_i x_i}{\sum m_i}$$

The **elastic modulus** of a material is defined as the ratio of stress to strain for that material. **Stress** is a quantity which is proportional to the force causing a deformation. **Strain** is a measure of the degree of deformation. *For a given material, there is an elastic modulus corresponding to each type of deformation:*
  (i) *resistance to change in length*
  (ii) *resistance to motion of parallel planes within a solid*
  (iii) *resistance to change in volume.*

$$\text{Elastic modulus} \equiv \frac{\text{stress}}{\text{strain}} \quad (12.5)$$

Young's modulus Y is a measure of the resistance of a body to elongation or compression, and is equal to the ratio of the tensile stress (the force per unit cross-sectional area) to the tensile strain (the fractional change in length.)

$$Y \equiv \frac{\text{tensile stress}}{\text{tensile strain}} = \frac{F/A}{\Delta L/L_i} \quad (12.6)$$

The **elastic limit** is the maximum stress from which a substance will recover to an initial length.

The **Shear modulus** S is a measure of the resistance of a solid to internal planes sliding past each other. This occurs when a force is applied to one face of an object while the opposite face is held fixed. The shear modulus equals the ratio of the shearing stress to the shear strain. *In Equation 12.7, Δx is the distance that the sheared face moves and h is the height of the object.*

$$S \equiv \frac{\text{shear stress}}{\text{shear strain}} = \frac{F/A}{\Delta x/h} \quad (12.7)$$

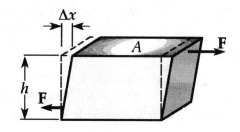

The **Bulk modulus** B is a parameter which characterizes the resistance of an object to a change in volume caused by a change in pressure. It is defined as the ratio of the volume stress (the pressure) to the volume strain $(\Delta V / V)$. *When the pressure increases (ΔP is positive), the volume decreases (ΔV is negative); the negative sign is included in Equation 12.8 so that the value of B will be positive.* The **compressibility** of a substance is the reciprocal of the bulk modulus.

$$B \equiv \frac{\text{volume stress}}{\text{volume strain}} = -\frac{\Delta P}{\Delta V / V} \qquad (12.8)$$

SUGGESTIONS, SKILLS, AND STRATEGIES

The following procedure is recommended when analyzing a body in static equilibrium under the action of several external forces:

- Make a diagram of the system under consideration.

- Draw a free-body diagram and label all external forces acting on each object in the system. Try to guess the correct direction for each force. If you select an incorrect direction that leads to a negative sign in your solution for a force, do not be alarmed; this merely means that the direction of the force is the opposite of what you assumed.

- Choose a convenient coordinate system and resolve all forces into rectangular components. Then apply the first condition for equilibrium along each coordinate direction. Remember to keep track of the signs of the various force components.

- Choose a convenient axis for calculating the net torque on the object. Remember that the choice of the origin for the torque equation is **arbitrary**; therefore, choose an origin that will simplify your calculation as much as possible. A force that acts along a direction that passes through the origin makes a zero contribution to the torque and can be ignored.

- The first and second conditions of equilibrium give a set of linear equations with several unknowns. All that is left is to solve the simultaneous equations for the unknowns in terms of the known quantities.

REVIEW CHECKLIST

You should be able to:

▷ Describe the two necessary conditions of equilibrium for a rigid body. (Section 12.1)

▷ Locate the center of gravity of a system of particles or a rigid body and understand the difference between center of gravity and center of mass. (Section 12.2)

▷ Analyze problems of rigid bodies in static equilibrium using the procedures presented in Section 12.3 of the text. (Section 12.3)

▷ Define and make calculations of stress, strain, and elastic modulus for each of the three types of deformation described in the chapter. (Section 12.4)

ANSWERS TO SELECTED CONCEPTUAL QUESTIONS

12. A ladder stands on the ground, leaning against a wall. Would you feel safer climbing up the ladder if you were told that the ground is frictionless but the wall is rough, or that the wall is frictionless but the ground is rough? Justify your answer.

Answer

The picture shows the forces on the ladder if both the wall and floor exert friction. If the floor is perfectly smooth, it can exert no frictional force to the right, to counterbalance the wall's normal force. Therefore a ladder on a smooth floor cannot stand in equilibrium. On the other hand, a smooth wall can still exert a normal force to hold the ladder in equilibrium against horizontal motion. The counterclockwise torque of this force prevents rotation about the foot of the ladder. So you should choose a rough floor.

☐   ☐   ☐   ☐

SOLUTIONS TO SELECTED END-OF-CHAPTER PROBLEMS

3. A uniform beam of mass m_b and length ℓ supports blocks with masses m_1 and m_2 at two positions, as in Figure P12.3. The beam rests on two knife edges. For what value of x will the beam be balanced at P such that the normal force at O is zero?

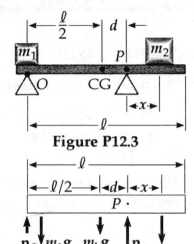

Figure P12.3

Solution

We use the rigid body in equilibrium model.

Refer to the free-body diagram, and take torques about point P.

$$\sum \tau_P = -n_0\left[\frac{\ell}{2}+d\right] + m_1 g\left[\frac{\ell}{2}+d\right] + m_b g d - m_2 g x = 0$$

We want to find x for which $n_0 = 0$. Let $n_0 = 0$ and solve for x.

$$m_1\left[\frac{\ell}{2}+d\right] + m_b d - m_2 x = 0 \qquad \text{so} \qquad x = \frac{m_1}{m_2}\left[\frac{\ell}{2}+d\right] + \frac{m_b}{m_2}d \qquad \diamond$$

---

7. Consider the following mass distribution: 5.00 kg at $(0,0)$ m, 3.00 kg at $(0, 4.00)$ m, and 4.00 kg at $(3.00, 0)$ m. Where should a fourth object of mass 8.00 kg be placed so that the center of gravity of the four-object arrangement will be at $(0,0)$?

Solution

$$\mathbf{r}_{CG} = \frac{\sum m_i \mathbf{r}_i}{\sum m_i} \qquad \mathbf{r}_{CG}\left(\sum m_i\right) = \sum m_i \mathbf{r}_i$$

We require the center of mass to be at the origin; this simplifies the situation, leaving

$$\sum m_i x_i = 0 \qquad \sum m_i y_i = 0$$

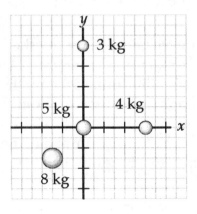

To find the x-coordinate:

$$[5.00 \text{ kg}][0 \text{ m}]+[3.00 \text{ kg}][0 \text{ m}]+[4.00 \text{ kg}][3.00 \text{ m}]+[8.00 \text{ kg}]x = 0$$

and $\qquad x = -1.50 \text{ m}$

Likewise, to find the y-coordinate, we solve:

$$[5.00 \text{ kg}][0 \text{ m}]+[3.00 \text{ kg}][4.00 \text{ m}]+[4.00 \text{ kg}][0 \text{ m}]+[8.00 \text{ kg}]y = 0$$

and $\qquad y = -1.50 \text{ m}$

Therefore, a fourth mass of 8.00 kg should be located at

$$\mathbf{r}_4 = \left(-1.50\hat{\mathbf{i}} - 1.50\hat{\mathbf{j}}\right) \text{ m} \qquad \Diamond$$

---

13. A 15.0-m uniform ladder weighing 500 N rests against a frictionless wall. The ladder makes a 60.0° angle with the horizontal. (a) Find the horizontal and vertical forces the ground exerts on the base of the ladder when an 800-N firefighter is 4.00 m from the bottom. (b) If the ladder is just on the verge of slipping when the firefighter is 9.00 m up, what is the coefficient of static friction between ladder and ground?

Solution

Conceptualize: Refer to the free-body diagram at right, as needed. Since the wall is frictionless, only the ground exerts an upward force on the ladder to oppose the combined weight of the ladder and firefighter, so $n_g = 1\,300$ N. Based on the angle of the ladder, $f < 1\,300$ N. The coefficient of friction is probably somewhere between 0 and 1.

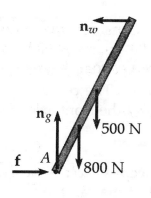

Categorize: Draw a free-body diagram, apply Newton's second law, and sum torques to find the unknown forces. Since this is a statics problem (no motion), both the net force and net torque are zero.

Analyze:

(a) $\quad \sum F_x = f - n_w = 0; \quad \sum F_y = n_g - 800 \text{ N} - 500 \text{ N} = 0$

so that $\qquad\qquad n_g = 1\,300 \text{ N} \quad \text{(upwards)} \qquad \Diamond$

Taking torques about an axis at the foot of the ladder, $\Sigma \tau_A = 0$:

$$-(800\ \text{N})(4.00\ \text{m})\sin 30° - (500\ \text{N})(7.50\ \text{m})\sin 30° + n_w(15.0\ \text{m})\cos 30° = 0$$

Solving the torque equation for n_w,

$$n_w = \frac{\left[(4.00\ \text{m})(800\ \text{N}) + (7.50\ \text{m})(500\ \text{N})\right]\tan 30.0°}{15.0\ \text{m}} = 267.5\ \text{N}$$

Next substitute this value into the F_x equation to find:

$$f = n_w = 268\ \text{N} \quad \text{(f is directed toward the wall)} \qquad \Diamond$$

(b) When the firefighter is 9.00 m up the ladder, the torque equation $\Sigma \tau_A = 0$ gives:

$$-(800\ \text{N})(9.00\ \text{m})\sin 30° - (500\ \text{N})(7.50\ \text{m})\sin 30° + n_w(15.0\ \text{m})\sin 60° = 0$$

Solving, $\qquad n_w = 421\ \text{N}$.

Since $\qquad f = n_w = 421\ \text{N}$ and $f = f_{max} = \mu_s n_g$,

$$\mu_s = \frac{f_{max}}{n_g} = \frac{421\ \text{N}}{1\,300\ \text{N}} = 0.324 \qquad \Diamond$$

Finalize: The calculated answers seem reasonable since they agree with our predictions. This problem would be more realistic if the wall were not frictionless, in which case an additional vertical force would be added. This more complicated problem could be solved if we knew at least one of the forces of friction.

---

17. A 1 500-kg automobile has a wheel base (the distance between the axles) of 3.00 m. The center of mass of the automobile is on the center line at a point 1.20 m behind the front axle. Find the force exerted by the ground on each wheel.

Solution

Conceptualize: Since the center of mass lies in the front half of the car, there should be more force on the front wheels than the rear ones, and the sum of the wheel forces must equal the weight of the car.

Categorize: Draw a free-body diagram, apply Newton's second law, and sum torques to find the unknown forces for this statics problem.

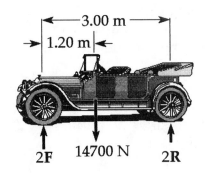

Analyze:

The car's weight is

$$F_g = mg = (1\,500 \text{ kg})(9.80 \text{ m}/\text{s}^2) = 14\,700 \text{ N}$$

Call **F** the force of the ground on each of the front wheels and **R** the normal force on each of the rear wheels. If we take torques around the front axle, with clockwise in the picture chosen as positive, the equations are as follows:

$$\sum F_x = 0: \quad 0 = 0$$

$$\sum F_y = 0: \quad 2R - 14\,700 \text{ N} + 2F = 0$$

$$\sum \tau = 0: \quad -2R(3.00 \text{ m}) + (14\,700 \text{ N})(1.20 \text{ m}) + 2F(0) = 0$$

The torque equation gives:

$$R = \frac{17\,640 \text{ N}\cdot\text{m}}{6.00 \text{ m}} = 2\,940 \text{ N} = 2.94 \text{ kN} \qquad \Diamond$$

Then, from the second force equation,

$$2(2.94 \text{ kN}) - 14.7 \text{ kN} + 2F = 0 \qquad \text{and} \qquad F = 4.41 \text{ kN} \qquad \Diamond$$

Finalize: As expected, the front wheels experience a greater force (about 50% more) than the rear wheels. Since the frictional force between the tires and road is proportional to this normal force, it makes sense that most cars today are built with front wheel drive so that the wheels under power are the ones with more traction (friction).

───────────────────

27. A 200-kg load is hung on a wire having a length of 4.00 m, cross-sectional area $0.200 \times 10^{-4} \text{ m}^2$, and Young's modulus $8.00 \times 10^{10} \text{ N}/\text{m}^2$. What is its increase in length?

Solution

Conceptualize: Since metal wire does not stretch very much, the length will probably not change by more than 1% (<4 cm in this case) unless it is stretched beyond its elastic limit.

Categorize: Apply the Young's Modulus strain equation to find the increase in length.

Analyze: Young's Modulus is $\quad Y = \dfrac{F/A}{\Delta L / L_i}$

The load force is $\qquad\qquad\qquad F = (200\text{ kg})(9.80\text{ m}/\text{s}^2) = 1\,960\text{ N}$

So $\qquad\qquad\qquad \Delta L = \dfrac{FL_i}{AY} = \dfrac{(1\,960\text{ N})(4.00\text{ m})(1\,000\text{ mm}/\text{m})}{(0.200 \times 10^{-4}\text{ m}^2)(8.00 \times 10^{10}\text{ N}/\text{m}^2)}$

$$\Delta L = 4.90\text{ mm} \qquad\qquad\qquad\qquad \Diamond$$

Finalize: The wire only stretched about 0.1% of its length, so this seems like a reasonable amount.

---

33. If the shear stress in steel exceeds 4.00×10^8 N/m^2, the steel ruptures. Determine the shearing force necessary to (a) shear a steel bolt 1.00 cm in diameter and (b) punch a 1.00-cm-diameter hole in a steel plate 0.500 cm thick.

Solution

(a) We do not need the equation $S = \text{stress}/\text{strain}$. Rather, we use just the definition of stress $\sigma = F/A$, where A is the area of one of the layers sliding over each other:

$$F = \sigma A = \pi\left(4.00 \times 10^8\text{ N}/\text{m}^2\right)\left(0.500 \times 10^{-2}\text{ m}\right)^2 = 31.4\text{ kN} \qquad\qquad \Diamond$$

(b) Now the area of the molecular layers sliding over each other is the curved lateral surface area of the cylinder punched out, a cylinder of radius 0.50 cm and height 0.50 cm. So,

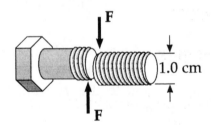

$$F = \sigma A = \sigma(h)(2\pi r)$$

$$F = 2\pi\left(4.00 \times 10^8\text{ N}/\text{m}^2\right)\left(0.500 \times 10^{-2}\text{ m}\right)\left(0.500 \times 10^{-2}\text{ m}\right) = 62.8\text{ kN} \qquad\qquad \Diamond$$

---

35. When water freezes, it expands by about 9.00%. What pressure increase would occur inside your automobile engine block if the water in it froze? (The bulk modulus of ice is 2.00×10^9 N/m².)

Solution

V represents the original volume; $0.0900V$ is the change in volume that happens if the block cracks open. Imagine squeezing the ice, with unstressed volume $1.09V$, back down ($\Delta V = -0.0900V$) to its previous volume, according to the definition of the bulk modulus (Equation 12.8):

$$\Delta P = -\frac{B(\Delta V)}{V_i} = -\frac{\left(2.00 \times 10^9 \text{ N}/\text{m}^2\right)\left(-0.0900V\right)}{1.09V} = 1.65 \times 10^8 \text{ N}/\text{m}^2 \qquad \Diamond$$

39. A bridge of length 50.0 m and mass 8.00×10^4 kg is supported on a smooth pier at each end as in Figure P12.39. A truck of mass 3.00×10^4 kg is located 15.0 m from one end. What are the forces on the bridge at the points of support?

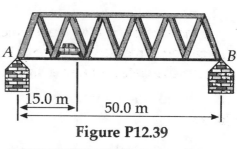

Figure P12.39

Solution

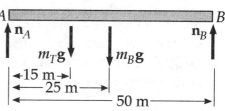

Let n_A and n_B be the normal forces at the points of support.

Choosing the origin at point A, we find:

$$\sum F_y = 0: \qquad n_A + n_B - \left(8.00 \times 10^4 \text{ kg}\right)g - \left(3.00 \times 10^4 \text{ kg}\right)g = 0$$

$$\sum \tau = 0: \qquad -\left(3.00 \times 10^4 \text{ kg}\right)(15.0 \text{ m})g - \left(8.00 \times 10^4 \text{ kg}\right)(25.0 \text{ m})g + n_B(50.0 \text{ m}) = 0$$

The equations combine to give

$$n_A = 5.98 \times 10^5 \text{ N}$$

and $\qquad n_B = 4.80 \times 10^5 \text{ N}$ $\qquad \Diamond$

45. A uniform sign of weight F_g and width $2L$ hangs from a light, horizontal beam, hinged at the wall and supported by a cable (Fig. P12.45). Determine (a) the tension in the cable and (b) the components of the reaction force exerted by the wall on the beam, in terms of F_g, d, L, and θ.

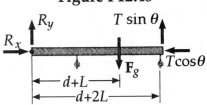

Figure P12.45

Solution

Choose the beam for analysis, and draw a free-body diagram as shown. We know that the direction of the force from the cable at the right end is along the cable, at an angle of θ above the horizontal.

Taking torques about the left end,

$$\sum F_x = 0: \quad +R_x - T\cos\theta = 0$$

$$\sum F_y = 0: \quad +R_y - F_g + T\sin\theta = 0$$

$$\sum \tau = 0: \quad R_y(0) + R_x(0) - F_g(d+L) + (0)(T\cos\theta) + (d+2L)(T\sin\theta) = 0$$

(a) The torque equation gives
$$T = \frac{F_g(d+L)}{(d+2L)\sin\theta} \qquad \Diamond$$

(b) Now from the force equations,
$$R_x = \frac{F_g(d+L)}{(d+2L)\tan\theta} \qquad \Diamond$$

and
$$R_y = F_g - \frac{F_g(d+L)}{d+2L} = \frac{F_g L}{d+2L} \qquad \Diamond$$

---

49. A 10 000-N shark is supported by a cable attached to a 4.00-m rod that can pivot at the base. Calculate the tension in the tie-rope between the rod and the wall if it is holding the system in the position shown in Figure P12.49. Find the horizontal and vertical forces exerted on the base of the rod. (Neglect the weight of the rod.)

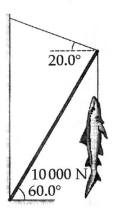

Figure P12.49

Solution

Conceptualize: Since the rod helps support the weight of the shark by exerting a vertical force, the tension in the upper portion of the cable must be less than 10 000 N. Likewise, the vertical force on the base of the rod should also be less than 10 kN.

Categorize: This is another statics problem where the sum of the forces and torques must be zero. To find the unknown forces, draw a free-body diagram of the rod, apply Newton's second law, and sum torques.

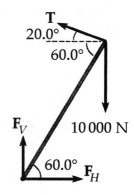

Analyze: From the diagram, the angle **T** makes with the rod is $\theta = 60.0° + 20.0° = 80.0°$ and the perpendicular component of **T** is $T\sin(80.0°)$.

Summing torques around the base of the rod, and applying Newton's second law in the horizontal and vertical directions,

$$\sum \tau = 0: \qquad -(4.00 \text{ m})(10\,000 \text{ N})\cos 60° + T(4.00 \text{ m})\sin(80°) = 0$$

$$T = \frac{(10\,000 \text{ N})\cos(60.0°)}{\sin(80.0°)} = 5.08 \times 10^3 \text{ N} \qquad \Diamond$$

$$\sum F_x = 0: \qquad F_H - T\cos(20.0°) = 0$$

$$F_H = T\cos(20.0°) = 4.77 \times 10^3 \text{ N} \qquad \Diamond$$

$$\sum F_y = 0: \qquad F_V + T\sin(20.0°) - 10\,000 \text{ N} = 0$$

$$F_V = (10\,000 \text{ N}) - T\sin(20.0°) = 8.26 \times 10^3 \text{ N} \qquad \Diamond$$

Finalize: The forces calculated are indeed less than 10 kN as predicted. That shark sure is a big catch; he weighs about a ton!

---

53. A force acts on a uniform rectangular cabinet weighing 400 N, as in Figure P12.53. (a) If the cabinet slides with constant speed when $F = 200$ N and $h = 0.400$ m, find the coefficient of kinetic friction and the position of the resultant normal force. (b) If $F = 300$ N, find the value of h for which the cabinet just begins to tip.

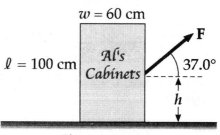

Figure P12.53

Solution

(a) Think of the normal force as acting at a distance x from the lower left corner. Moving with constant speed, the cabinet is in equilibrium:

$$\sum F_x = 0: \quad -f + (200 \text{ N})\cos(37.0°) = 0$$

$$\sum F_y = 0: \quad -400 \text{ N} + n + (200 \text{ N})\sin(37.0°) = 0$$

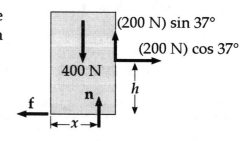

These equations tell us that $f = 160$ N and $n = 280$ N,

so $\mu_k = f/n = 0.571$ ◊

Take torques about the lower left corner; $\Sigma\tau = 0$ gives

$-(400 \text{ N})(30 \text{ cm}) + nx + (200 \text{ N})(60 \text{ cm})\sin(37.0°) - (200 \text{ N})(40 \text{ cm})\cos(37.0°) = 0$

Substituting $n = 280$ N (and converting cm to m, and back) gives

$x = \dfrac{120 \text{ N·m} - 72.2 \text{ N·m} + 63.9 \text{ N·m}}{280 \text{ N}} = 39.9 \text{ cm}$ ◊

(b) When the cabinet is just about to tip, the normal force is located at the lower right corner, and $\Sigma\tau = 0$ is still true.

Because most of the forces are directed through the lower right corner, we choose to take torques about that point. This leaves only two forces to deal with.

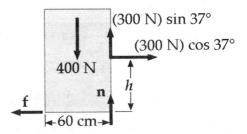

$\sum\tau = 0:$ $-(300 \text{ N})(h)\cos(37.0°) + (400 \text{ N})(30.0 \text{ cm}) = 0$

Solving for h, $h = \dfrac{120 \text{ N·m}}{240 \text{ N}} = 50.1 \text{ cm}$ ◊

---

55. A uniform beam of mass m is inclined at an angle θ to the horizontal. Its upper end produces a ninety-degree bend in a very rough rope tied to a wall, and its lower end rests on a rough floor (Fig. P12.55). (a) If the coefficient of static friction between beam and floor is μ_s, determine an expression for the maximum mass M that can be suspended from the top before the beam slips. (b) Determine the magnitude of the reaction force at the floor and the magnitude of the force exerted by the beam on the rope at P in terms of m, M, and μ_s.

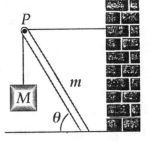

Figure P12.55

Solution

Conceptualize: The solution to this problem is not so obvious as some other problems because there are three independent variables that affect the maximum mass M. We could at least expect that more mass can be supported for higher coefficients of friction (μ_s), larger angles (θ), and a more massive beam (m).

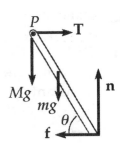

Categorize: Draw a free-body diagram, apply Newton's second law, and sum torques to find the unknown forces for this statics problem.

Analyze:

(a) Use $\Sigma F_x = \Sigma F_y = \Sigma \tau = 0$ and choose the axis at the point of contact with the floor to simplify the torque analysis. Since the rope is described as very rough, we will assume that it will never slip on the end of the beam. First, let us determine what frictional force at the floor is necessary to put the system in equilibrium; then we can check whether that frictional force can be obtained.

$$\sum F_x = 0: \qquad\qquad T - f = 0$$

$$\sum F_y = 0: \qquad\qquad n - Mg - mg = 0$$

$$\sum \tau = 0: \qquad\qquad Mg(\cos\theta)L + mg(\cos\theta)\frac{L}{2} - T(\sin\theta)L = 0$$

Substituting for T, we get $\quad f = \left(M + \frac{1}{2}m\right)g\cot\theta$

In order for the beam not to slip, we need $f \le \mu_s n$. Substituting for n and f, we obtain the requirement

$$\mu_s \ge \left[\frac{M + \frac{1}{2}m}{M + m}\right]\!\cot\theta$$

The factor in brackets is always less than one, so if $\mu_s \ge \cot\theta$ then M can be increased without limit. In this case there is no maximum mass! ◊

Otherwise, if $\mu_s < \cot\theta$, on the verge of slipping the equality will apply, and solving for M yields

$$M = \frac{m}{2}\left[\frac{2\mu_s\sin\theta - \cos\theta}{\cos\theta - \mu_s\sin\theta}\right] \qquad\qquad ◊$$

(b) If $\mu_s > \cot\theta$, then the frictional force will never reach its maximum, and the forces at the ends of the beam can be expressed in terms of θ but not in terms of μ_s as requested. ◊

Otherwise, if $\mu_s < \cot\theta$, at the floor we see that the normal force is in the y direction and frictional force is in the $-x$ direction. The reaction force then is

$$R = \sqrt{n^2 + (\mu_s n)^2} = g(M + m)\sqrt{1 + \mu_s^2} \qquad\qquad ◊$$

At point P, the force of the beam on the rope is

$$F = \sqrt{T^2 + (Mg)^2} = g\sqrt{M^2 + \mu_s^2(M + m)^2} \qquad\qquad ◊$$

Finalize: In our answer to part (a), notice that this result does not depend on L, which is reasonable since the lever arm of the beam's weight is proportional to the length of the beam.

The answer to this problem is certainly more complex than most problems. We can see that the maximum mass M that can be supported is proportional to m, but it is not clear from the solution that M increases proportionally to μ_s, and θ as predicted. To further examine the solution to part (a), we could graph or calculate the ratio M/m as a function of θ for several reasonable values of μ_s, ranging from 0.5 to 1. This would show that M does increase with increasing μ_s and θ, as predicted. The complex requirements for stability explain why we don't encounter this precarious configuration very often.

---

57. A stepladder of negligible weight is constructed as shown in Figure P12.57. A painter of mass 70.0 kg stands on the ladder 3.00 m from the bottom. Assuming the floor is frictionless, find (a) the tension in the horizontal bar connecting the two halves of the ladder, (b) the normal forces at A and B, and (c) the components of the reaction force at the single hinge C that the left half of the ladder exerts on the right half. (**Suggestion:** Treat the ladder as a single object; but also each half of the ladder separately.)

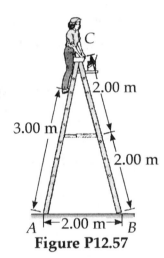

Figure P12.57

Solution

If we think of the whole ladder, we can solve part (b).

(b) The painter is 3/4 of the way up the ladder, so the lever arm of her weight about A is

$$\frac{3}{4}(1.00 \text{ m}) = 0.750 \text{ m}$$

$$\sum F_x = 0: \quad 0 = 0$$

$$\sum F_y = 0: \quad n_A - 686 \text{ N} + n_B = 0$$

$$\sum \tau_A = 0: \quad n_A(0) - (686 \text{ N})(0.750 \text{ m}) + n_B(2.00 \text{ m}) = 0$$

Thus, $n_B = 257 \text{ N}$ ◊

and $n_A = 686 \text{ N} - 257 \text{ N} = 429 \text{ N}$ ◊

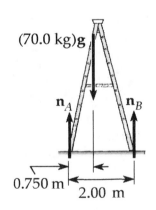

Now consider the left half of the ladder. We know the direction of the bar tension, and we make guesses for the directions of the components of the hinge force. If a guess is wrong, the answer will be negative.

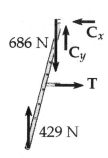

The side rails make an angle with the horizontal

$$\theta = \cos^{-1}(1/4) = 75.5°$$

Taking torques about the top of the ladder, we have

$$\sum \tau_c = 0: \quad (-429 \text{ N})(1.00 \text{ m}) + T(2.00 \text{ m})\sin 75.5° + (686 \text{ N})(0.250 \text{ m}) = 0$$

$$\sum F_x = 0: \quad T - C_x = 0$$

$$\sum F_y = 0: \quad 429 \text{ N} - 686 \text{ N} + C_y = 0$$

(a) From the torque equation,
$$T = \frac{257 \text{ N} \cdot \text{m}}{1.94 \text{ m}} = 133 \text{ N} \qquad \diamond$$

(b) From the force equations,
$$C_x = T = 133 \text{ N}$$

$$C_y = 686 \text{ N} - 429 \text{ N} = 257 \text{ N up}$$

The force that the left half exerts on the right half has opposite components:

133 N to the right, and 257 N down. $\qquad \diamond$

---

61. A wire of length L, Young's modulus Y, and cross-sectional area A is stretched elastically by an amount ΔL. By Hooke's law (Section 7.4) the restoring force is $-k\Delta L$. (a) Show that $k = YA/L$. (b) Show that the work done in stretching the wire by an amount ΔL is $W = YA(\Delta L)^2/2L$.

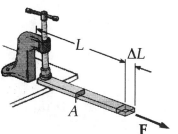

Figure P12.61

Solution

(a) According to Hooke's law
$$|\mathbf{F}| = k\,\Delta L$$

Young's modulus is defined as
$$Y = \frac{F/A}{\Delta L/L} = k\frac{L}{A}$$

or
$$k = \frac{YA}{L} \qquad \diamond$$

(b) Since we have determined that the wire stretches like a spring, we can determine the work done by integrating the force $F = -kx$ over the distance we stretched the wire.

$$W = -\int_0^{\Delta L} F\,dx = -\int_0^{\Delta L} (-kx)dx = \frac{YA}{L}\int_0^{\Delta L} x\,dx = \left[\frac{YA}{L}\left(\frac{1}{2}x^2\right)\right]_{x=0}^{x=\Delta L}$$

Therefore, $\quad W = \dfrac{YA}{2L}\Delta L^2$ ◊

---

65. (a) Estimate the force with which a karate master strikes a board if the hand's speed at time of impact is 10.0 m/s, decreasing to 1.00 m/s during a 0.002 00-s time-of-contact with the board. The mass of his hand and arm is 1.00 kg. (b) Estimate the shear stress if this force is exerted on a 1.00-cm-thick pine board that is 10.0 cm wide. (c) If the maximum shear stress a pine board can support before breaking is 3.60×10^6 N/m^2, will the board break?

Solution

The impulse-momentum theorem describes the force of the board on his hand:

$$Ft = mv_f - mv_i$$

$$F(0.002\,00\text{ s}) = (1.00\text{ kg})(1.00\text{ m/s} - 10.0\text{ m/s})$$

and $\quad F = -4\,500$ N

(a) Therefore, the force of his hand on board is 4 500 N. ◊

(b) That force produces a shear stress on the area that is exposed when the board snaps:

$$\text{Stress} = \frac{F}{A} = \frac{4\,500\text{ N}}{(10.0^{-1}\text{ m})(10.0^{-2}\text{ m})} = 4.50 \times 10^6 \text{ N/m}^2 \qquad ◊$$

(c) **Yes**; this suffices to break the board. ◊

Chapter 13
UNIVERSAL GRAVITATION

EQUATIONS AND CONCEPTS

Newton's law of universal gravitation states that every particle in the Universe attracts every other particle with a force that is directly proportional to the product of their masses and inversely proportional to the square of the distance between them. *This one example of a force form referred to as an inverse square law.*

$$F_g = G\frac{m_1 m_2}{r^2} \qquad (13.1)$$

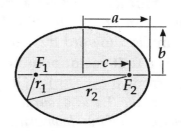

The constant G is called the **universal gravitational constant.**

$$G = 6.673 \times 10^{-11} \text{ N} \cdot \text{m}^2/\text{kg}^2 \qquad (13.2)$$

The **acceleration due to gravity,** g, decreases with increasing altitude. In Equation 13.6, $r = R_E + h$ where h is measured from the Earth's surface. *At the surface of the earth, $h = 0$ and as $r \to \infty$, the weight of an object approaches zero.*

$$g = \frac{GM_E}{r^2} = \frac{GM_E}{(R_E + h)^2} \qquad (13.6)$$

Kepler's first law states that all planets move in elliptical orbits with the Sun at one focus (see figure at right). *This law is a direct result of the inverse square nature of the gravitational force. The semimajor axis has length a and the semiminor axis has length b.*

Kepler's second law states that the radius vector from the Sun to any planet sweeps out equal areas in equal times. *This result applies in the case of any central force and implies conservation of angular momentum.*

$$\frac{dA}{dt} = \frac{L}{2M_P} = \text{constant} \qquad (13.7)$$

Kepler's third law states that the square of the orbital period of any planet is proportional to the cube of the semimajor axis of the elliptical orbit. *Kepler's third law is valid for circular and elliptical orbits and the constant of proportionality is independent of the mass of the planet orbiting the Sun.* In the case of a satellite of a planet (e.g. the moon orbiting the earth), M_S is replaced by the mass of the planet and the constant K_E would have a different value.

$$T^2 = \left(\frac{4\pi^2}{GM_S} \right) a^3 = K_S a^3 \qquad (13.8)$$

$$K_S = 2.97 \times 10^{-19} \ s^2 / m^3$$

The **gravitational field of the Earth** at a point in space, **g**, equals the gravitational force $\mathbf{F}_g$ experienced by a test particle when placed at that point divided by the mass of the test particle. *The Earth creates the gravitational field; the presence of the test particle is not necessary for the existence of the field.*

$$\mathbf{g} \equiv \frac{\mathbf{F}_g}{m} \qquad (13.9)$$

The **gravitational field at a distance r from the center of the Earth** points radially inward toward the center of the Earth ($\hat{\mathbf{r}}$ is a unit vector pointing radially outward from the Earth). *Over a small region near the Earth's surface, **g** is an approximately uniform downward field.*

$$\mathbf{g} \equiv \frac{\mathbf{F}_g}{m} = -\frac{GM_E}{r^2} \hat{\mathbf{r}} \qquad (13.10)$$

The **gravitational potential energy function** of an Earth-particle system when the particle is a distance r from the center of the earth varies as $1/r$. *Recall that the force varies as $1/r^2$.*

$$U(r) = -\frac{GM_E m}{r} \qquad (13.13)$$

The gravitational potential energy for any pair of particles follows from Equation 13.13 and varies as $1/r$. *In Equations 13.13 and 13.14, it is assumed that the potential is zero at infinity; the negative sign indicates that the force is attractive. An external agent must do positive work to increase the separation of the particles.*

$$U(r) = -\frac{Gm_1 m_2}{r} \qquad (13.14)$$

The total energy of a system in which an object of mass m is in orbit around a body of mass M (where $M \gg m$) equals the sum of the kinetic energy of m (taking the massive body to be at rest) and the potential energy of the system. *E may be positive, negative or zero depending on the value of v.*

$$E = K + U$$

$$E = \frac{1}{2}mv^2 - \frac{GMm}{r} \qquad (13.16)$$

The total mechanical energy is negative for both circular and elliptical orbits. The kinetic energy is positive and equal to half the absolute value of the potential energy. *The total energy and the total angular momentum of a gravitationally bound, two-object system are constants of the motion.*

$$E = -\frac{GMm}{2r} \quad \text{(circular orbits)} \qquad (13.18)$$

$$E = -\frac{GMm}{2a} \quad \text{(elliptical orbits)} \qquad (13.19)$$

The escape velocity is defined as the minimum velocity an object must have, when projected from the Earth, in order to escape the Earth's gravitational field (that is, to just reach $r = \infty$ with zero speed). *Note that v_{esc} does not depend on the mass of the projected body.*

$$v_{esc} = \sqrt{\frac{2GM_E}{R_E}} \qquad (13.22)$$

REVIEW CHECKLIST

You should be able to:

▷ Calculate the net gravitational force on an object due to one or more nearby masses. (Section 13.1)

▷ Calculate the free-fall acceleration at given heights above the surface of a planet. (Section 13.3)

▷ State Kepler's three laws of planetary motion, describe the properties of an ellipse, and make calculations using the second and third laws. (Section 13.4)

▷ Determine the gravitational field (magnitude and direction) at a point in the vicinity of a system of masses. (Section 13.5)

▷ Calculate the potential energy of an Earth-mass system and the energy required to move a mass between two points in the Earth's gravitational field. (Section 13.6)

▷ Calculate the total energy of a satellite moving in a circular orbit about a large body (e.g. planet orbiting the Sun or an earth satellite) located at the center of motion. Note that the total energy is negative, as it must be for any closed orbit. (Section 13.7)

▷ Obtain the expression for v_{esc} using the principle of conservation of energy and calculate the escape velocity from a planet. (Section 13.7)

ANSWERS TO SELECTED CONCEPTUAL QUESTIONS

10. Explain why it takes more fuel for a spacecraft to travel from the Earth to the Moon than for the return trip. Estimate the difference.

Answer The mass and radius of the Earth and Moon, and the distance between the two are

$$M_E = 5.98 \times 10^{24} \text{ kg}, \quad R_E = 6.37 \times 10^6 \text{ m},$$

$$M_M = 7.36 \times 10^{22} \text{ kg}, \quad R_M = 1.75 \times 10^6 \text{ m}, \quad \text{and} \quad d = 3.84 \times 10^8 \text{ m}$$

To travel between the Earth and the Moon, a distance d, a rocket engine must boost the spacecraft over the point of zero total gravitational field in between. Call x the distance of this point from Earth. To cancel, the Earth and Moon must here produce equal fields:

$$\frac{GM_E}{x^2} = \frac{GM_M}{(d-x)^2}$$

Isolating x:
$$\frac{M_E}{M_M}(d-x)^2 = x^2 \quad \text{and} \quad \sqrt{\frac{M_E}{M_M}}(d-x) = x$$

Thus,
$$x = \frac{d\sqrt{M_E}}{\sqrt{M_M} + \sqrt{M_E}} = 3.46 \times 10^8 \text{ m}$$

and
$$(d-x) = \frac{d\sqrt{M_M}}{\sqrt{M_M} + \sqrt{M_E}} = 3.83 \times 10^7 \text{ m}$$

In general, we can ignore the gravitational pull of the far planet when we're close to the other; our results will still be approximately correct. Thus the approximate energy difference between point x and the Earth's surface, is:

$$\frac{\Delta E_{E \to x}}{m} = \frac{-GM_E}{x} - \frac{-GM_E}{R_E}$$

Similarly, flying from the moon:
$$\frac{\Delta E_{M \to x}}{m} = \frac{-GM_M}{d-x} - \frac{-GM_M}{R_M}$$

Taking the ratio of the energy terms,
$$\frac{\Delta E_{E \to x}}{\Delta E_{M \to x}} = \left(\frac{M_E}{x} - \frac{M_E}{R_E}\right) \bigg/ \left(\frac{M_M}{d-x} - \frac{M_M}{R_M}\right) \approx 23.0$$

This would also be the minimum fuel ratio, if the spacecraft was driven by a method other than rocket exhaust. If rockets were used, the total fuel ratio would be much larger.

□   □   □   □

12. Why don't we put a geosynchronous weather satellite in orbit around the 45th parallel? Wouldn't this be more useful in the United States than one in orbit around the equator?

Answer   While a satellite in orbit above the 45th parallel might be more useful, it isn't possible. The center of a satellite orbit must be the center of the Earth, since that is the force center for the gravitational force. If the satellite is north of the plane of the equator for part of its orbit, it must be south of the equatorial plane for the rest.

□   □   □   □

16. At what position in its elliptical orbit is the speed of a planet a maximum? At what position is the speed a minimum?

Answer   At the planet's closest approach to the Sun—its perihelion—the system's potential energy is at its minimum (most negative), so that the planet's kinetic energy and speed take their maximum values. At the planet's greatest separation from the Sun—its aphelion—the gravitational energy is maximum, and the speed is minimum.

□   □   □   □

19. In his 1798 experiment, Cavendish was said to have "weighed the Earth." Explain this statement.

Answer   The Earth creates a gravitational field at its surface according to $g = GM_E/R_E^2$. The factors g and R_E were known, so as soon as Cavendish measured G, he could compute the mass of the Earth.

□   □   □   □

SOLUTIONS TO SELECTED END-OF-CHAPTER PROBLEMS

7. In introductory physics laboratories, a typical Cavendish balance for measuring the gravitational constant G uses lead spheres of masses of 1.50 kg and 15.0 g whose centers are separated by about 4.50 cm. Calculate the gravitational force between these spheres, treating each as a particle located at the center of the sphere.

Solution

$$F = \frac{Gm_1m_2}{r^2} = \frac{\left(6.67 \times 10^{-11}\ \text{N} \cdot \text{m}^2/\text{kg}^2\right)(1.50\ \text{kg})(0.015\ 0\ \text{kg})}{\left(4.50 \times 10^{-2}\ \text{m}\right)^2}$$

$$F = 7.41 \times 10^{-10}\ \text{N} = 741\ \text{pN} \quad \text{toward the other sphere.} \qquad \lozenge$$

This is the force that each sphere has exerted on it by the other. In the space literally "between the spheres," no force acts because no object is there to feel a force.

9. When a falling meteoroid is at a distance above the Earth's surface of 3.00 times the Earth's radius, what is its acceleration due to the Earth's gravitation?

Solution The acceleration of gravity, $g = GM_E/r^2$, follows an inverse-square law. At the surface, a distance of one Earth-radius (R_E) from the center, it is 9.80 m/s^2.

At an altitude $3.00R_E$ above the surface (at distance $4.00R_E$ from the center), the acceleration of gravity will be $4.00^2 = 16.0$ times smaller:

$$g = \frac{GM_E}{(4.00R_E)^2} = \frac{GM_E}{16.0R_E{}^2} = \frac{9.80 \text{ m}/\text{s}^2}{16.0} = 0.612 \text{ m}/\text{s}^2 \text{ down} \qquad \Diamond$$

10. The free-fall acceleration on the surface of the Moon is about one sixth of that on the surface of the Earth. If the radius of the Moon is about $0.250R_E$, find the ratio of their average densities, $\rho_{\text{Moon}}/\rho_{\text{Earth}}$.

Solution The gravitational field at the surface of the Earth or Moon is given by

$$g = \frac{GM}{R^2}$$

The expression for density is

$$\rho = \frac{M}{V} = \frac{M}{\frac{4}{3}\pi R^3}$$

so

$$M = \frac{4}{3}\pi\rho R^3$$

and

$$g = \frac{G\frac{4}{3}\pi\rho R^3}{R^2} = G\frac{4}{3}\pi\rho R$$

Noting that this equation applies to both the Moon and the Earth, and dividing the two equations,

$$\frac{g_M}{g_E} = \frac{G\frac{4}{3}\pi\rho_M R_M}{G\frac{4}{3}\pi\rho_E R_E} = \frac{\rho_M R_M}{\rho_E R_E}$$

Substituting,

$$\frac{1}{6} = \frac{\rho_M}{\rho_E}\left(\frac{1}{4}\right) \quad \text{and} \quad \frac{\rho_M}{\rho_E} = \frac{4}{6} = \frac{2}{3} \qquad \Diamond$$

13. Plaskett's binary system consists of two stars that revolve in a circular orbit about a center of mass midway between them. This means that the masses of the two stars are equal (Figure P13.13). Assume the orbital speed of each star is 220 km/s and the orbital period of each is 14.4 days. Find the mass M of each star. (For comparison, the mass of our Sun is 1.99×10^{30} kg.)

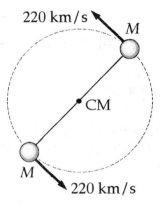

220 km/s

M

CM

M

220 km/s

Figure P13.13

Solution

Conceptualize: From the given data, it is difficult to estimate a reasonable answer to this problem without actually working through the details to actually solve it. A reasonable guess might be that each star has a mass equal to or larger than our Sun because fourteen days is short compared to the periods of all the Sun's planets.

Categorize: The only force acting on the two stars is the central gravitational force of attraction which results in a centripetal acceleration. When we solve Newton's 2nd law, we can find the unknown mass in terms of the variables given in the problem.

Analyze: Applying Newton's 2nd Law, $\Sigma F = ma$ yields $F_g = ma_c$ for each star:

$$\frac{GMM}{(2r)^2} = \frac{Mv^2}{r} \qquad \text{so} \qquad M = \frac{4v^2r}{G}$$

We can write r in terms of the period, T, by considering the time and distance of one complete cycle. The distance traveled in one orbit is the circumference of the stars' common orbit, so $2\pi r = vT$.

Therefore, $\quad M = \dfrac{4v^2r}{G} = \left(\dfrac{4v^2}{G}\right)\left(\dfrac{vT}{2\pi}\right) = \dfrac{2v^3T}{\pi G}$

$$M = \frac{2(220 \times 10^3 \text{ m / s})^3(14.4 \text{ d})(86\ 400 \text{ s / d})}{\pi\left(6.67 \times 10^{-11} \text{ N} \cdot \text{m}^2/\text{ kg}^2\right)} = 1.26 \times 10^{32} \text{ kg} \qquad \lozenge$$

Finalize: The mass of each star is about 63 solar masses, much more than our initial guess! A quick check in an astronomy book reveals that stars over 8 solar masses are considered to be **heavyweight** stars, and astronomers estimate that the maximum theoretical limit is about 100 solar masses before a star becomes unstable. So these two stars are exceptionally massive!

15. Io, a moon of Jupiter, has an orbital period of 1.77 days and an orbital radius of 4.22×10^5 km. From these data, determine the mass of Jupiter.

Solution

We use the particle under a net force and the particle in uniform circular motion models. The gravitational force of Jupiter on Io provides the centripetal acceleration of Io.

$$\sum F_{Io} = M_{Io} a: \quad \frac{GM_J M_{Io}}{r^2} = \frac{M_{Io} v^2}{r} = \frac{M_{Io}}{r}\left(\frac{2\pi r}{T}\right)^2 = \frac{4\pi^2 r M_{Io}}{T^2}$$

Thus,
$$M_J = \frac{4\pi^2 r^3}{GT^2} = \frac{4\pi^2 (4.22 \times 10^8 \text{ m})^3}{(6.67 \times 10^{-11} \text{ N} \cdot \text{m}^2 / \text{kg}^2)(1.77 \text{ d})^2}\left(\frac{1 \text{ d}}{86\,400 \text{ s}}\right)^2\left(\frac{\text{N} \cdot \text{s}^2}{\text{kg} \cdot \text{m}}\right)$$

and
$$M_J = 1.90 \times 10^{27} \text{ kg} \qquad \diamond$$

---

19. A synchronous satellite, which always remains above the same point on a planet's equator, is put in orbit around Jupiter to study the famous red spot. Jupiter rotates about its axis once every 9.84 h. Use the data of Table 13.2 to find the altitude of the satellite.

Solution   Jupiter's rotational period, in seconds, is

$$T = (9.84 \text{ h})(3\,600 \text{ s}/\text{h}) = 35\,424 \text{ s}$$

Jupiter's gravitational force is the central force on the satellite:

$$\frac{GM_s M_J}{r^2} = \frac{M_s v^2}{r} = \left(\frac{M_s}{r}\right)\left(\frac{2\pi r}{T}\right)^2 \quad \text{so} \quad GM_J T^2 = 4\pi^2 r^3$$

$$r = \sqrt[3]{\frac{GM_J T^2}{4\pi^2}} = \left(\frac{(6.67 \times 10^{-11} \text{ N} \cdot \text{m}^2 / \text{kg}^2)(1.90 \times 10^{27} \text{ kg})(3.54 \times 10^4 \text{ s})^2}{4\pi^2}\right)^{1/3}$$

$$r = 15.9 \times 10^7 \text{ m}$$

Thus, we can calculate the altitude of the synchronous satellite to be

$$\text{Altitude} = \left(15.91 \times 10^7 \text{ m}\right) - \left(6.99 \times 10^7 \text{ m}\right) = 8.92 \times 10^7 \text{ m} \qquad \diamond$$

25. Compute the magnitude and direction of the gravitational field at a point P on the perpendicular bisector of the line joining two objects of equal mass separated by a distance $2a$ as shown in Figure P13.25.

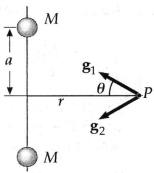

Figure P13.25

Solution

We must add the vector fields created by each mass. In equation form, $\mathbf{g} = \mathbf{g}_1 + \mathbf{g}_2$ where

$$\mathbf{g}_1 = \frac{GM}{r^2 + a^2} \qquad \text{to the left and upward at } \theta$$

and

$$\mathbf{g}_2 = \frac{GM}{r^2 + a^2} \qquad \text{to the left and downward at } \theta$$

Therefore,

$$\mathbf{g} = \frac{GM}{r^2 + a^2}\cos\theta(-\hat{\mathbf{i}}) + \frac{GM}{r^2 + a^2}\sin\theta(\hat{\mathbf{j}}) + \frac{GM}{r^2 + a^2}\cos\theta(-\hat{\mathbf{i}}) + \frac{GM}{r^2 + a^2}\sin\theta(-\hat{\mathbf{j}})$$

$$\mathbf{g} = \frac{2GM}{r^2 + a^2}\frac{r}{\sqrt{r^2 + a^2}}(-\hat{\mathbf{i}}) + 0\hat{\mathbf{j}} = \frac{-2GMr}{(r^2 + a^2)^{3/2}}\hat{\mathbf{i}} \qquad \Diamond$$

---

29. After our Sun exhausts its nuclear fuel, its ultimate fate may be to collapse to a **white dwarf** state, in which it has approximately the same mass as it has now, but a radius equal to the radius of the Earth. Calculate (a) the average density of the white dwarf, (b) the free-fall acceleration, and (c) the gravitational potential energy of a 1.00-kg object at its surface.

Solution

(a) $\rho = \dfrac{M_S}{V} = \dfrac{M_S}{\frac{4}{3}\pi R_E^3} = \dfrac{1.99 \times 10^{30} \text{ kg}}{\frac{4}{3}\pi(6.37 \times 10^6 \text{ m})^3} = 1.84 \times 10^9 \text{ kg / m}^3$ $\qquad \Diamond$

(This is on the order of 1 million times the density of concrete!)

(b) For an object of mass m on its surface, $mg = GM_S m / R_E^2$. Thus,

$$g = \frac{GM_S}{R_E^2} = \frac{(6.67 \times 10^{-11} \text{ N} \cdot \text{m}^2/ \text{kg}^2)(1.99 \times 10^{30} \text{ kg})}{(6.37 \times 10^6 \text{ m})^2} = 3.27 \times 10^6 \text{ m / s}^2 \qquad \Diamond$$

(This acceleration is on the order of 1 million times more than g_{Earth}.)

(c)   Relative to $U_g = 0$ at infinity,

$$U_g = \frac{-GM_S m}{R_E} = \frac{(-6.67 \times 10^{-11} \text{ N} \cdot \text{m}^2/\text{kg}^2)(1.99 \times 10^{30} \text{ kg})(1 \text{ kg})}{(6.37 \times 10^6 \text{ m})} = -2.08 \times 10^{13} \text{ J} \quad \Diamond$$

(Such a large potential energy could yield a big gain in kinetic energy with even small changes in height. For example, dropping the 1.00-kg object from 1.00 m would result in a final velocity of 2 560 m/s.)

---

33.   A space probe is fired as a projectile from the Earth's surface with an initial speed of 2.00×10^4 m/s. What will its speed be when it is very far from the Earth? Ignore friction and the rotation of the Earth.

Solution

We apply the work-kinetic energy theorem between an initial point just after the payload is fired off at the Earth's surface, and a final point when it is coasting along far away.

$$K_i + U_i + W_{\text{other forces}} - f_k \Delta r = K_f + U_f$$

$$\frac{1}{2} m_{\text{probe}} v_i^2 - \frac{GM_{\text{Earth}} m_{\text{probe}}}{r_i} = \frac{1}{2} m_{\text{probe}} v_f^2 - \frac{GM_E m_{\text{probe}}}{r_f}$$

The reciprocal of the final distance is negligible compared with the reciprocal of the original distance from the Earth's center.

$$\frac{1}{2} v_i^2 - \frac{GM_{\text{Earth}}}{R_{\text{Earth}}} = \frac{1}{2} v_f^2 - 0$$

We solve for the final speed, and substitute values tabulated on the endpapers:

$$v_f^2 = v_i^2 - \frac{2GM_{\text{Earth}}}{R_{\text{Earth}}}$$

$$v_f^2 = (2.00 \times 10^4 \text{ m/s})^2 - \frac{2(6.67 \times 10^{-11} \text{ N} \cdot \text{m}^2/\text{kg}^2)(5.98 \times 10^{24} \text{ kg})}{6.37 \times 10^6 \text{ m}}$$

$$v_f^2 = 2.75 \times 10^8 \text{ m}^2/\text{s}^2 \qquad \text{and} \qquad v_f = 1.66 \times 10^4 \text{ m/s} \qquad \Diamond$$

35. A "treetop satellite" (Fig. P13.35) moves in a circular orbit just above the surface of a planet, assumed to offer no air resistance. Show that its orbital speed v and the escape speed from the planet are related by the expression $v_{esc} = \sqrt{2}\,v$.

Solution Call M the mass of the planet and R its radius. For the orbiting "treetop satellite,"

$$\sum F = ma \quad \text{becomes} \quad \frac{GMm}{R^2} = \frac{mv^2}{R} \quad \text{or} \quad v = \sqrt{\frac{GM}{R}}$$

If the object is launched with escape velocity, applying conservation of energy to the object-Earth system gives

$$\frac{1}{2}mv_{esc}^2 - \frac{GMm}{R} = 0 \quad \text{or} \quad v_{esc} = \sqrt{\frac{2GM}{R}}$$

Thus, $\qquad\qquad\qquad v_{esc} = \sqrt{2}\,v$ ◊

61. Two hypothetical planets of masses m_1 and m_2 and radii r_1 and r_2, respectively, are nearly at rest when they are an infinite distance apart. Because of their gravitational attraction, they head toward each other on a collision course. (a) When their center-to-center separation is d, find expressions for the speed of each planet and for their relative speed. (b) Find the kinetic energy of each planet just before they collide if $m_1 = 2.00 \times 10^{24}$ kg, $m_2 = 8.00 \times 10^{24}$ kg, $r_1 = 3.00 \times 10^6$ m, and $r_2 = 5.00 \times 10^6$ m. (**Note:** Both energy and momentum of the system are conserved.)

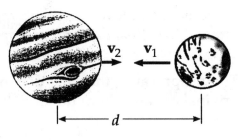

Solution We use both the energy version and the momentum version of the isolated system model.

(a) At infinite separation, $U = 0$; and at rest, $K = 0$. Since energy of the two-planet system is conserved, we have

$$0 = \frac{1}{2}m_1v_1^2 + \frac{1}{2}m_2v_2^2 - \frac{Gm_1m_2}{d} \qquad\qquad [1]$$

The initial momentum of the system is zero and momentum is conserved.

Therefore, $\qquad\qquad 0 = m_1v_1 - m_2v_2 \qquad\qquad [2]$

Combine Equations [1] and [2] to find

$$v_1 = m_2\sqrt{\frac{2G}{d(m_1 + m_2)}} \quad \text{and} \quad v_2 = m_1\sqrt{\frac{2G}{d(m_1 + m_2)}}$$

Relative velocity $\quad v_r = v_1 - (-v_2) = \sqrt{\dfrac{2G(m_1 + m_2)}{d}}$ ◊

(b) Substitute the given numerical values into the equation found for v_1 and v_2 in part (a) to find

$$v_1 = 1.03 \times 10^4 \text{ m / s} \quad \text{and} \quad v_2 = 2.58 \times 10^3 \text{ m / s}.$$

Therefore, $\quad K_1 = \frac{1}{2}m_1v_1^2 = 1.07 \times 10^{32} \text{ J}; \quad K_2 = \frac{1}{2}m_2v_2^2 = 2.67 \times 10^{31} \text{ J}$ ◊

---

69. Two stars of masses M and m, separated by a distance d, revolve in circular orbits about their center of mass (Fig. P13.69). Show that each star has a period given by

$$T^2 = \frac{4\pi^2}{G(M + m)}d^3$$

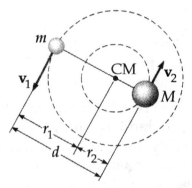

Figure P13.69

Proceed as follows: Apply Newton's second law to each star. Note that the center-of-mass condition requires that $Mr_2 = mr_1$, where $r_1 + r_2 = d$.

Solution For the star of mass M and orbital radius r_2,

$$\sum F = ma \quad \text{gives} \quad \frac{GMm}{d^2} = \frac{Mv_2^2}{r_2} = \frac{M}{r_2}\left(\frac{2\pi r_2}{T}\right)^2$$

For the star of mass m, $\quad \sum F = ma \quad$ gives $\quad \dfrac{GMm}{d^2} = \dfrac{mv_1^2}{r_1} = \dfrac{m}{r_1}\left(\dfrac{2\pi r_1}{T}\right)^2$

Clearing fractions, we then obtain simultaneous equations:

$$GmT^2 = 4\pi^2 d^2 r_2 \quad \text{and} \quad GMT^2 = 4\pi^2 d^2 r_1$$

Adding, we find $\quad G(M + m)T^2 = 4\pi^2 d^2(r_1 + r_2) = 4\pi^2 d^3$

$$T^2 = \frac{4\pi^2 d^3}{G(M + m)}$$ ◊

In a visual binary star system T, d, r_1, and r_2 can sometimes be measured, so the mass of each component can be computed.

Chapter 14

FLUID MECHANICS

EQUATIONS AND CONCEPTS

The **density of a homogeneous substance** is defined as mass per unit volume. *Density is characteristic of a particular type of material and independent of the total quantity of material in the sample.*

$$\rho \equiv \frac{m}{V}$$

The **SI units of density** are kilograms per cubic meter.

$$1\,g/cm^3 = 1\,000\,kg/m^3$$

Pressure is defined as the magnitude of the normal force per unit area acting on a surface. *Pressure is a scalar quantity.*

$$P \equiv \frac{F}{A} \qquad (14.1)$$

The **SI units of pressure** are newtons per square meter, or pascal (Pa).

$$1\,Pa \equiv 1\,N/m^2 \qquad (14.3)$$

Atmospheric pressure is often expressed in other units: atmospheres, mm of mercury (Torr), or pounds per square inch.

$$1\,atm = 1.013 \times 10^5\,Pa$$

$$1\,Torr = 133.3\,Pa$$

$$1\,lb/in^2 = 6\,895\,Pa$$

The **absolute pressure**, P, at a depth h below the surface of a liquid, which is open to the atmosphere, is greater than atmospheric pressure, P_0, by an amount which depends on the depth below the surface. The quantity $P - P_0 = \rho g h$ is called the gauge pressure and P is the absolute pressure. *The pressure has the same value at all points at a given depth and does not depend on the shape of the container.*

$$P = P_0 + \rho g h \qquad (14.4)$$

$$P_0 = 1.013 \times 10^5\,Pa = 1\,atm$$

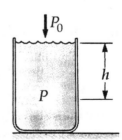

218

Pascal's law states that a change in the pressure applied to an enclosed fluid (liquid or gas) is transmitted undiminished to every point within the fluid and to the walls of the container.

Archimedes's principle states that when an object is partially or fully immersed in a fluid, the fluid exerts an upward buoyant force on the object. The magnitude of the buoyant force equals the weight of the fluid displaced by the object. The weight of the displaced fluid depends on the density, ρ_{fluid}, and volume of the displaced fluid, V.

$$B = \rho_{\text{fluid}} g V \qquad (14.5)$$

For a **floating object**, the fraction of the volume that is below the fluid surface is equal to the ratio of the density of the object to that of the fluid. *In Equation 14.6, V_{fluid} is the volume of the displaced fluid, and is therefore the volume of the object that is submerged.*

$$\frac{V_{\text{fluid}}}{V_{\text{obj}}} = \frac{\rho_{\text{obj}}}{\rho_{\text{fluid}}} \qquad (14.6)$$

The **ideal fluid model** is based on the following four assumptions:

- **nonviscous** — internal friction between adjacent fluid layers is negligible.
- **incompressible** — the density of the fluid is constant throughout the fluid.
- **steady flow** — the velocity at each point in the fluid remains constant.
- **irrotational** (nonturbulent) — there are no eddy currents within the fluid; each element of the fluid has zero angular momentum about its center of mass.

The **equation of continuity for fluids** states that the flow rate (product of area and speed of flow) of an incompressible fluid (ρ = constant) is constant at every point along a pipe.

$$A_1 v_1 = A_2 v_2 = \text{constant} \qquad (14.7)$$

Bernoulli's equation states that the sum of pressure, kinetic energy per unit volume, and potential energy per unit volume remains constant along a streamline of an ideal fluid. *The equation is a statement of the law of conservation of mechanical energy as applied to an ideal fluid.*

$$P + \frac{1}{2}\rho v^2 + \rho g y = \text{constant} \qquad (14.9)$$

REVIEW CHECKLIST

You should be able to:

▷ Calculate pressure below the surface of a liquid and determine the total force exerted on a surface due to hydrostatic pressure. (Section 14.2)

▷ Convert among the several different units commonly used to express pressure. (Section 14.3)

▷ Determine the buoyant force on a floating or submerged object and determine the fraction of a floating object that is below the surface of a fluid. (Section 14.4)

▷ Calculate the density of an object or a fluid using Archimedes's principle (Section 14.4)

▷ State the simplifying assumptions of an ideal fluid moving with streamline flow. (Section 14.5)

▷ Make calculations using the equation of continuity and Bernoulli's equation. (Sections 14.5 and 14.6)

ANSWERS TO SELECTED CONCEPTUAL QUESTIONS

1. Two drinking glasses having equal weights but different shapes and different cross-sectional areas are filled to the same level with water. According to the expression $P = P_0 + \rho g h$, the pressure is the same at the bottom of both glasses. In view of this, why does one weigh more than the other?

Answer For the cylindrical container shown, the weight of the water is equal to the gauge pressure at the bottom multiplied by the area of the bottom. For the narrow-necked bottle, on the other hand, the weight of the fluid is much less than PA, the bottom pressure times the area. The water does exert the large PA force downward on the bottom of the bottle, but the water also exerts an upward force nearly as large on the ring-shaped horizontal area surrounding the neck. The net vector force that the water exerts on the glass is equal to the small weight of the water.

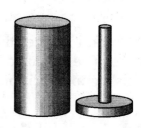

☐   ☐   ☐   ☐

6. A fish rests on the bottom of a bucket of water while the bucket is being weighed on a scale. When the fish begins to swim around, does the scale reading change?

Answer In either case, the scale is supporting the container, the water, and the fish. Therefore the weight will remains the same. The reading on the scale, however, can change if the net center of mass accelerates in the vertical direction, as when the fish jumps out of the water. In that case, the scale, which registers force, will also show an additional force caused by Newton's law, $F = ma$.

☐   ☐   ☐   ☐

10. The water supply for a city is often provided from reservoirs built on high ground. Water flows from the reservoir, through pipes, and into your home when you turn the tap on your faucet. Why is the water flow more rapid out of a faucet on the first floor of a building than in an apartment on a higher floor?

Answer   The water supplied to the building flows through a pipe connected to the water tower. Near the earth, the water pressure is greater because the pressure increases with increasing depth beneath the surface of the water. The penthouse apartment is not as far below the water surface; hence the water flow will not be as rapid as on a lower floor.

☐   ☐   ☐   ☐

17.   A barge is carrying a load of gravel along a river. It approaches a low bridge and the captain realizes that the top of the pile of gravel is not going to make it under the bridge. The captain orders the crew to quickly shovel gravel from the pile into the water. Is this a good decision?

Answer   It turns out that this is not a good decision. Imagine a given shovel full of gravel. According to Archimedes's principle, the portion of the total buoyant force which can be associated with keeping this shovel-full of gravel afloat on the barge is due to a portion of water whose weight is equal to the weight of the gravel. Since water is less dense than gravel, the volume of water associated with keeping a shovel-full of gravel afloat is larger than the volume of the gravel. When the gravel is thrown overboard, the barge rises, in association with a volume of water larger than the volume of the gravel removed from the pile. Thus, as the gravel is shoveled overboard, the barge rises by an amount which is larger than the amount by which the height of the pile of gravel decreases. The better approach would be to **add** gravel to the barge from a riverside source, causing it to sink deeper into the water at a greater rate than the height of the pile of gravel increases.

☐   ☐   ☐   ☐

29.   An unopened can of diet cola floats when placed in a tank of water, whereas a can of regular cola of the same brand sinks in the tank. What do you suppose could explain this behavior?

Answer   Regular cola is sugar syrup. Its density is higher than the density of diet cola, which is nearly pure water. The low-density air inside the can has a bigger effect than the thin aluminum shell, so the can of diet soda floats.

☐   ☐   ☐   ☐

SOLUTIONS TO SELECTED END-OF-CHAPTER PROBLEMS

1.   Calculate the mass of a solid iron sphere that has a diameter of 3.00 cm.

Solution   The definition of density $\rho = m/V$ is often written as $m = \rho V$.

Here $V = \frac{4}{3}\pi r^3$ so $m = \rho V = \rho\left(\frac{4}{3}\pi(d/2)^3\right)$

Thus, $m = \left(7.86 \times 10^3 \text{ kg}/\text{m}^3\right)\left(\frac{4}{3}\right)\pi\left((1.50 \text{ cm})^3\right)\left(10^{-6} \text{ m}^3/\text{cm}^3\right) = 111 \text{ g}$ ◊

3.   A 50.0-kg woman balances on one heel of a pair of high-heeled shoes. If the heel is circular and has a radius of 0.500 cm, what pressure does she exert on the floor?

Solution

The area of the circular base of the heel is

$$\pi r^2 = \pi (0.500 \text{ cm})^2\left(\frac{1 \text{ m}^2}{10\,000 \text{ cm}^2}\right) = 7.85 \times 10^{-5} \text{ m}^2$$

The force she exerts is her weight,

$$mg = (50.0 \text{ kg})(9.80 \text{ m}/\text{s}^2) = 490 \text{ N}$$

Then $P = \dfrac{F}{A} = \dfrac{490 \text{ N}}{7.85 \times 10^{-5} \text{ m}^2} = 6.24 \text{ MPa}$ ◊

7.   The spring of the pressure gauge shown in Figure 14.2 has a force constant of $1\,000$ N/m, and the piston has a diameter of 2.00 cm. As the gauge is lowered into water, what change in depth causes the piston to move in by 0.500 cm?

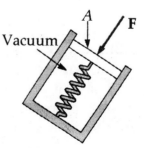

Figure 14.2

Solution

$$F_{\text{spring}} = F_{\text{fluid}} \qquad \text{or} \qquad kx = \rho g h A$$

$$h = \frac{kx}{\rho g A} = \frac{(1\,000 \text{ N}/\text{m}^2)(0.005\,00 \text{ m})}{(1\,000 \text{ kg}/\text{m}^3)(9.80 \text{ m}/\text{s}^2)(0.010\,0 \text{ m})^2 \pi} = 1.62 \text{ m}$$ ◊

9. What must be the contact area between a suction cup (completely exhausted) and a ceiling if the cup is to support the weight of an 80.0-kg student?

Solution

Conceptualize: The suction cups used by burglars seen in movies are about 10 cm in diameter, and it seems reasonable that one of these might be able to support the weight of an 80-kg student. The face area of a 10-cm cup is approximately:

$$A = \pi r^2 \approx 3(0.05 \text{ m})^2 \approx 0.008 \text{ m}^2$$

Categorize: "Suction" is not a new kind of force. Familiar forces hold the cup in equilibrium, one of which is the atmospheric pressure acting over the area of the cup. This problem is simply another application of Newton's 2nd law.

Analyze: The vacuum between cup and ceiling exerts no force on either. The atmospheric pressure of the air below the cup pushes up on it with a force $(P_{atm})(A)$. If the cup barely supports the student's weight, then the normal force of the ceiling is approximately zero, and

$$\sum F_y = 0 + (P_{atm})(A) - mg = 0: \qquad A = \frac{mg}{P_{atm}} = \frac{784 \text{ N}}{1.013 \times 10^5 \text{ N} / \text{m}^2} = 7.74 \times 10^{-3} \text{ m}^2 \qquad \lozenge$$

Finalize: This calculated area agrees with our prediction and corresponds to a suction cup that is 9.93 cm in diameter (Our 10 cm estimate was right on — a lucky guess, considering that a burglar would probably use at least two suction cups, not one.) As an aside, the suction cup we have drawn, above, appears to be about 30 cm in diameter, plenty big enough to support the weight of the student.

---

17. Blaise Pascal duplicated Torricelli's barometer using a red Bordeaux wine, of density 984 kg/m³, as the working liquid (Fig. P14.17). What was the height h of the wine column for normal atmospheric pressure? Would you expect the vacuum above the column to be as good as for mercury?

Solution

In Bernoulli's equation, $P_1 + \frac{1}{2}\rho v_1^2 + \rho g y_1 = P_2 + \frac{1}{2}\rho v_2^2 + \rho g y_2$

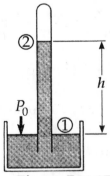

Figure P14.17

Take point 1 at the wine surface in the pan, where $P_1 = P_{atm}$, and point 2 at the wine surface up in the tube. Here we approximate $P_2 = 0$, although some alcohol and water will evaporate. The vacuum is not so good as with mercury. Unless you are careful, a lot of dissolved oxygen or carbon dioxide may come bubbling out.

Now, since the speed of the fluid at both points is zero, we can rearrange the terms to arrive essentially at Equation 14.4:

$$P_1 = P_2 + \rho g(y_2 - y_1)$$

$$1 \, \text{atm} = 0 + (984 \, \text{kg}/\text{m}^3)(9.80 \, \text{m}/\text{s}^2)(y_2 - y_1)$$

$$y_2 - y_1 = \frac{1.013 \times 10^5 \, \text{N}/\text{m}^2}{9643 \, \text{N}/\text{m}^3} = 10.5 \, \text{m} \qquad \Diamond$$

A water barometer in a stairway of a three-story building is a nice display. Red wine makes the fluid level easier to see.

===

23. A Ping-Pong ball has a diameter of 3.80 cm and average density of 0.0840 g/cm³. What force is required to hold it completely submerged under water?

Solution

Conceptualize: According to Archimedes's Principle, the buoyant force acting on the submerged ball will be equal to the weight of the water the ball will displace. The ball has a volume of about 30 cm³, so the weight of this water is approximately:

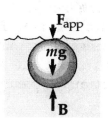

$$B = F_g = \rho V g \approx (1 \, \text{g}/\text{cm}^3)(30 \, \text{cm}^3)(10 \, \text{m}/\text{s}^2) = 0.3 \, \text{N}$$

Since the ball is much less dense than the water, the applied force will approximately equal this buoyant force.

Categorize: Apply Newton's 2nd law to find the applied force.

Analyze: At equilibrium, $\quad \Sigma F = 0 \quad$ or $\quad -F_{app} - mg + B = 0$

where the buoyant force is $\quad B = \rho_w V g \quad$ and $\quad \rho_w = 1\,000 \, \text{kg}/\text{m}^3$

The applied force is then $\quad F_{app} = \rho_w V g - mg$

Using $m = \rho_{ball}V$ to eliminate the unknown mass of the ball, this becomes

$$F_{app} = Vg(\rho_w - \rho_{ball}) = \tfrac{4}{3}\pi r^3 g(\rho_w - \rho_{ball})$$

$$F_{app} = \tfrac{4}{3}\pi(1.90 \times 10^{-2}\ \text{m})^3(9.80\ \text{m}/\text{s}^2)(1\,000\ \text{kg}/\text{m}^3 - 84\ \text{kg}/\text{m}^3)$$

$$F_{app} = 0.258\ \text{N} \qquad \Diamond$$

Finalize: The force is approximately what we expected, so our result is reasonable. If a force greater than 0.258 N were to be applied, the ball would accelerate down until it hit the bottom (which would then provide a normal force directed upwards).

---

29. A cube of wood having an edge dimension of 20.0 cm and a density of 650 kg/m³ floats on water. (a) What is the distance from the horizontal top surface of the cube to the water level? (b) How much lead weight has to be placed on top of the cube so that its top is just level with the water?

Solution

Set h equal to the distance from the top of the cube to the water level.

(a) According to Archimedes's principle,

$$B = \rho_w Vg = (1.00\ \text{g}/\text{cm}^3)\big[(20.0\ \text{cm})^2(20.0\ \text{cm} - h\ \text{cm})\big]g$$

But $B = \text{Weight of block} = mg = \rho_{wood}V_{wood}g = (0.650\ \text{g}/\text{cm}^3)(20.0\ \text{cm})^3 g$

Setting these two equations equal,

$$(0.650\ \text{g}/\text{cm}^3)(20.0\ \text{cm})^3 g = (1.00\ \text{g}/\text{cm}^3)(20.0\ \text{cm})^2(20.0\ \text{cm} - h\ \text{cm})g$$

$$20.0\ \text{cm} - h = 20.0(0.650)\ \text{cm}$$

$$h = 20.0(1.00\ \text{cm} - 0.650\ \text{cm}) = 7.00\ \text{cm} \qquad \Diamond$$

(b) $B = mg + Mg$ where M=mass of lead

$$(1.00\ \text{g}/\text{cm}^3)(20.0\ \text{cm})^3 g = (0.650\ \text{g}/\text{cm}^3)(20.0\ \text{cm})^3 g + Mg$$

$$M = (20.0\ \text{cm})^3(1.00\ \text{g}/\text{cm}^3 - 0.650\ \text{g}/\text{cm}^3) = (20.0\ \text{cm})^3(0.350\ \text{g}/\text{cm}^3) = 2.80\ \text{kg} \quad \Diamond$$

---

33. How many cubic meters of helium are required to lift a balloon with a 400-kg payload to a height of 8 000 m? (Take $\rho_{He} = 0.180$ kg/m³.) Assume that the balloon maintains a constant volume and that the density of air decreases with the altitude z according to the expression $\rho_{air} = \rho_0 e^{-z/8\,000}$, where z is in meters and $\rho_0 = 1.25$ kg/m³ is the density of air at sea level.

Solution   At $z = 8\,000$ m, the density of air is

$$\rho_{air} = \rho_0 e^{-z/8\,000} = (1.25 \text{ kg}/\text{m}^3)e^{-1} = (1.25 \text{ kg}/\text{m}^3)(0.368) = 0.460 \text{ kg}/\text{m}^3$$

Think of the balloon reaching equilibrium at this height.

The weight of its payload is $\qquad\qquad$ $Mg = (400 \text{ kg})(9.80 \text{ m}/\text{s}^2) = 3\,920$ N

The weight of the helium in it is $\qquad$ $mg = \rho_{He}Vg$

$\sum F_y = 0$ becomes $\qquad\qquad$ $+\rho_{air}Vg - Mg - \rho_{He}Vg = 0$

Solving, $\qquad\qquad\qquad\qquad\qquad$ $(\rho_{air} - \rho_{He})V = M$

and $\quad V = \dfrac{M}{\rho_{air} - \rho_{He}} = \dfrac{400 \text{ kg}}{(0.460 - 0.18)\text{ kg}/\text{m}^3} = 1.43 \times 10^3$ m³ $\qquad\qquad$ ◊

---

35. A plastic sphere floats in water with 50.0 percent of its volume submerged. This same sphere floats in glycerin with 40.0 percent of its volume submerged. Determine the densities of the glycerin and the sphere.

Solution

The forces on the ball are its weight $\qquad$ $F_g = mg = \rho_{plastic}V_{ball}g$

and the buoyant force of the liquid $\qquad$ $B = \rho_{fluid}V_{immersed}g$

When floating in water, $\sum F_y = 0$: $\qquad$ $-(\rho_{plastic})(V_{ball})g + (\rho_{water})(0.500\,V_{ball})g = 0$

$\qquad\qquad\qquad\qquad\qquad\qquad\qquad$ $\rho_{plastic} = 0.500\,\rho_{water} = 500 \text{ kg}/\text{m}^3$ $\qquad$ ◊

When floating in glycerin, $\sum F_y = 0$: $\quad$ $-(\rho_{plastic})(V_{ball})g + (\rho_{glycerin})(0.400\,V_{ball})g = 0$

$\qquad\qquad\qquad\qquad\qquad\qquad\qquad$ $\rho_{plastic} = 0.400\,\rho_{glycerin}$

$\qquad\qquad\qquad\qquad\qquad$ $\rho_{glycerin} = \dfrac{500 \text{ kg}/\text{m}^3}{0.400} = 1\,250 \text{ kg}/\text{m}^3$ $\qquad$ ◊

This glycerin would sink in water.

---

39. A large storage tank, open at the top and filled with water, develops a small hole in its side at a point 16.0 m below the water level. If the rate of flow from the leak is equal to 2.50×10^{-3} m^3/min, determine (a) the speed at which the water leaves the hole and (b) the diameter of the hole.

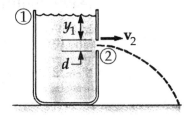

Solution Take point 1 at the top surface and point 2 at the exiting stream. Assuming the top is open to the atmosphere, then $P_1 = P_0$. At point 2, by Newton's third law, the water must push on the air just as strongly as the air pushes on the water, so $P_2 = P_0$.

(a) $P_1 + \frac{1}{2}\rho v_1^2 + \rho g y_1 = P_2 + \frac{1}{2}\rho v_2^2 + \rho g y_2$

$A_1 \gg A_2$, so $\qquad\qquad\qquad v_1 \ll v_2$

With the simplifications $v_1 \approx 0$ and $P_1 = P_2 = P_0$,

$$v_2 = \sqrt{2g(y_1 - y_2)} = \sqrt{2(9.80 \text{ m} / \text{s}^2)(16 \text{ m})} = 17.7 \text{ m} / \text{s} \quad \Diamond$$

(b) From the flow rate, we find that

$$A_2 v_2 = 2.50 \times 10^{-3} \text{ m}^3 / \text{min}$$

$$\left(\frac{\pi d^2}{4}\right)(17.7 \text{ m} / \text{s})(60 \text{ s} / \text{min}) = 2.50 \times 10^{-3} \text{ m}^3 / \text{min}$$

Thus, $\qquad\qquad\qquad d = 1.73 \times 10^{-3} \text{ m} = 1.73 \text{ mm} \qquad\qquad\qquad \Diamond$

---

41. Water flows through a fire hose of diameter 6.35 cm at a rate of 0.012 0 m^3/s. The fire hose ends in a nozzle of inner diameter 2.20 cm. What is the speed with which the water exits the nozzle?

Solution Take point 1 inside the hose and point 2 at the surface of the water stream leaving the nozzle. The volume flow rate is constant:

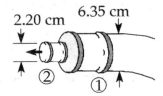

$$0.012\,0 \text{ m}^3 / \text{s} = A_1 v_1 = A_2 v_2$$

$$v_1 = \frac{0.012\,0 \text{ m}^3 / \text{s}}{\pi (6.35 \times 10^{-2} \text{ m} / 2)^2} = 3.79 \text{ m} / \text{s}$$

and $\qquad\qquad v_2 = \dfrac{0.012\,0 \text{ m}^3 / \text{s}}{\pi (1.10 \times 10^{-2} \text{ m})^2} = 31.6 \text{ m} / \text{s} \qquad\qquad \Diamond$

Related Calculation: Find the pressure at both points.

At point 2, the water pressure is 101.3 kPa. The air exerts this much force per unit area on the water. By Newton's third law, the water exerts the same force on the air. If we have a nonviscous incompressible fluid in steady nonturbulent flow, and the height at both points is the same,

$$P_1 + \frac{1}{2}\rho v_1^2 + \rho g h_1 = P_2 + \frac{1}{2}\rho v_2^2 + \rho g h_2$$

$$P_1 = P_2 + \frac{1}{2}\rho\left(v_2^2 - v_1^2\right)$$

$$P_1 = 1.01\times10^5 \ \frac{N}{m^2} + \frac{1}{2}\left(1\,000 \ \frac{kg}{m^3}\right)\left(\left(31.6 \ \frac{m}{s}\right)^2 - \left(3.79 \ \frac{m}{s}\right)^2\right)$$

$$P_1 = 101 \text{ kPa} + 491 \text{ kPa} = 592 \text{ kPa}$$ ◊

---

57. The true weight of an object can be measured in a vacuum, where buoyant forces are absent. An object of volume V is weighed in air on a balance with the use of weights of density ρ. If the density of air is ρ_{air} and the balance reads F_g', show that the true weight F_g is

$$F_g = F_g' + \left(V - \frac{F_g'}{\rho g}\right)\rho_{air}g$$

Solution

We use the rigid body in equilibrium model. The "balanced" condition is one in which the net torque on the balance is zero. Since the balance has lever arms of equal length, the total force on each pan is equal. Applying $\Sigma\tau = 0$ around the pivot gives us this in equation form:

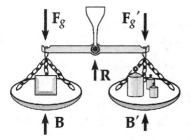

$$F_g - B = F_g' - B'$$

where B and B' are the buoyant forces on the body and weights respectively.

The buoyant force experienced by an object of volume V in air is

$$B = V\rho_{air}g \ .$$

So for the test mass and for the weights, respectively,

$$B = V\rho_{air}g \qquad \text{and} \qquad B' = V'\rho_{air}g$$

Since the volume of the weights is not given explicitly, we must use the density equation to eliminate it:

$$V' = \frac{m'}{\rho} = \frac{m'g}{\rho g} = \frac{F_g'}{\rho g}$$

With this substitution, the buoyant force on the weights is $B' = (F_g' / \rho g)\rho_{air}g$:

Therefore, $\qquad F_g = F_g' + \left(V - \frac{F_g'}{\rho g}\right)\rho_{air}g$ ◊

Related Comment: We can now answer the popular riddle: Which weighs more, a pound of feathers or a pound of bricks? Like in the problem above, the feathers feel a greater buoyant force than the bricks, so if they "weigh" the same on a scale as a pound of bricks, then the feathers must have more mass and therefore a greater "true weight."

---

61. Review problem. With reference to Figure 14.5, show that the total torque exerted by the water behind the dam about a horizontal axis through O is $\frac{1}{6}\rho gwH^3$. Show that the effective line of action of the total force exerted by the water is at a distance $\frac{1}{3}H$ above O.

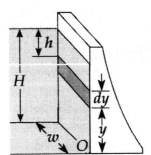

Figure 14.5

Solution The torque is calculated from the equation

$$\tau = \int d\tau = \int r\, dF$$

From Figure 14.5, we have

$$\tau = \int y\, dF = \int_0^H y[\rho g(H - y)w]\,dy = \frac{1}{6}\rho gwH^3 \qquad ◊$$

$$F = \int dF = \int_0^H [\rho g(H - y)w]\,dy = \frac{1}{2}\rho gwH^2$$

If this were applied at a height y_{eff} such that the torque remains unchanged,

$$\frac{1}{6}\rho gwH^3 = y_{eff}\left[\frac{1}{2}\rho gwH^2\right]$$

and

$$y_{eff} = \frac{1}{3}H \qquad ◊$$

---

65. In 1983 the United States began coining the cent piece out of copper-clad zinc rather than pure copper. The mass of the old copper penny is 3.083 g, while that of the new cent is 2.517 g. Calculate the percentage of zinc (by volume) in the new cent. The density of copper is 8.960 g/cm^3 and that of zinc is 7.133 g/cm^3. The new and old coins have the same volume.

Solution

Let f represent the fraction of the volume V occupied by zinc in the new coin.

We have $m = \rho V$ for both coins:

$$3.083 \text{ g} = \left(8.960 \text{ g} / \text{cm}^3\right)V$$

$$2.517 \text{ g} = \left(7.133 \text{ g} / \text{cm}^3\right)(fV) + \left(8.960 \text{ g} / \text{cm}^3\right)(1-f)V$$

By substitution,

$$2.517 \text{ g} = \left(7.133 \text{ g} / \text{cm}^3\right)fV + 3.083 \text{ g} - \left(8.960 \text{ g} / \text{cm}^3\right)fV$$

$$fV = \frac{3.083 \text{ g} - 2.517 \text{ g}}{8.960 \text{ g} / \text{cm}^3 - 7.133 \text{ g} / \text{cm}^3}$$

$$f = \frac{0.566 \text{ g}}{1.827 \text{ g} / \text{cm}^3}\left(\frac{8.960 \text{ g} / \text{cm}^3}{3.083 \text{ g}}\right) = 0.900\,4 = 90.04\%$$

◊

Chapter 15
OSCILLATORY MOTION

EQUATIONS AND CONCEPTS

Hooke's law gives the force exerted by a spring that is stretched or compressed beyond the equilibrium position. The force is proportional in magnitude to the displacement x from equilibrium position, shown in the figure as $x=0$. *The negative sign means that the force is a restoring force, always directed toward the equilibrium position. The constant of proportionality k is the elastic constant or the spring constant; it is always positive and has a value which corresponds to the relative stiffness of the elastic medium.*

$$F_s = -kx \qquad (15.1)$$

Simple harmonic motion is one form of periodic motion and occurs when the force exerted on a mass has the form given by Equation 15.1.

The **acceleration of an object in simple harmonic motion** is proportional to the displacement from equilibrium and oppositely directed. *Equation 15.2 is the result of applying Newton's second law to a mass, m, where the force is given by Equation 15.1.*

$$a_x = -\left(\frac{k}{m}\right)x \qquad (15.2)$$

The **mathematical representation** of a mass in simple harmonic motion along the x axis is a second order differential equation. The ratio k/m from Equation 15.2 is denoted by the factor ω^2 in Equation 15.5.

$$\frac{d^2x}{dt^2} = -\omega^2 x \qquad (15.5)$$

For an **ideal oscillator** in harmonic motion along the x axis, the following constants of the motion are found in expressions for position, velocity and acceleration:

Amplitude, A, represents the maximum displacement along either positive or negative x.

Angular frequency, ω, has units of radians per second and is a measure of how rapidly the oscillations are occurring.

Phase constant, ϕ, is determined by the position and velocity of the oscillating particle when $t=0$. The quantity $(\omega t + \phi)$ is called the phase.

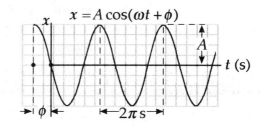

$$x = A\cos(\omega t + \phi)$$

For the displacement shown in the figure:

$$A = 4$$

$$\omega = \frac{1\ \text{rev}}{2\pi\ \text{s}} = \frac{2\pi\ \text{rad}}{2\pi\ \text{s}} = 1\ \text{rad / s}$$

and

$$\phi = +\pi / 2\ \text{rad}$$

The **position as a function of time** describes the displacement from equilibrium for a particle moving in simple harmonic motion along the x axis.

$$x(t) = A\cos(\omega t + \phi) \tag{15.6}$$

$$\text{where } \omega = \sqrt{\frac{k}{m}} \tag{15.9}$$

The **velocity as a function of time** and **acceleration as a function of time** are found by taking successive derivatives of Equation 15.6. Taking the first derivative of x with respect to time gives the velocity and the time derivative of the velocity (or the second derivative of the position) gives the acceleration.

$$v = \frac{dx}{dt} = -\omega A \sin(\omega t + \phi) \tag{15.15}$$

$$a = \frac{d^2 x}{dt^2} = -\omega^2 A \cos(\omega t + \phi) \tag{15.16}$$

The **maximum magnitudes of velocity and acceleration** can be found directly from Equations 15.15 and 15.16.

$$v_{\text{max}} = \omega A = \sqrt{\frac{k}{m}}\, A \tag{15.17}$$

or

$$a_{\text{max}} = \omega^2 A = \left(\frac{k}{m}\right) A \tag{15.18}$$

The **period of motion** T equals the time required for a particle to complete one full cycle of its motion. *The values of x and v at any time t equals the values of x and v at a later time $(t + T)$.*

$$T = \frac{2\pi}{\omega} \tag{15.10}$$

$$T = 2\pi\sqrt{\frac{m}{k}} \tag{15.13}$$

The **frequency of the motion,** f, numerically equals the inverse of the period and represents the number of oscillations per unit time. T is measured in seconds; f is measured in cycles/s or hertz (Hz). *The period and frequency depend only on the values of m and k.*

$$f = \frac{1}{T} = \frac{\omega}{2\pi} \tag{15.11}$$

$$f = \frac{1}{2\pi}\sqrt{\frac{k}{m}} \tag{15.14}$$

The **kinetic and potential energies** of a harmonic oscillator are always positive quantities.

$$E = \frac{1}{2}mv^2 + \frac{1}{2}kx^2$$

The **total mechanical energy** of a simple harmonic oscillator is a constant of the motion and is proportional to the square of the amplitude. *Equation 15.21 is obtained by substituting for v and x from Equations 15.6 and 15.15.*

$$E = \frac{1}{2}kA^2 \tag{15.21}$$

An expression for **velocity as a function of position** can be found from the principle of conservation of energy. *The magnitude of the velocity of an object in simple harmonic motion is a maximum at $x = 0$; the velocity is zero when the mass is at the points of maximum displacement $(x = \pm A)$.*

$$v = \pm\sqrt{\frac{k}{m}\left(A^2 - x^2\right)} \tag{15.22}$$

$$v = \pm\omega\sqrt{\left(A^2 - x^2\right)}$$

The **equation of motion for the simple pendulum** assumes a small angular displacement θ so that $\sin\theta \approx \theta$. *This small angle approximation is valid when the angle θ is measured in radians.*

$$\frac{d^2\theta}{dt^2} = -\frac{g}{L}\theta \tag{15.24}$$

The **period and frequency of a simple pendulum** depend only on the length of the supporting string and the value of the acceleration due to gravity.

$$\omega = \sqrt{\frac{g}{L}}$$ (15.25)

$$T = \frac{2\pi}{\omega} = 2\pi\sqrt{\frac{L}{g}}$$ (15.26)

The **period of a physical pendulum** depends on the moment of inertia, I, and the distance, d, between the pivot point and the center of mass. *This result can be used to measure the moment of inertia of a flat rigid body when the location of the center of mass is known.*

$$T = \frac{2\pi}{\omega} = 2\pi\sqrt{\frac{I}{mgd}}$$ (15.28)

The **period of a torsional pendulum** depends on the moment of inertia of the oscillating body and the torsion constant, κ, of the supporting wire. *The parameter, κ, is a measure of the restoring torque when a wire or long rod is twisted through an angle θ.*

$$T = 2\pi\sqrt{\frac{I}{\kappa}}$$ (15.30)

The **position as a function of time for a damped oscillator** includes an amplitude that decreases exponentially with time. *For this expression the damping force is assumed to be proportional to the velocity, and the damping coefficient, b, is assumed to be small.*

$$x = Ae^{-bt/2m}\cos(\omega t + \phi)$$ (15.32)

The **angular frequency of a damped oscillator** can be expressed in terms of ω_0, the natural frequency of the system. *The manner in which the amplitude of a given damped oscillator approaches zero depends on the value of the damping coefficient, b, and the natural frequency of the system, ω_0. The damping modes are called underdamped, critically damped, and overdamped.* See Figure 15.23 in the textbook.

$$\omega = \sqrt{\frac{k}{m} - \left(\frac{b}{2m}\right)^2}$$ (15.33)

$$\text{where } (\omega_0)^2 = \frac{k}{m}$$

SUGGESTIONS, SKILLS, AND STRATEGIES

SIMILAR STRUCTURE OF SEVERAL SYSTEMS IN SIMPLE HARMONIC MOTION

You should note the similar structure of the following equations and the corresponding expressions for period, T:

Oscillating system	Equation of motion	Period
Harmonic oscillator along x axis	$\dfrac{d^2x}{dt^2} = -\left(\dfrac{k}{m}\right)x$	$T = 2\pi\sqrt{\dfrac{m}{k}}$
Simple Pendulum	$\dfrac{d^2\theta}{dt^2} = -\left(\dfrac{g}{L}\right)\theta$	$T = 2\pi\sqrt{\dfrac{L}{g}}$
Physical pendulum	$\dfrac{d^2\theta}{dt^2} = -\left(\dfrac{mgd}{I}\right)\theta$	$T = 2\pi\sqrt{\dfrac{I}{mgd}}$
Torsion pendulum	$\dfrac{d^2\theta}{dt^2} = -\left(\dfrac{\kappa}{I}\right)\theta$	$T = 2\pi\sqrt{\dfrac{I}{\kappa}}$

In each case, when the equation of motion is in the form

$$\frac{d^2(\text{displacement})}{dt^2} = -C(\text{displacement})$$

the period is

$$T = 2\pi\sqrt{\frac{1}{C}}$$

COMPARISON OF SIMPLE HARMONIC MOTION AND UNIFORM CIRCULAR MOTION

Problems involving simple harmonic motion can often be solved more easily by treating simple harmonic motion along a straight line as the projection of uniform circular motion along a diameter of a reference circle. This technique is described in Section 15.4 of the text.

REVIEW CHECKLIST

You should be able to:

▷   Describe the general characteristics of simple harmonic motion, and the significance of the various parameters which appear in the expression for the displacement versus time, $x = A\cos(\omega t + \phi)$. (Sections 15.1 and 15.2)

▷   Start with the general expression for the displacement versus time for a simple harmonic oscillator, and obtain expressions for the velocity and acceleration as functions of time. (Section 15.2)

▷   Determine the frequency, period, amplitude, phase constant, and position at a specified time of a simple harmonic oscillator given an equation for $x(t)$. (Section 15.2)

▷   Start with an expression for $x(t)$ for an oscillator and determine the maximum speed, maximum acceleration, and the total distance traveled in a specified time. (Section 15.2)

▷   Describe the phase relations among displacement, velocity, and acceleration for simple harmonic motion. Given a curve of one of these variables versus time, sketch curves showing the time dependence of the other two. (Section 15.2)

▷   Calculate the mechanical energy, maximum velocity, and maximum acceleration of a mass-spring system given values for amplitude, spring constant, and mass. (Section 15.3)

▷   Describe the relationship between uniform circular motion and simple harmonic motion. (Section 15.4)

▷   Make calculations using the expressions for the period of simple, physical and torsion pendulums. (Section 15.5)

▷   Describe the conditions that give rise to the different types of damped motion. (Section 15.6)

ANSWERS TO SELECTED CONCEPTUAL QUESTIONS

2. If the coordinate of a particle varies as $x = -A\cos\omega t$, what is the phase constant in Equation 15.6? At what position is the particle at $t = 0$?

Answer Equation 15.6 says $x = A\cos(\omega t + \phi)$. Since negating a cosine wave is equivalent to changing the phase by 180°, we can say that

$$\phi = 180°\left(\frac{\pi\,\text{rad}}{180°}\right) = \pi\,\text{rad}$$

At time $t = 0$, the particle is at a position of $x = -A$.

□   □   □   □

4. Determine whether or not the following quantities can be in the same direction for a simple harmonic oscillator: (a) position and velocity, (b) velocity and acceleration, (c) position and acceleration.

Answer In a simple harmonic oscillator, the velocity follows the position by 1/4 of a cycle, and the acceleration follows the position by 1/2 of a cycle. Referring to Figure 15.1 of this study guide, it can be noted that there exist times (a) when both the position and the velocity are positive, and therefore in the same direction. (b) There also exist times when both the velocity and the acceleration are positive, and therefore in the same direction. On the other hand, (c) when the position is positive, the acceleration is always negative, and therefore position and acceleration always have opposite signs.

□   □   □   □

16. Is it possible to have damped oscillations when a system is at resonance? Explain.

Answer Yes. At resonance, the amplitude of a damped oscillator will remain constant. If the system were not damped, the amplitude would increase without limit at resonance.

□   □   □   □

SOLUTIONS TO SELECTED END-OF-CHAPTER PROBLEMS

3. The position of a particle is given by the expression $x = (4.00 \text{ m}) \cos(3.00\pi t + \pi)$, where x is in meters and t is in seconds. Determine (a) the frequency and period of the motion, (b) the amplitude of the motion, (c) the phase constant, and (d) the position of the particle at $t = 0.250$ s.

Solution

We use the particle in simple harmonic motion model.

The particular position function $x = (4.00 \text{ m}) \cos(3.00\pi t + \pi)$ and the general one, $x = A \cos(\omega t + \phi)$ have a specially powerful kind of equality called functional equality. They must give the same x value for all values of the variable t. This requires, then, that all parts be the same:

(a) $\omega = 3.00\pi \text{ rad} / \text{s} = (2\pi \text{ rad} / 1 \text{ cycle})f$ or $f = 1.50$ Hz

$$T = \frac{1}{f} = 0.667 \text{ s} \qquad \Diamond$$

(b) $A = 4.00 \text{ m}$ $\qquad \Diamond$

(c) $\phi = \pi$ rad $\qquad \Diamond$

(d)   At $t = 0.250$ s, $x = (4.00 \text{ m}) \cos(1.75\pi \text{ rad}) = (4.00 \text{ m}) \cos(5.50 \text{ rad})$

Note that 5.50 rad is **not** 5.50°.

Instead, $x = (4.00 \text{ m}) \cos(5.50 \text{ rad}) = (4.00 \text{ m}) \cos(315°) = 2.83 \text{ m}$ $\qquad \Diamond$

5. A particle moving along the x axis in simple harmonic motion starts from its equilibrium position, the origin, at $t = 0$ and moves to the right. The amplitude of its motion is 2.00 cm, and the frequency is 1.50 Hz. (a) Show that the position of the particle is given by $x = (2.00 \text{ cm}) \sin(3.00\pi t)$. Determine (b) the maximum speed and the earliest time $(t > 0)$ at which the particle has this speed, (c) the maximum acceleration and the earliest time $(t > 0)$ at which the particle has this acceleration, and (d) the total distance traveled between $t = 0$ and $t = 1.00$ s.

Solution

(a) At $t=0$, $x=0$ and v is positive (to the right). The sine function is zero and the cosine is positive at $\theta=0$, so this situation corresponds to $x = A\sin\omega t$ and $v = v_i\cos\omega t$.

Since $f=1.50$ Hz, $\qquad\qquad\qquad\qquad\qquad \omega = 2\pi f = 3\pi\,\text{s}^{-1}$

Also, $\qquad\qquad\qquad\qquad\qquad\qquad A = 2.00$ cm

so that $\qquad\qquad\qquad\qquad\qquad\quad x = (2.00\ \text{cm})\sin(3\pi t)$ ◊

(b) This is equivalent to writing $\qquad x = A\cos(\omega t + \phi)$

with $\qquad\qquad\qquad\qquad A=2.00$ cm, $\omega=3.00\,\pi\,\text{s}^{-1}$ and $\phi = -90° = -\dfrac{\pi}{2}$

Note also that $\qquad\qquad\qquad\qquad T = \dfrac{1}{f} = 0.667$ s

The velocity is $\qquad\qquad\qquad\qquad v = \dfrac{dx}{dt} = (2.00)(3.00\,\pi)\cos(3.00\,\pi t)\ \text{cm}/\text{s}$

The maximum speed is $\qquad\qquad v_{max} = v_i = A\omega = (2.00)(3.00\,\pi)\ \text{cm}/\text{s} = 18.8\ \text{cm}/\text{s}$ ◊

This speed occurs at $t=0$, when $\cos(3.00\,\pi t) = +1$

and next at $\ t = \dfrac{T}{2} = 0.333$ s, $\quad$ when $\quad \cos\left[(3.00\pi\,\text{s}^{-1})(0.333\ \text{s})\right] = -1$ ◊

(c) Again, $\qquad a = \dfrac{dv}{dt} = (-2.00\ \text{cm})(3.00\,\pi\ \text{s}^{-1})^2\sin(3.00\,\pi t)$

Its maximum value is

$$a_{max} = A\omega^2 = (2.00\ \text{cm})(3.00\,\pi\ \text{s}^{-1})^2 = 178\ \text{cm}/\text{s}^2$$

The acceleration has this positive value for the first time

at $\qquad\qquad\qquad t = \dfrac{3T}{4} = 0.500$ s ◊

when $\qquad a = -(2.00\ \text{cm})(3.00\,\pi\ \text{s}^{-1})^2\sin\left[(3.00\,\pi\ \text{s}^{-1})(0.500\ \text{s})\right] = 178\ \text{cm}/\text{s}^2$ ◊

(d) Since $A=2.00$ cm, the particle will travel 8.00 cm in one period, $T = \dfrac{2}{3}$ s.

Hence, in $\qquad\qquad\qquad\qquad\qquad (1.00\ \text{s}) = \dfrac{3}{2}T$

the particle will travel $\qquad\qquad\qquad \dfrac{3}{2}(8.00\ \text{cm}) = 12.0\ \text{cm}$ ◊

9. A 7.00-kg object is hung from the bottom end of a vertical spring fastened to an overhead beam. The object is set into vertical oscillations having a period of 2.60 s. Find the force constant of the spring.

Solution An object hanging from a vertical spring moves with simple harmonic motion just like an object moving without friction attached to a horizontal spring.

We are given the period; by Equation 15.10, $\dfrac{2\pi}{\omega} = T = 2.60$ s

Solving for the angular frequency, $\omega = \dfrac{2\pi \text{ rad}}{2.60 \text{ s}} = 2.42 \text{ rad} / \text{s}$

However, $\omega = \sqrt{\dfrac{k}{m}}$, so $k = \omega^2 m = (2.42 \text{ rad} / \text{s})^2 (7.00 \text{ kg})$

Thus we solve for the force constant: $k = 40.9 \text{ kg} / \text{s}^2 = 40.9 \text{ N} / \text{m}$ ◊

---

11. A 0.500-kg object attached to a spring with a force constant of 8.00 N/m vibrates in simple harmonic motion with an amplitude of 10.0 cm. Calculate (a) the maximum value of its speed and acceleration, (b) the speed and acceleration when the object is 6.00 cm from the equilibrium position, and (c) the time interval required for the object to move from $x = 0$ to $x = 8.00$ cm.

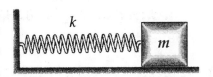

Solution $\omega = \sqrt{\dfrac{k}{m}} = \sqrt{\dfrac{8.00 \text{ N} / \text{m}}{0.500 \text{ kg}}} = 4.00 \text{ s}^{-1}$

Therefore, position is given by $x = (10.0 \text{ cm}) \sin \left[(4.00 \text{ s}^{-1}) t \right]$

(a) From this we find that $v = \dfrac{dx}{dt} = (40.0 \text{ cm} / \text{s}) \cos(4.00 t)$

$v_{max} = 40.0 \text{ cm} / \text{s}$ ◊

$a = \dfrac{dv}{dt} = -(160 \text{ cm} / \text{s}^2) \sin(4.00 t)$

$a_{max} = 160 \text{ cm} / \text{s}^2$ ◊

(b)
$$t = \frac{1}{4.00}\sin^{-1}\left(\frac{x}{10.0 \text{ cm}}\right)$$

When $x = 6.00$ cm, $\quad t = 0.161$ s

and we find that $\quad v = (40.0 \text{ cm/s})\cos\left[(4.00 \text{ s}^{-1})(0.161 \text{ s})\right] = 32.0 \text{ cm/s}$ ◊

$\quad a = -(160 \text{ cm/s}^2)\sin\left[(4.00 \text{ s}^{-1})(0.161 \text{ s})\right] = -96.0 \text{ cm/s}^2$ ◊

(c) Using
$$t = \frac{1}{4}\sin^{-1}\left(\frac{x}{10.0 \text{ cm}}\right)$$

When $x = 0$, $t = 0$ and when $x = 8.00$ cm, $t = 0.232$ s. Therefore, $\quad \Delta t = 0.232$ s ◊

---

17. An automobile having a mass of 1 000 kg is driven into a brick wall in a safety test. The bumper behaves like a spring of force constant 5.00×10^6 N/m and compresses 3.16 cm as the car is brought to rest. What was the speed of the car before impact, assuming that no mechanical energy is lost during impact with the wall?

Solution

Conceptualize: If the bumper is only compressed 3 cm, the car is probably not permanently damaged, so v is most likely less than 10 mph (~ 5 m/s).

Categorize: Assuming no energy is lost during impact with the wall, the initial energy (kinetic) equals the final energy (elastic potential).

Analyze: Energy conservation gives $\quad K_i = U_f \quad$ or $\quad \frac{1}{2}mv^2 = \frac{1}{2}kx^2$

Solving for the velocity, $\quad v = x\sqrt{\frac{k}{m}} = (3.16 \times 10^{-2} \text{ m})\sqrt{\frac{5.00 \times 10^6 \text{ N/m}}{1\,000 \text{ kg}}}$

Thus, $\quad v = 2.23$ m/s ◊

Finalize: The speed is less than 5 m/s as predicted, so the answer seems reasonable. If the speed of the car were sufficient to compress the bumper beyond its elastic limit, then some of the initial kinetic energy would be lost to deforming the front of the car. In that case, some other procedure would have to be used to estimate the car's initial speed.

---

23. A particle executes simple harmonic motion with an amplitude of 3.00 cm. At what position does its speed equal half its maximum speed?

Solution

Conceptualize: If we consider the speed of the particle along its path as shown in the sketch, we can see that the particle is at rest momentarily at one endpoint while being accelerated toward the middle by an elastic force that decreases as the particle approaches the equilibrium position. When it reaches the midpoint, the direction of acceleration changes so that the particle slows down until it stops momentarily at the opposite endpoint. From this analysis, we can estimate that $v = v_{max}/2$ somewhere in the outer half of the travel (since this is the region where the speed is changing most rapidly): $1.50 < \pm x < 3.00$

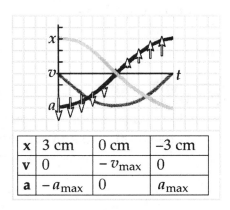

x	3 cm	0 cm	−3 cm
v	0	$-v_{max}$	0
a	$-a_{max}$	0	a_{max}

Categorize: One way to analyze this problem is to start with the equation for linear SHM, take the first derivative with respect to time to find $v(t)$, and solve this equation for x when $v = v_{max}/2$.

Analyze: For linear SHM, $x = A\cos\omega t$

noting that the negative indicates direction, $dx/dt = v = -A\omega\sin\omega t$

Since $A\omega$ is a constant, $v = \frac{1}{2}v_{max}$ when $\sin\omega t = \pm\frac{1}{2}$

or $\omega t = \sin^{-1}\left(\pm\frac{1}{2}\right) = \pm\frac{1}{6}\pi$ or $\pm\frac{5}{6}\pi$

Thus, in our displacement equation, $\cos\omega t = \cos\left(\pm\frac{\pi}{6}\right) = \frac{\sqrt{3}}{2}$

or $\cos\omega t = \cos\left(\pm\frac{5\pi}{6}\right) = -\frac{\sqrt{3}}{2}$

Substituting $A = 3.00$ cm, $x = \pm\frac{A\sqrt{3}}{2} = \pm\frac{(3.00 \text{ cm})\sqrt{3}}{2}$

Solving, $x = \pm 2.60$ cm   ◊

We could get the same answer from $v = \pm\omega\sqrt{A^2 - x^2}$

Finalize: The calculated position is in the outer half of the travel as predicted, and is in fact very close to the endpoints. This means that the speed of the particle changes remarkably little until the particle reaches the ends of its travel, where it experiences the maximum restoring force of the spring, which is proportional to x.

31. A simple pendulum has a mass of 0.250 kg and a length of 1.00 m. It is displaced through an angle of 15.0° and then released. What are (a) the maximum speed, (b) the maximum angular acceleration, and (c) the maximum restoring force? **What if?** Solve this problem by using the simple harmonic motion model for the motion of the pendulum, and then solve the problem more precisely by using more general principles.

Solution

METHOD ONE:

Since 15.0° is small enough that (in radians) $\sin\theta \approx \theta$ within 1%, we may model the motion as simple harmonic motion. The constant angular frequency characterizing the motion is

$$\omega = \sqrt{\frac{g}{L}} = \sqrt{\frac{9.80 \text{ m/s}^2}{1.00 \text{ m}}} = 3.13 \text{ rad/s}$$

The amplitude as a distance is $\quad A = L\theta = (1.00 \text{ m})(0.262 \text{ rad}) = 0.262 \text{ m}$

(a) The maximum linear speed is $\quad v_{max} = \omega A = (3.13 \text{ s}^{-1})(0.262 \text{ m}) = 0.820 \text{ m/s}$ ◊

(b) Similarly, $\quad a_{max} = \omega^2 A = (3.13 \text{ s}^{-1})^2(0.262 \text{ m}) = 2.57 \text{ m/s}^2$

This implies maximum angular acceleration

$$\alpha = \frac{a}{r} = \frac{2.57 \text{ m/s}^2}{1.00 \text{ m}} = 2.57 \text{ rad/s}^2$$ ◊

(c) $\qquad\qquad \sum F = ma = (0.250 \text{ kg})(0.257 \text{ m/s}^2) = 0.641 \text{ N}$ ◊

METHOD TWO:

We may work out slightly more precise answers by using the energy version of the isolated system model and the particle under a net force model. At release, the pendulum has height above its equilibrium position.

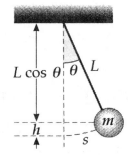

$$h = L - L\cos 15.0° = (1.00 \text{ m})(1 - \cos 15.0°) = 0.034\,1 \text{ m}$$

(a) The energy of the pendulum-Earth system is conserved as the pendulum swings down:

$$(K + U)_{top} = (K + U)_{bottom}$$

$$0 + mgh = \frac{1}{2}mv_{max}^2 + 0$$

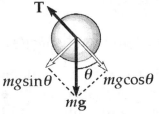

$$v_{max} = \sqrt{2gh} = \sqrt{2(9.80 \text{ m/s}^2)(0.034\,1 \text{ m})} = 0.817 \text{ m/s}$$ ◊

(c)   The restoring force at release is

$$mg\sin 15.0° = (0.250\text{ kg})(9.80\text{ m}/\text{s}^2)(\sin 15.0°) = 0.634\text{ N} \qquad \Diamond$$

(b)   This produces linear acceleration $a = \dfrac{\sum F}{m} = \dfrac{0.634\text{ N}}{0.250\text{ kg}} = 2.54\text{ m}/\text{s}^2$ $\Diamond$

and angular acceleration $\alpha = \dfrac{a}{r} = \dfrac{2.54\text{ m}/\text{s}^2}{1.00\text{ m}} = 2.54\text{ rad}/\text{s}^2$ $\Diamond$

---

33.   A particle of mass m slides without friction inside a hemispherical bowl of radius R. Show that, if it starts from rest with a small displacement from equilibrium, the particle moves in simple harmonic motion with an angular frequency equal to that of a simple pendulum of length R. That is, $\omega = \sqrt{g/R}$.

Solution

Locate the center of curvature C of the bowl. We can measure the excursion of the object from equilibrium by the angle θ between the radial line to C and the vertical. The distance the object moves from equilibrium is $s=R\theta$.

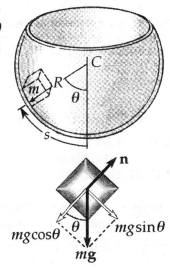

$\sum F_s = ma$ becomes $-mg\sin\theta = m\dfrac{d^2s}{dt^2}$

For small angles $\sin\theta \approx \theta$

so by substitution, $-mg\theta = m\dfrac{d^2s}{dt^2}$; $-mg\dfrac{s}{R} = m\dfrac{d^2s}{dt^2}$

Isolating the derivative, $\dfrac{d^2s}{dt^2} = -\left(\dfrac{g}{R}\right)s$

By the form of this equation, we can see that the acceleration is proportional to the position and in the opposite direction, so we have SHM. $\Diamond$

We identify its angular frequency by comparing our equation to Eq. 15.5: $\dfrac{d^2x}{dt^2} = -\omega^2 x$

Now x and s both measure position, so $\omega^2 = \dfrac{g}{R}$ and $\omega = \sqrt{\dfrac{g}{R}}$ $\Diamond$

---

35. A physical pendulum in the form of a planar body moves in simple harmonic motion with a frequency of 0.450 Hz. If the pendulum has a mass of 2.20 kg and the pivot is located 0.350 m from the center of mass, determine the moment of inertia of the pendulum about the pivot point.

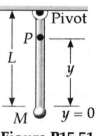

Solution $f = 0.450$ Hz; $d = 0.350$ m; $m = 2.20$ kg

By Eq. 15.28, $T = 2\pi \sqrt{\dfrac{I}{mgd}}$ and $T^2 = \dfrac{4\pi^2 I}{mgd}$

$$I = \frac{T^2 mgd}{4\pi^2} = \left(\frac{1}{f}\right)^2 \frac{mgd}{4\pi^2} = \frac{(2.20 \text{ kg})(9.80 \text{ m}/\text{s}^2)(0.350 \text{ m})}{(0.450 \text{ s}^{-1})^2(4\pi^2)} = 0.944 \text{ kg} \cdot \text{m}^2 \quad \lozenge$$

51. A small ball of mass M is attached to the end of a uniform rod of equal mass M and length L that is pivoted at the top (Fig. P15.51). (a) Determine the tensions in the rod at the pivot and at the point P when the system is stationary. (b) Calculate the period of oscillation for small displacements from equilibrium, and determine this period for $L = 2.00$ m. (**Suggestions:** Model the object at the end of the rod as a particle and use Eq. 15.28.)

Figure P15.51

Solution

Conceptualize: The tension in the rod at the pivot is the weight of the rod plus the weight of the mass M, so at the pivot point $T = 2Mg$. The tension at point P should be slightly less since the portion of the rod between P and the pivot does not contribute to the tension.

Categorize: The tension can be found from applying Newton's Second Law. The period of this physical pendulum can be found by analyzing its moment of inertia and using Equation 15.28.

Analyze:

(a) When the pendulum is stationary, the tension at any point in the rod is simply the weight of everything below that point. This conclusion comes from applying $\Sigma F_y = ma_y = 0$ to everything below that point.

Thus, at the pivot the tension is $F_T = F_{g,\text{ball}} + F_{g,\text{rod}} = Mg + Mg = 2Mg \quad \lozenge$

At point P, $F_T = F_{g,\text{ball}} + F_{g,\text{rod below } P}$

$$F_T = Mg + Mg\left(\frac{y}{L}\right) = Mg\left(1 + \frac{y}{L}\right) \quad \lozenge$$

(b) For a physical pendulum where I is the moment of inertia about the pivot, $m = 2M$, and d is the distance from the pivot to the center of mass,

the period of oscillation is: $\qquad\qquad T = 2\pi\sqrt{I/(mgd)}$

Relative to the pivot, $\qquad\qquad\qquad I_{total} = I_{rod} + I_{ball} = \frac{1}{3}ML^2 + ML^2 = \frac{4}{3}ML^2$

The center of mass distance is $\qquad d = \dfrac{\sum m_i x_i}{\sum m_i} = \dfrac{(ML/2 + ML)}{(M + M)} = \dfrac{3L}{4}$

so we have $\quad T = 2\pi\sqrt{\dfrac{I}{mgd}} \quad$ or $\quad T = 2\pi\sqrt{\dfrac{(4ML^2/3)}{(2M)g(3L/4)}} = \dfrac{4\pi}{3}\sqrt{\dfrac{2L}{g}}$ ◊

For $L = 2.00$ m, $\qquad\qquad\qquad T = \dfrac{4\pi}{3}\sqrt{\dfrac{2(2.00\ \text{m})}{9.80\ \text{m}/\text{s}^2}} = 2.68$ s ◊

Finalize: In part (a), the tensions agree with the initial predictions. In part (b) we found that the period is slightly less (by about 6%) than a simple pendulum of length L. It is interesting to note that we were able to calculate a value for the period despite not knowing the mass value. This is because the period of any pendulum depends on the **location** of the center of mass and not on the **size** of the mass.

---

53. A large block P executes horizontal simple harmonic motion as it slides across a frictionless surface with a frequency $f = 1.50$ Hz. Block B rests on it, as shown in Figure P15.53, and the coefficient of static friction between the two is $\mu_s = 0.600$. What maximum amplitude of oscillation can the system have if block B is not to slip?

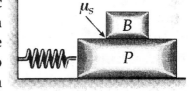

Figure P15.53

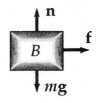

Solution If the block B does not slip, it undergoes simple harmonic motion (SHM) with the same amplitude and frequency as those of P, and with its acceleration caused by the static friction force exerted on it by P. Think of the block when it is just ready to slip at a turning point in its motion:

$\sum F = ma \qquad$ becomes $\qquad f_{max} = \mu_s n = \mu_s mg = ma_{max} = mA\omega^2$

Then $\qquad\qquad\qquad A = \dfrac{\mu_s g}{\omega^2} = \dfrac{0.600(9.80\ \text{m}/\text{s}^2)}{\left[2\pi(1.50\ \text{s}^{-1})\right]^2} = 6.62$ cm ◊

---

59. A pendulum of length L and mass M has a spring of force constant k connected to it at a distance h below its point of suspension (Fig. P15.59). Find the frequency of vibration of the system for small values of the amplitude (small θ). Assume the vertical suspension of length L is rigid, but ignore its mass.

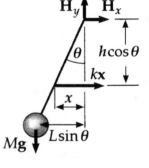

Figure P15.59

Solution

Conceptualize: The frequency of vibration should be greater than that of a simple pendulum since the spring adds an additional restoring force:

$$f > \frac{1}{2\pi}\sqrt{\frac{g}{L}}$$

Categorize: We can find the frequency of oscillation from the angular frequency, ω, which is found in the equation for angular SHM: $d^2\theta/dt^2 = -\omega^2\theta$. The angular acceleration can be found from analyzing the torques acting on the pendulum.

Analyze: For the pendulum (see sketch)

$$\sum \tau = I\alpha \quad \text{and} \quad d^2\theta/dt^2 = -\alpha$$

The negative sign appears because positive θ is measured clockwise in the picture. We take torque around the point of suspension:

$$\sum \tau = MgL\sin\theta + kxh\cos\theta = I\alpha$$

For small amplitude vibrations, use the approximations:

$$\sin\theta \approx \theta, \quad \cos\theta \approx 1, \quad \text{and} \quad x = h\tan\theta \approx h\theta$$

Therefore, with $I = mL^2$,

$$\frac{d^2\theta}{dt^2} = -\left[\frac{MgL + kh^2}{I}\right]\theta = -\left[\frac{MgL + kh^2}{ML^2}\right]\theta$$

This is of the SHM form

$$\frac{d^2\theta}{dt^2} = -\omega^2\theta$$

with angular frequency,

$$\omega = \sqrt{\frac{MgL + kh^2}{ML^2}} = 2\pi f$$

The ordinary frequency is

$$f = \frac{\omega}{2\pi} = \frac{1}{2\pi}\sqrt{\frac{MgL + kh^2}{ML^2}} \qquad \lozenge$$

Finalize: The frequency is greater than for a simple pendulum as we expected. In fact, the additional contribution inside the square root looks like the frequency of a mass on a spring scaled by h/L since the spring is connected to the rod and not directly to the mass. So we can think of the solution as:

$$f^2 = \frac{1}{4\pi^2}\left(\frac{MgL + kh^2}{ML^2}\right) = f^2_{pendulum} + \frac{h^2}{L^2}f^2_{spring}$$

63. A simple pendulum with a length of 2.23 m and a mass of 6.74 kg is given an initial speed of 2.06 m/s at its equilibrium position. Assume it undergoes simple harmonic motion, and determine its (a) period, (b) total energy, and (c) maximum angular displacement.

Solution

(a) The period is

$$T = \frac{2\pi}{\omega}$$

$$T = 2\pi\sqrt{\frac{L}{g}} = 2\pi\sqrt{\frac{2.23\ m}{9.80\ m/s^2}} = 3.00\ s \qquad \lozenge$$

(b) The total energy is

$$E = \tfrac{1}{2}mv^2_{max}$$

$$E = \tfrac{1}{2}(6.74\ kg)(2.06\ m/s)^2 = 14.3\ J \qquad \lozenge$$

(c) At angular displacement θ_{max},

$$mgh = \tfrac{1}{2}mv^2_{max}$$

and

$$h = \frac{v^2_{max}}{2g} = 0.217\ m$$

By geometry,

$$h = L - L\cos\theta_{max} = L(1 - \cos\theta_{max})$$

Solving for θ_{max},

$$\cos\theta_{max} = 1 - \frac{h}{L}$$

and

$$\theta_{max} = 25.5° \qquad \lozenge$$

Alternatively, we could write $v_{max} = \omega A$

$$A = \frac{v_{max}}{\omega} = \frac{2.06 \text{ m / s}}{2.10 \text{ / s}} = 0.983 \text{ m}$$

and

$$\theta = \frac{A}{L} = \frac{0.983 \text{ m}}{2.23 \text{ m}} = 0.441 \text{ rad} = 25.2°$$

Our two answers are not precisely equal because the pendulum does not move with precisely simple harmonic motion.

67. A ball of mass m is connected to two rubber bands of length L, each under tension T, as in Figure P15.67. The ball is displaced by a small distance y perpendicular to the length of the rubber bands. Assuming that the tension does not change, show that (a) the restoring force is $-(2T/L)y$ and (b) the system exhibits simple harmonic motion with an angular frequency $\omega = \sqrt{2T/mL}$.

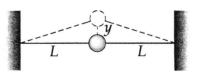

Figure P15.67

Solution

(a) $\sum \mathbf{F} = -2T\sin\theta\,\hat{\mathbf{j}}$ where $\theta = \tan^{-1}(y/L)$

Since for a small displacement, $\sin\theta \approx \tan\theta = \dfrac{y}{L}$

and the resultant force is $\sum \mathbf{F} = \left(-\dfrac{2Ty}{L}\right)\hat{\mathbf{j}}$ ◊

(b) Since there is a restoring force that is proportional to the position, it causes the system to move with simple harmonic motion like a block-spring system.

Thus, $\sum \mathbf{F} = -k\mathbf{x}$ becomes $\sum F = -\left(\dfrac{2T}{L}\right)y$

Therefore, $\omega = \sqrt{\dfrac{k}{m}} = \sqrt{\dfrac{2T}{mL}}$ ◊

69. A smaller disk of radius r and mass m is attached rigidly to the face of a second larger disk of radius R and mass M as shown in Figure P15.69. The center of the small disk is located at the edge of the large disk. The large disk is mounted at its center on a frictionless axle. The assembly is rotated through a small angle θ from its equilibrium position and released. (a) Show that the speed of the center of the small disk as it passes through the equilibrium position is

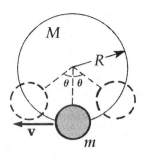

$$v = 2\sqrt{\frac{Rg(1-\cos\theta)}{(M/m)+(r/R)^2+2}}$$

Figure P15.69

(b) Show that the period of the motion is

$$T = 2\pi\sqrt{\frac{(M+2m)R^2+mr^2}{2mgR}}$$

Solution

(a) $\Delta K + \Delta U = 0$; thus $K_{top} + U_{top} = K_{bot} + U_{bot}$

where $K_{top} = U_{bot} = 0$

Therefore, $mgh = \frac{1}{2}I\omega^2$

where $h = R - R\cos\theta = R(1-\cos\theta)$

$\omega = v/R$

and $I = \frac{1}{2}MR^2 + \frac{1}{2}mr^2 + mR^2$

Substituting, we find $mgR(1-\cos\theta) = \frac{1}{2}\left(\frac{1}{2}MR^2 + \frac{1}{2}mr^2 + mR^2\right)\frac{v^2}{R^2}$

$$mgR(1-\cos\theta) = \left(\frac{1}{4}M + \frac{1}{4}\frac{r^2}{R^2}m + \frac{1}{2}m\right)v^2$$

and $v^2 = \dfrac{4gR(1-\cos\theta)}{(M/m)+\left(r^2/R^2\right)+2}$ so $v = 2\sqrt{\dfrac{Rg(1-\cos\theta)}{(M/m)+(r/R)^2+2}}$ ◊

(b) $T = 2\pi\sqrt{\dfrac{I}{M_T gd}}$ with $M_T = m + M$ and $d = \dfrac{mR + M(0)}{m+M}$

Therefore, $T = 2\pi\sqrt{\dfrac{\frac{1}{2}MR^2 + \frac{1}{2}mr^2 + mR^2}{mgR}} = 2\pi\sqrt{\dfrac{(M+2m)R^2+mr^2}{2mgR}}$ ◊

Chapter 16
WAVE MOTION

EQUATIONS AND CONCEPTS

The **wave function** $y(x,t)$ represents the y coordinate of an element or small segment of a medium as a wave pulse travels along the x direction. The value of y depends on two variables (position and time) and is read "y as a function of x and t".) *When t has a fixed value, $y = y(x)$ and defines the shape of the wave pulse at an instant in time.*

$$y(x,t) = f(x - vt) \qquad (16.1)$$

pulse traveling right

$$y(x,t) = f(x + vt) \qquad (16.2)$$

pulse traveling left

A **sinusoidal wave (periodic wave)** form, $y = f(x)$, is shown in the figure at right.

Wavelength, λ — distance between identical points on adjacent crests,

Amplitude, A — maximum displacement from equilibrium of an element of the medium, and

Period, T — time required for two identical points of adjacent wave crests to pass a given point in the medium.

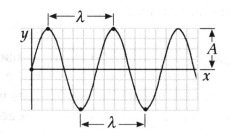

The **frequency, f, of a periodic wave** is the inverse of the period. *It is the same as the frequency of the harmonic oscillation of one element of the medium through which the wave is moving.*

$$f = \frac{1}{T} \qquad (16.3)$$

The **displacement of a sinusoidal wave** repeats itself when x is increased by an integral multiple of λ. *If the wave moves to the left, the quantity $(x - vt)$ is replaced with $(x + vt)$.*

$$y(x,t) = A\sin\left(\frac{2\pi}{\lambda}(x - vt)\right) \qquad (16.5)$$

The **wave speed** can be expressed in terms of the wavelength and period. *A traveling wave moves a distance $x = \lambda$ in a time $t = T$.*

$$v = \frac{\lambda}{T} \qquad (16.6)$$

The **periodic nature of y** as seen in Equation 16.7, found by combining Equations 16.5 and 16.6.

$$y = \sin\left[2\pi\left(\frac{x}{\lambda} - \frac{t}{T}\right)\right] \qquad (16.7)$$

Characteristic wave quantities:
   wave number k,
   angular frequency ω.

$$k \equiv \frac{2\pi}{\lambda} \qquad (16.8)$$

$$\omega \equiv \frac{2\pi}{T} \qquad (16.9)$$

The **wave function** can be expressed in a more compact form in terms of the parameters defined above.

$$y = A\sin(kx - \omega t) \qquad (16.10)$$

The **general expression for a sinusoidal wave** includes a phase constant ϕ. The phase constant is necessary in cases when the transverse displacement is not zero at $x = 0$ and $t = 0$.

$$y = A\sin(kx - \omega t + \phi) \qquad (16.13)$$

The **wave speed v** or phase velocity can also be expressed in alternative forms.

$$v = \frac{\omega}{k} \qquad (16.11)$$

$$v = \lambda f \qquad (16.12)$$

The **transverse speed v_y** and **transverse acceleration a_y** of a point on a sinusoidal wave on a string are out of phase by $\pi/2$ radians. *Do not confuse transverse speed with wave speed; v_y and a_y describe the motion of an element of the string that moves perpendicular to the direction of propagation of the wave.* Note the **maximum values of transverse speed and acceleration.**

$$v_y = -\omega A\cos(kx - \omega t) \qquad (16.14)$$

$$a_y = -\omega^2 A\sin(kx - \omega t) \qquad (16.15)$$

$$v_{y,\text{max}} = \omega A \qquad (16.16)$$

$$a_{y,\text{max}} = \omega^2 A \qquad (16.17)$$

A **wave on a stretched string** has a wave speed that depends on the tension in the string T and the linear density μ of the string (mass per unit length).

$$v = \sqrt{\frac{T}{\mu}}$$

(16.18)

The **power** (rate of energy transfer) transmitted by a sinusodial wave on a string is proportional to the square of the angular frequency and to the square of the amplitude, where μ is the mass per unit length of the string.

$$\mathcal{P} = \frac{1}{2}\mu\omega^2 A^2 v$$

(16.21)

The **linear wave equation** is satisfied by any wave function having the form $y = f(x \pm vt)$.

$$\frac{\partial^2 y}{\partial x^2} = \frac{1}{v^2}\frac{\partial^2 y}{\partial t^2}$$

(16.27)

REVIEW CHECKLIST

You should be able to:

▷ Recognize whether or not a given function is a possible description of a traveling wave and identify the direction (+ or −) in which the wave is travelling. (Section 16.1).

▷ Express a given sinusoidal wave function in several alternative forms involving different combinations of the wave parameters: wavelength, period, phase velocity, wave number, angular frequency, and sinusoidal frequency. (Section 16.2)

▷ Obtain values for the characteristic wave parameters: A, ω, k, λ, f, and ϕ if given a specific wave function for a sinusoidal wave. (Section 16.2)

▷ Calculate the speed of a transverse pulse travelling on a string. (Section 16.3)

▷ Describe the reflection of a traveling wave from a boundary between two media. (Section 16.4)

▷ Calculate the rate at which energy is transported by sinusoidal waves in a string. (Section 16.5)

ANSWERS TO SELECTED CONCEPTUAL QUESTIONS

2. How would you create a longitudinal wave in a stretched spring? Would it be possible to create a transverse wave in a spring?

Answer A longitudinal wave can be set up in a stretched spring by compressing the coils in a small region, and releasing the compressed region. The disturbance will proceed to propagate as a longitudinal pulse. It is quite possible to set up a transverse wave in a spring, simply by displacing the spring in a direction perpendicular to its length.

□   □   □   □

7. A vibrating source generates a sinusoidal wave on a string under constant tension. If the power delivered to the string is doubled, by what factor does the amplitude change? Does the wave speed change under these circumstances?

Answer Power is always proportional to the square of the amplitude, so if power doubles, amplitude increases by $\sqrt{2}$ times, a factor of 1.41. The wave speed does not depend on amplitude, but stays constant.

□   □   □   □

13. If you stretch a rubber hose and pluck it, you can observe a pulse traveling up and down the hose. What happens to the speed of the pulse if you stretch the hose more tightly? What happens to the speed if you fill the hose with water?

Answer If you stretch the hose tighter, you increase the tension, and increase the speed of the wave. If the hose elongates at all, then you decrease the linear density, which also increases the speed of the wave. If you fill it with water, you increase the linear density of the hose, and decrease the speed of the wave.

□   □   □   □

SOLUTIONS TO SELECTED END-OF-CHAPTER PROBLEMS

1.   At $t=0$, a transverse pulse in a wire is described by the function

$$y = \frac{6}{x^2 + 3}$$

where x and y are in meters. Write the function $y(x, t)$ that describes this pulse if it is traveling in the positive x direction with a speed of 4.50 m/s.

Solution $y(x, t)$ must be a function of both x and t, but must become $y = 6/(x^2+3)$ when $t=0$. To guarantee the same form, we substitute the term $x = x' + ut$, and solve for y:

$$y = \frac{6}{(x' + ut)^2 + 3}$$

Note that as t increases, x' must decrease by $u\Delta t$; aside from that, the equation remains the same. We first define that at $t=0$, $x=0$.

So $x' + u(0) = x = 0$ [1]

In order to cause the wave to appear to move to the right, we need to force our reference point (x) to the left. Therefore, one second later, at $t=1$, the wave has moved 4.50 m in the $+x$ direction, but x moves 4.50 m in the $-x$ direction.

So $x' + u(1\ \text{s}) = x = -4.50\ \text{m}$ [2]

Subtracting equations [1] and [2], $u = -4.50$ m/s, and our new equation is:

$$y(x,\ t) = \frac{6}{(x - 4.50t)^2 + 3} \qquad \Diamond$$

In general, we can cause any waveform to move along the x axis at a velocity v_x by substituting $(x - v_x t)$ for x. The same principle applies to motion in other directions.

7.   A sinusoidal wave is traveling along a rope. The oscillator that generates the wave completes 40.0 vibrations in 30.0 s. Also, a given maximum travels 425 cm along the rope in 10.0 s. What is the wavelength?

Solution $f = \dfrac{40.0\ \text{waves}}{30.0\ \text{s}} = 1.33\ \text{s}^{-1}$ and $v = \dfrac{425\ \text{cm}}{10.0\ \text{s}} = 42.5\ \text{cm/s}$

Since $v = \lambda t$, $\lambda = \dfrac{v}{f} = \dfrac{42.5\ \text{cm/s}}{1.33\ \text{s}^{-1}} = 0.319\ \text{m}$ $\Diamond$

9. A wave is described by $y = (2.00 \text{ cm})\sin(kx - \omega t)$, where $k = 2.11 \text{ rad/m}$, $\omega = 3.62 \text{ rad/s}$, x is in meters, and t is in seconds. Determine the amplitude, wavelength, frequency, and speed of the wave.

Solution

Given the general sinusoidal wave equation, $y = A\sin(kx - \omega t + \phi)$

By comparison, the amplitude is $A = 2.00 \text{ cm}$ ◊

The angular wave number $k = 2.11 \text{ rad/m}$: $\lambda = \dfrac{2\pi}{k} = 2.98 \text{ m}$ ◊

The angular frequency $\omega = 3.62 \text{ rad/s}$: $f = \dfrac{\omega}{2\pi} = 0.576 \text{ Hz}$ ◊

The speed is $v = f\lambda$, so $v = (0.576 \text{ s}^{-1})(2.98 \text{ m}) = 1.72 \text{ m/s}$ ◊

---

15. (a) Write the expression for y as a function of x and t for a sinusoidal wave traveling along a rope in the **negative** x direction with the following characteristics: $A = 8.00 \text{ cm}$, $\lambda = 80.0 \text{ cm}$, $f = 3.00 \text{ Hz}$, and $y(0, t) = 0$ at $t = 0$.   (b) **What if?** Write the expression for y as a function of x and t for the wave in part (a) assuming that $y(x, 0) = 0$ at the point $x = 10.0 \text{ cm}$.

Solution   The amplitude is $A = y_{max} = 8.00 \text{ cm} = 0.0800 \text{ m}$

$$k = \frac{2\pi}{\lambda} = \frac{2\pi}{0.800 \text{ m}} = 7.85 \text{ m}^{-1}$$

$$\omega = 2\pi f = 2\pi(3.00 \text{ s}^{-1}) = 6.00\pi \text{ rad/s}$$

(a)   Since $\phi = 0$, $y = A\sin(kx + \omega t + \phi)$

   becomes $y = (0.0800 \text{ m})\sin(7.85x + 6.00\pi t)$ ◊

(b)   In general, $y = (0.0800 \text{ m})\sin(7.85x + 6.00\pi t + \phi)$

   If $y(x,0) = 0$ at $x = 0.100 \text{ m}$, $0 = (0.0800 \text{ m})\sin(0.785 + \phi)$

   and $\phi = -0.785 \text{ rad}$

   Therefore, $y = (0.0800 \text{ m})\sin(7.85x + 6.00\pi t - 0.785)$ ◊

---

27. Transverse waves travel with a speed of 20.0 m/s in a string under a tension of 6.00 N. What tension is required for a wave speed of 30.0 m/s in the same string?

Solution

Conceptualize: Since $v \propto \sqrt{T}$, the new tension must be about twice as much as the original to achieve a 50% increase in the wave speed.

Categorize: The equation for the speed of a transverse wave on a string under tension can be used if we assume that the linear density of the string is constant. Then the ratio of the two wave speeds can be used to find the new tension.

Analyze: The two wave speeds can be written as:

$$v_1 = \sqrt{T_1/\mu} \qquad \text{and} \qquad v_2 = \sqrt{T_2/\mu}$$

Dividing, $\dfrac{v_2}{v_1} = \sqrt{\dfrac{T_2}{T_1}}$ so $T_2 = \left(\dfrac{v_2}{v_1}\right)^2 T_1 = \left(\dfrac{30.0 \text{ m/s}}{20.0 \text{ m/s}}\right)^2 (6.00 \text{ N}) = 13.5 \text{ N}$ ◊

Finalize: The new tension is slightly more than twice the original, so the result agrees with our initial prediction and is therefore reasonable.

---

31. A 30.0-m steel wire and a 20.0-m copper wire, both with 1.00-mm diameters, are connected end to end and stretched to a tension of 150 N. How long does it take a transverse wave to travel the entire length of the two wires?

Solution The total time of travel is the sum of the two times.

In each wire, $\qquad t = \dfrac{L}{v} = L\sqrt{\dfrac{\mu}{T}} \qquad$ where $\qquad \mu = \dfrac{m}{L} = \rho \dfrac{V}{L} = \rho A = \dfrac{\pi \rho d^2}{4}$

so $\qquad t = L\sqrt{\dfrac{\pi \rho d^2}{4T}}$

For copper, $\qquad t_1 = (20.0 \text{ m})\sqrt{\dfrac{\pi(8920 \text{ kg/m}^3)(0.00100 \text{ m})^2}{4(150 \text{ kg}\cdot\text{m/s}^2)}} = 0.137 \text{ s}$

For steel, $\qquad t_2 = (30.0 \text{ m})\sqrt{\dfrac{\pi(7860 \text{ kg/m}^3)(0.00100 \text{ m})^2}{4(150 \text{ kg}\cdot\text{m/s}^2)}} = 0.192 \text{ s}$

The total time is $\qquad (0.137 \text{ s}) + (0.192 \text{ s}) = 0.329 \text{ s}$ ◊

---

37. Sinusoidal waves 5.00 cm in amplitude are to be transmitted along a string that has a linear mass density of 4.00×10^{-2} kg/m. If the source can deliver a maximum power of 300 W and the string is under a tension of 100 N, what is the highest frequency at which the source can operate?

Solution The wave speed is $\quad v = \sqrt{\dfrac{T}{\mu}} = \sqrt{\dfrac{100 \text{ N}}{4.00 \times 10^{-2} \text{ kg / m}}} = 50.0 \text{ m / s}$

From $\mathcal{P} = \frac{1}{2}\mu\omega^2 A^2 v$, $\qquad \omega^2 = \dfrac{2\mathcal{P}}{\mu A^2 v} = \dfrac{2(300 \text{ N} \cdot \text{m / s})}{(4.00 \times 10^{-2} \text{ kg/m})(5.00 \times 10^{-2} \text{ m})^2 (50.0 \text{ m/s})}$

Solving, $\qquad\qquad\qquad\qquad \omega = 346.4 \text{ rad / s} \quad$ and $\quad f = \dfrac{\omega}{2\pi} = 55.1 \text{ Hz} \qquad\qquad \diamond$

39. A sinusoidal wave on a string is described by the equation

$$y = (0.15 \text{ m})\sin(0.80x - 50t)$$

where x and y are in meters and t is in seconds. If the mass per unit length of this string is 12.0 g/m, determine (a) the speed of the wave, (b) the wavelength, (c) the frequency, and (d) the power transmitted to the wave.

Solution

Compare the given wave function, $\qquad y = (0.15 \text{ m})\sin(0.80x - 50t)$

with the general wave equation, $\qquad y = A\sin(kx - \omega t)$

(a) $\quad v = f\lambda = \dfrac{\omega}{k} = \dfrac{50 \text{ rad / s}}{0.80 \text{ rad / m}} = 62.5 \text{ m / s}$ $\qquad\qquad\qquad\qquad\qquad \diamond$

(b) $\quad \lambda = \dfrac{2\pi}{k} = \dfrac{2\pi \text{ rad}}{0.80 \text{ rad / m}} = 7.85 \text{ m}$ $\qquad\qquad\qquad\qquad\qquad \diamond$

(c) $\quad f = \dfrac{\omega}{2\pi} = \dfrac{50 \text{ rad / s}}{2\pi \text{ rad}} = 7.96 \text{ Hz}$ $\qquad\qquad\qquad\qquad\qquad\qquad \diamond$

(d) $\quad \mathcal{P} = \frac{1}{2}\mu\omega^2 A^2 v = \frac{1}{2}(0.012\,0 \text{ kg / m})(50.0 \text{ s}^{-1})^2 (0.150 \text{ m})^2 (62.5 \text{ m / s}) = 21.1 \text{ W}$ $\qquad \diamond$

45. Show that the wave function $y = \ln[b(x - vt)]$ is a solution to Equation 16.27, where b is a constant.

Solution   The important thing to remember with partial derivatives is that **you treat all variables as constants, except the single variable of interest.** Keeping this in mind, we must apply two common rules of derivation to the function $y = \ln[b(x - vt)]$:

$$\frac{\partial}{\partial x}[\ln f(x)] = \left(\frac{1}{f(x)}\right)\frac{\partial(f(x))}{\partial x} \qquad [1]$$

$$\frac{\partial}{\partial x}\left[\frac{1}{f(x)}\right] = \frac{\partial}{\partial x}[f(x)]^{-1} = (-1)[f(x)]^{-2}\frac{\partial(f(x))}{\partial x} = -\left(\frac{1}{[f(x)]^2}\right)\frac{\partial(f(x))}{\partial x} \qquad [2]$$

Applying [1], $\dfrac{\partial y}{\partial x} = \left(\dfrac{1}{b(x - vt)}\right)\dfrac{\partial(bx - bvt)}{\partial x} = \left(\dfrac{1}{b(x - vt)}\right)(b) = \dfrac{1}{x - vt}$

Applying [2], $\dfrac{\partial^2 y}{\partial x^2} = -\dfrac{1}{(x - vt)^2}$

In a similar way, $\dfrac{\partial y}{\partial t} = \dfrac{-v}{(x - vt)}$ and $\dfrac{\partial^2 y}{\partial t^2} = -\dfrac{v^2}{(x - vt)^2}$

From the second-order partial derivatives, we see that $\dfrac{\partial^2 y}{\partial x^2} = \dfrac{1}{v^2}\dfrac{\partial^2 y}{\partial t^2}$ ◊

---

49.   The wave function for a traveling wave on a taut string is (in SI units)

$$y(x,t) = (0.350 \text{ m})\sin(10\pi t - 3\pi x + \pi/4)$$

(a) What are the speed and direction of travel of the wave?   (b) What is the vertical position of an element of the string at $t = 0$, $x = 0.100$ m? (c) What are the wavelength and frequency of the wave? (d) What is the maximum magnitude of the transverse speed of the string?

Solution   We use the traveling wave model.

We compare the given equation with $y = A\sin(kx - \omega t + \phi)$

We note that $\sin(\theta) = -\sin(-\theta) = \sin(-\theta + \pi)$

and find that $k = 3\pi \text{ rad/m}$ and $\omega = 10\pi \text{ rad/s}$

(a) The speed and direction of the wave can be defined in terms of the velocity:

$$\mathbf{v} = f\lambda\hat{\mathbf{i}} = \frac{\omega}{k}\hat{\mathbf{i}} = \frac{10\pi \text{ rad / s}}{3\pi \text{ rad / m}}\hat{\mathbf{i}} = 3.33\hat{\mathbf{i}} \text{ m / s}$$ ◊

(b) Substituting $t = 0$ and $x = 0.100$ m,

$$y = (0.350 \text{ m}) \sin (-0.300\pi + 0.250\pi) = -0.0548 \text{ m} = -5.48 \text{ cm}$$ ◊

Note that when you take the sine of a quantity with no units, it is not in degrees, but in radians.

(c) $$\lambda = \frac{2\pi \text{ rad}}{k} = \frac{2\pi \text{ rad}}{3\pi \text{ rad / m}} = 0.667 \text{ m}$$ ◊

and $$f = \frac{\omega}{2\pi \text{ rad}} = \frac{10\pi \text{ rad / s}}{2\pi \text{ rad}} = 5.00 \text{ Hz}$$ ◊

(d) $$v_y = \frac{\partial y}{\partial t} = (0.350 \text{ m})(10\pi \text{ rad / s}) \cos (10\pi t - 3\pi x + \pi/4)$$

The maximum occurs when the cosine term is 1:

$$v_{y,\text{max}} = (10\pi \text{ rad / s})(0.350 \text{ m / s}) = 11.0 \text{ m / s}$$ ◊

Note the large difference between the maximum particle speed and the wave speed found in part (a).

═══════════════════════════════

59. A rope of total mass m and length L is suspended vertically. Show that a transverse wave pulse travels the length of the rope in a time interval $\Delta t = 2\sqrt{L/g}$. (**Suggestion:** First find an expression for the wave speed at any point a distance x from the lower end by considering the tension in the rope as resulting from the weight of the segment below that point.)

Solution

We define $x = 0$ at the bottom of the rope and $x = L$ at the top of the rope. The tension in the rope at any point is the weight of the rope below that point. We can thus write the tension in the rope at each point x as $T = \mu x g$, where μ is the mass per unit length of the rope.

The speed of the wave pulse at each point along the rope's length is therefore

$$v = \sqrt{\frac{T}{\mu}} \qquad \text{or} \qquad v = \sqrt{gx}$$

But at each point x, the wave phase progresses at a rate of $v = \dfrac{dx}{dt}$

So we can then substitute for v, and generate the differential equation:

$$\frac{dx}{dt} = \sqrt{gx} \qquad \text{or} \qquad dt = \frac{dx}{\sqrt{gx}}$$

Integrating both sides, $\qquad \Delta t = \frac{1}{\sqrt{g}} \int_0^L \frac{dx}{\sqrt{x}} = \frac{\left[2\sqrt{x}\right]_0^L}{\sqrt{g}} = 2\sqrt{\frac{L}{g}}$ ◊

---

61. It is stated in Problem 59 that a wave pulse travels from the bottom to the top of a hanging rope of length L in a time interval $\Delta t = 2\sqrt{L/g}$. Use this result to answer the following questions. (It is not necessary to set up any new integrations.) (a) How long does it take for a pulse to travel halfway up the rope? Give your answer as a fraction of the quantity $2\sqrt{L/g}$. (b) A pulse starts traveling up the rope. How far has it traveled after a time $\sqrt{L/g}$?

Solution

Conceptualize: The wave pulse travels faster as it goes up the rope because the tension higher in the rope is greater (to support the weight of the rope below it). Therefore it should take more than half the total time Δt for the wave to travel halfway up the rope. Likewise, the pulse should travel less than halfway up the rope in time $\Delta t/2$.

Categorize: By using the time relationship given in the problem and making suitable substitutions, we can find the required time and distance.

Analyze:

(a) From the equation given, the time for a pulse to travel any distance, d, up from the bottom of a rope is $\Delta t_d = 2\sqrt{d/g}$. So the time for a pulse to travel a distance $L/2$ from the bottom is

$$\Delta t_{L/2} = 2\sqrt{\frac{L}{2g}} = 0.707\left(2\sqrt{\frac{L}{g}}\right)$$ ◊

(b) Likewise, the distance a pulse travels from the bottom of a rope in a time Δt_d is $d = g\Delta t_d^2/4$. So the distance traveled by a pulse after a time $\Delta t_d = \sqrt{L/g}$ is

$$d = \frac{g(L/g)}{4} = \frac{L}{4}$$ ◊

Finalize: As expected, it takes the pulse more than 70% of the total time to cover 50% of the distance. In half the total trip time, the pulse has climbed only 1/4 of the total length.

---

63. An aluminum wire is clamped at each end under zero tension at room temperature. The tension in the wire is increased by reducing the temperature, which results in a decrease in the wire's equilibrium length. What strain ($\Delta L/L$) results in a transverse wave speed of 100 m/s? Take the cross-sectional area of the wire to be 5.00×10^{-6} m^2, the density to be 2.70×10^3 kg/m^3, and Young's modulus to be 7.00×10^{10} N/m^2.

Solution

The expression for the elastic modulus,
$$Y = \frac{F/A}{\Delta L/L}$$

becomes an equation for strain
$$\frac{\Delta L}{L} = \frac{F/A}{Y} \qquad \text{[1]}$$

Substitute into the equation for the wave speed
$$v = \sqrt{T/\mu} = \sqrt{F/\mu}$$

the definition for μ
$$\mu = \frac{m}{L} = \frac{\rho(AL)}{L} = \rho A$$

and solve,
$$v^2 = \frac{F}{\mu} = \frac{1}{\rho}\left(\frac{F}{A}\right) \quad \text{or} \quad \left(\frac{F}{A}\right) = \rho v^2 \qquad \text{[2]}$$

Substituting [2] into [1],
$$\frac{\Delta L}{L} = \frac{\rho v^2}{Y} = \frac{(2.70 \times 10^3 \text{ kg/m}^3)(100 \text{ m/s})^2}{7.00 \times 10^{10} \text{ N/m}^2}$$

$$\left(\frac{\Delta L}{L}\right) = 3.86 \times 10^{-4} \qquad \Diamond$$

Chapter 17
SOUND WAVES

EQUATIONS AND CONCEPTS

The **speed of sound** depends on the bulk modulus B and equilibrium density ρ of the medium in which the wave (a compression wave) is propagating.

$$v = \sqrt{\frac{B}{\rho}} \qquad (17.1)$$

The **Bulk Modulus** is equal to the negative ratio of the pressure variation to the fractional change in volume of the medium.

$$B = -\frac{\Delta P}{\Delta V / V}$$

The **displacement from equilibrium** of an element of medium in which a harmonic sound wave is propagating has a sinusoidal variation in time. The maximum amplitude of the displacement is s_{max}. *The displacement is parallel to the direction of the propagation of the wave (longitudinal wave).*

$$s(x,\, t) = s_{max} \cos(kx - \omega t) \qquad (17.2)$$

The **pressure variations** in a medium conducting a sound wave vary harmonically in time and are out of phase with the displacements by $\pi/2$ radians (compare Equations 17.2 and 17.3).

$$\Delta P = \Delta P_{max} \sin(kx - \omega t) \qquad (17.3)$$

The **maximum change in pressure** from the equilibrium value is ΔP_{max}. A sound wave may be considered as either a displacement wave or a pressure wave. *The pressure amplitude is proportional to the displacement amplitude.*

$$\Delta P_{max} = \rho v \omega s_{max} \qquad (17.4)$$

The **intensity of a periodic sound wave** is proportional to the square of the source frequency and to the square of the amplitude of the displacement. The intensity is also proportional to the square of the pressure amplitude.

$$I = \frac{\mathcal{P}}{A} = \frac{1}{2}\rho v (\omega s_{max})^2$$

or

$$I = \frac{\Delta P_{max}^2}{2\rho v} \qquad (17.6)$$

The **intensity I of a spherical wave** decreases as the square of the distance from the source.

$$I = \frac{\mathcal{P}_{av}}{A} = \frac{\mathcal{P}_{av}}{4\pi r^2} \qquad (17.7)$$

Sound levels are measured on a logarithmic scale. β is measured in decibels (dB) and I is the corresponding intensity measured in W/m². I_0 *is a reference intensity corresponding to the threshold of hearing.*

$$\beta \equiv 10 \log\left(\frac{I}{I_0}\right) \qquad (17.8)$$

$$I_0 = 10^{-12} \ \text{W} / \text{m}^2$$

The **Doppler effect (apparent shift in frequency)** is observed whenever there is relative motion between a source and an observer. *Equation 17.13 is a general Doppler shift expression; it applies to relative motion of source and observer toward or away from each other with correct use of algebraic signs.* For each of v_0 and v_s, a positive value is used when the corresponding object moves toward the other object; negative values are used when the object moves away from the other. In Equation 17.13, v_0 and v_s are measured **relative to the medium in which the sound travels.**

$$f' = \left(\frac{v + v_0}{v - v_s}\right)f \qquad (17.13)$$

where

v = speed of sound in a medium,
v_0 = velocity of the observer, and
v_s = velocity of the source.

Shock waves are produced when a sound source moves through a medium with a speed which is greater than the wave speed in that medium. The shock wave front has a conical shape with a half angle which depends on the Mach number of the source, defined as the ratio v_s/v.

$$\sin\theta = \frac{v}{v_s}$$

SUGGESTIONS, SKILLS, AND STRATEGIES

SOUND INTENSITY AND THE DECIBEL SCALE

When making calculations using Equation 17.8 which defines the intensity of a sound wave on the decibel scale, the properties of logarithms must be kept clearly in mind.

In order to determine the decibel level corresponding to two sources sounded simultaneously, you must first find the intensity, I, of each source in W/m^2; add these values, and then convert the resulting intensity to the decibel scale. As an illustration of this technique, determine the dB level when two sounds with intensities of 40 dB and 45 dB are sounded together.

Solve Eq. 17.8, $\beta = 10\log(I / I_0)$ to find $I = I_0 10^{\beta/10}$

Then, for $\beta_1 = 40$ dB, $I_1 = \left(10^{-12} \text{ W / m}^2\right)10^4 = 1.00 \times 10^{-8} \text{ W / m}^2$

For $\beta_2 = 45$ dB, $I_2 = \left(10^{-12} \text{ W / m}^2\right)10^{4.5} = 3.16 \times 10^{-8} \text{ W / m}^2$

and $I_{\text{total}} = 4.16 \times 10^{-8} \text{ W / m}^2$

Again using Eq. 17.8, $\beta_{\text{total}} = 10\log(I_{\text{total}} / I_0)$

So $\beta_{\text{total}} = 10\log\left(\dfrac{4.16 \times 10^{-8} \text{ W / m}^2}{1.00 \times 10^{-12} \text{ W / m}^2}\right) = 10\log\left(4.16 \times 10^4\right) = 46.19$ dB

The intensity level of the combined sources is 46.2 dB (not 85 dB).

DOPPLER EFFECT

Equation 17.13 is the generalized equation for the Doppler effect:

$$f' = \left(\frac{v + v_0}{v - v_s}\right)f \quad \text{where:}$$

f = source frequency, v_0 = speed of observer

f' = observed frequency, v_s = speed of source

v = speed of sound in medium

The most likely error in using Equation 17.13 is using the incorrect algebraic sign for the velocity of either the observer or the source.

When the **relative motion** of source and observer is:

either or both toward the other: Enter v_s and v_0 with + signs.
either or both away from the other: Enter v_s and v_0 with − signs.

Remember in Equation 17.13 stated above, the algebraic signs (the plus sign in the numerator and the minus sign in the denominator) are part of the structure of the equation. These signs remain and correct signs, stated above, must be entered along with the values of v_0 and v_s.

Consider the following examples given a source frequency of 300 Hz and the speed of sound in air of 343 m/s.

(1) Source moving with a speed of 40 m/s toward a fixed observer.
In this case, $v_0 = 0$ and $v_s = +40$ m/s. Substituting into Equation 17.13,

$$f' = \left(\frac{v + v_0}{v - v_s}\right)f = \left(\frac{343 \text{ m/s} + 0}{343 \text{ m/s} - 40 \text{ m/s}}\right)(300 \text{ Hz}) = 340 \text{ Hz}$$

(2) Source moving away from a fixed observer with a speed of 40 m/s.
In this case, $v_0 = 0$ and $v_s = -40$ m/s; Substituting into Equation 17.13,

$$f' = \left(\frac{v + v_0}{v - v_s}\right)f = \left(\frac{343 \text{ m/s} + 0}{343 \text{ m/s} - (-40 \text{ m/s})}\right)(300 \text{ Hz}) = 269 \text{ Hz}$$

(3) Observer moving with speed of 40 m/s toward a fixed source:
$v_0 = +40$ m/s and $v_s = 0$. Substituting into Equation 17.13,

$$f' = \left(\frac{v + v_0}{v - v_s}\right)f = \left(\frac{343 \text{ m/s} + 40 \text{ m/s}}{343 \text{ m/s} - 0}\right)(300 \text{ Hz}) = 335 \text{ Hz}$$

(4) Observer moving with speed 40 m/s away from a fixed source:
$v_0 = -40$ m/s and $v_s = 0$. Substituting into Equation 17.13,

$$f' = \left(\frac{v + v_0}{v - v_s}\right)f = \left(\frac{343 \text{ m/s} + (-40 \text{ m/s})}{343 \text{ m/s} - 0}\right)(300 \text{ Hz}) = 265 \text{ Hz}$$

As a check on your calculated value of f', remember:

- When the relative motion is source or observer toward the other, $f' > f$.
- When the relative motion is source or observer away from the other, $f' < f$.

REVIEW CHECKLIST

You should be able to:

▷   Calculate the speed of sound in various media in terms of the appropriate elastic properties of a particular medium (including bulk modulus, Young's modulus, and the pressure-volume relationships of an ideal gas), and the corresponding inertial properties (usually mass density). (Section 17.1)

▷   Describe the harmonic displacement and pressure variation as functions of time and position for a harmonic sound wave. Relate the displacement amplitude to the pressure amplitude for a harmonic sound wave and calculate the wave intensity from each of these parameters. (Sections 17.2 and 17.3)

▷   Determine the total intensity due to one or several sound sources whose individual decibel levels are known. Calculate the decibel level due to some combination of sources whose individual intensities are known. (Section 17.3)

▷   Make calculations for the various situations under which the Doppler effect is observed. **Pay particular attention to the correct use of algebraic signs.** (Section 17.4)

ANSWERS TO SELECTED CONCEPTUAL QUESTIONS

2.   If an alarm clock is placed in a good vacuum and then activated, no sound is heard. Explain.

Answer   We assume that a perfect vacuum surrounds the clock. The sound waves require a medium for them to travel to your ear. The hammer on the alarm will strike the bell, and the vibration will spread as sound waves through the body of the clock. If a bone of your skull were in contact with the clock, you would hear the bell. However, in the absence of a surrounding medium like air or water, no sound can be radiated away.

What happens to the sound energy? Here is the answer: As the sound wave travels through the steel and plastic, traversing joints and going around corners, its energy is converted into additional internal energy, raising the temperature of the materials. After the sound has died away, the clock will glow very slightly brighter in the infrared spectrum. For a larger-scale example of the same effect: Colossal storms raging on the Sun are deathly still for us.

□   □   □   □

8. If the distance from a point source is tripled, by what factor does the intensity decrease?

Answer We suppose that a point source has no structure, and radiates sound equally in all directions (isotropically). The sound wavefronts are expanding spheres, so the area over which the sound energy spreads increases according to $A = 4\pi r^2$. Thus, if the distance is tripled, the area increases by a factor of nine, and the new intensity will be one ninth of the old intensity. This answer according to the inverse-square law applies if the medium is uniform and unbounded.

For contrast, suppose that the sound is confined to move in a horizontal layer. (Thermal stratification in an ocean can have this effect on sonar "pings.") Then the area over which the sound energy is dispersed will only increase according to the circumference of an expanding circle: $A = 2\pi r h$, and so three times the distance will result in one third the intensity.

In the case of an entirely enclosed speaking tube (such as a ship's telephone), the area perpendicular to the energy flow stays the same, and increasing the distance will not change the intensity appreciably.

□   □   □   □

15. How can an object move with respect to an observer so that the sound from it is not shifted in frequency?

Answer For the sound from a source not to shift in frequency, the radial velocity of the source relative to the observer must be zero; that is, the source must not be moving toward or away from the observer.

This can happen if the source and observer are not moving at all; if they have equal velocities relative to the medium; or, it can happen if the source moves around the observer in a circular pattern of constant radius. Even if the source accelerates along the circle, decelerates, or stops, the frequency heard will equal the frequency emitted by the source.

□   □   □   □

SOLUTIONS TO SELECTED END-OF-CHAPTER PROBLEMS

1. Suppose that you hear a clap of thunder 16.2 s after seeing the associated lightning stroke. The speed of sound waves in air is 343 m/s, and the speed of light is 3.00×10^8 m/s. How far are you from the lightning stroke?

Solution

Conceptualize: There is a common rule of thumb that lightning is about a mile away for every 5 seconds of delay between the flash and thunder (or ~3 s/km). Therefore, this lightning strike is about 3 miles (~5 km) away.

Categorize: The distance can be found from the speed of sound and the elapsed time. The time for the light to travel to the observer will be much less than the sound delay, so the speed of light can be taken as ∞.

Analyze: Assuming that the speed of sound is constant through the air between the lightning strike and the observer,

$$v_s = \frac{d}{\Delta t}$$

or $\quad d = v_s \Delta t = (343 \text{ m / s})(16.2 \text{ s}) = 5.56 \text{ km}$ ◊

Finalize: Our calculated answer is consistent with our initial estimate, but we should check the validity of our assumption that the speed of light could be ignored. The time delay for the light is

$$t_{light} = \frac{d}{c_{air}} = \frac{5\ 560 \text{ m}}{3.00 \times 10^8 \text{ m / s}} = 1.85 \times 10^{-5} \text{ s}$$

and $\quad \Delta t = t_{sound} - t_{light} = 16.2 \text{ s} - 1.85 \times 10^{-5} \text{ s} \approx 16.2 \text{ s}$

Since the travel time for the light is much smaller than the uncertainty in the time of 16.2 s, t_{light} can be ignored without affecting the distance calculation. However, our assumption of a constant speed of sound in air is probably not valid due to local variations in air temperature during a storm. We must assume that the given speed of sound in air is an accurate **average** value for the conditions described.

13. Write an expression that describes the pressure variation as a function of position and time for a sinusoidal sound wave in air, if $\lambda = 0.100$ m and $\Delta P_{max} = 0.200$ Pa.

Solution

We write the pressure variation as $\qquad \Delta P = \Delta P_{max} \sin(kx - \omega t)$

Noting that $k = \dfrac{2\pi}{\lambda}$, $\qquad\qquad\qquad k = \dfrac{2\pi \text{ rad}}{0.100 \text{ m}} = 62.8 \text{ rad} / \text{m}$

Likewise, $\quad \omega = \dfrac{2\pi v}{\lambda}$, so $\qquad\qquad \omega = \dfrac{(2\pi \text{ rad})(343 \text{ m} / \text{s})}{0.100 \text{ m}} = 2.16 \times 10^4 \text{ rad} / \text{s}$

We now can create our equation: $\qquad \Delta P = (0.200 \text{ Pa}) \sin(62.8x - 21600t)$ $\qquad$ ◊

---

15. An experimenter wishes to generate in air a sound wave that has a displacement amplitude of 5.50×10^{-6} m. The pressure amplitude is to be limited to 0.840 N/m^2. What is the minimum wavelength the sound wave can have?

Solution $\quad$ We are given $\quad s_{max} = 5.50 \times 10^{-6}$ m and $\Delta P_{max} = 0.840$ Pa

The pressure amplitude is $\quad \Delta P_{max} = \rho v \omega s_{max} = \rho v \left(\dfrac{2\pi v}{\lambda} \right) s_{max}$

or $\lambda_{min} = \dfrac{2\pi \rho v^2 s_{max}}{\Delta P_{max}}$: $\qquad \lambda_{min} = \dfrac{2\pi (1.20 \text{ kg} / \text{m}^3)(343 \text{ m} / \text{s})^2 (5.50 \times 10^{-6} \text{ m})}{0.840 \text{ Pa}} = 5.81$ m ◊

---

19. Calculate the sound level in decibels of a sound wave that has an intensity of 4.00 μW/m^2.

Solution

We use the equation $\quad \beta = 10 \log(I / I_0)$, $\qquad$ where $\qquad I_0 = 10^{-12}$ W / m^2.

$$\beta = 10 \log \left(\dfrac{4.00 \times 10^{-6} \text{ W} / \text{m}^2}{10^{-12} \text{ W} / \text{m}^2} \right) = 66.0 \text{ dB} \qquad ◊$$

---

25. A family ice show is held at an enclosed arena. The skaters perform to music with level 80.0 dB. This is too loud for your baby, who yells at 75.0 dB. (a) What total sound intensity engulfs you? (b) What is the combined sound level?

Solution

$$\beta = 10 \ \log\left(\frac{I}{10^{-12} \ W \ / \ m^2}\right)$$

so

$$I = \left[10^{\beta/10}\right]10^{-12} \ W \ / \ m^2$$

(a) For your baby,

$$I_b = \left(10^{75.0/10}\right)\left(10^{-12} \ W \ / \ m^2\right) = 3.16 \times 10^{-5} \ W \ / \ m^2$$

For the music,

$$I_m = \left(10^{80.0/10}\right)\left(10^{-12} \ W \ / \ m^2\right) = 10.0 \times 10^{-5} \ W \ / \ m^2$$

The combined intensity is $I_{total} = I_m + I_b$

$$I_{total} = 10.0 \times 10^{-5} \ W \ / \ m^2 + 3.16 \times 10^{-5} \ W \ / \ m^2$$

$$I_{total} = 13.2 \times 10^{-5} \ W \ / \ m^2 \qquad \Diamond$$

(b) The combined sound level is then

$$\beta_{total} = 10 \ \log\left(\frac{I_{total}}{10^{-12} \ W \ / \ m^2}\right) = 10 \ \log\left(\frac{1.32 \times 10^{-4} \ W \ / \ m^2}{10^{-12} \ W \ / \ m^2}\right) = 81.2 \ dB \qquad \Diamond$$

29. A firework charge is detonated many meters above the ground. At a distance of 400 m from the explosion, the acoustic pressure reaches a maximum of 10.0 N/m². Assume that the speed of sound is constant at 343 m/s throughout the atmosphere over the region considered, that the ground absorbs all the sound falling on it, and that the air absorbs sound energy as described by the rate 7.00 dB/km. What is the sound level (dB) at 4.00 km from the explosion?

Solution

Conceptualize: At a distance of 4 km, an explosion should be audible, but probably not extremely loud. So based on the data in Table 17.2, we might expect the sound level to be somewhere between 40 and 80 dB.

Categorize: From the sound pressure data given in the problem, we can find the intensity, which is used to find the sound level in dB. The sound intensity will decrease with increased distance from the source and from the absorption of the sound by the air.

Analyze: At a distance of 400 m from the explosion, $\Delta P_{max} = 10.0\ \text{Pa}$. At this point,

$$I = \frac{\Delta P_{max}^2}{2\rho v} = \frac{\left(10.0\ \text{N} / \text{m}^2\right)^2}{2\left(1.20\ \text{kg} / \text{m}^3\right)(343\ \text{m} / \text{s})} = 0.121\ \text{W} / \text{m}^2$$

Therefore, the sound level is

$$\beta = 10\ \log\left(\frac{I}{I_0}\right) = 10\ \log\left(\frac{0.121\ \text{W} / \text{m}^2}{1.00 \times 10^{-12}\ \text{W} / \text{m}^2}\right) = 111\ \text{dB}$$

From Equations 17.8 and 17.7, we can calculate the intensity and decibel level (due to distance alone) 4 km away:

$$I' = I(400\ \text{m})^2 / (4\,000\ \text{m})^2 = 1.21 \times 10^{-3}\ \text{W} / \text{m}^2$$

and

$$\beta = 10\ \log\left(\frac{I'}{I_0}\right) = \left(\frac{1.21 \times 10^{-3}\ \text{W} / \text{m}^2}{1.00 \times 10^{-12}\ \text{W} / \text{m}^2}\right) = 90.8\ \text{dB}$$

At a distance of 4 km from the explosion, absorption from the air will have decreased the sound level by an additional

$$\Delta\beta = (7.00\ \text{dB} / \text{km})(3.60\ \text{km}) = 25.2\ \text{dB}$$

So at 4 km, the sound level will be

$$\beta_f = \beta - \Delta\beta = 90.8\ \text{dB} - 25.2\ \text{dB} = 65.6\ \text{dB} \qquad \Diamond$$

Finalize: This sound level falls within our expected range. Evidently, this explosion is rather loud (about the same as a vacuum cleaner) even at a distance of 4 km from the source. It is interesting to note that the distance and absorption effects each reduce the sound level by about the same amount ($\sim 20\ \text{dB}$). If the explosion were at ground level, the sound level would be further reduced by reflection and absorption from obstacles between the source and observer, and the calculation would be much more complicated (if not impossible).

---

33. The sound level at a distance of 3.00 m from a source is 120 dB. At what distance will the sound level be (a) 100 dB and (b) 10.0 dB?

Solution $\quad \beta = 10\ \log\left(\frac{I}{10^{-12}\ \text{W} / \text{m}^2}\right) \qquad$ so $\qquad I = \left[10^{\beta/10}\right]10^{-12}\ \text{W} / \text{m}^2$

$$I_{120} = 1\ \text{W} / \text{m}^2 \qquad I_{100} = 10^{-2}\ \text{W} / \text{m}^2 \qquad I_{10} = 10^{-11}\ \text{W} / \text{m}^2$$

(a) The power passing through any sphere around the source is $\mathcal{P} = 4\pi r^2 I$, so conservation of energy requires that $r_{120}^2 I_{120} = r_{100}^2 I_{100} = r_{10}^2 I_{10}$.

and $\quad r_{100} = r_{120}\sqrt{\dfrac{I_{120}}{I_{100}}} = (3.00\text{ m})\sqrt{\dfrac{1\text{ W}/\text{m}^2}{10^{-2}\text{ W}/\text{m}^2}} = 30.0\text{ m}$ ◊

(b) $\quad r_{10} = r_{120}\sqrt{\dfrac{I_{120}}{I_{10}}} = (3.00\text{ m})\sqrt{\dfrac{1\text{ W}/\text{m}^2}{10^{-11}\text{ W}/\text{m}^2}} = 9.49\times10^5\text{ m}$ ◊

39. Standing at a crosswalk, you hear a frequency of 560 Hz from the siren of an approaching ambulance. After the ambulance passes, the observed frequency of the siren is 480 Hz. Determine the ambulance's speed from these observations.

Solution

Conceptualize: We can assume that a ambulance with its siren on is in a hurry to get somewhere, and is probably traveling between 20 and 100 mi/h (~10 m/s to 50 m/s), depending on the driving conditions.

Categorize: We can use the equation for the Doppler effect to find the speed of the car.

Analyze: Let v_s represent the magnitude of the velocity of the ambulance.

Approaching car: $\quad f' = \left(\dfrac{v}{v - v_s}\right)f \qquad$ Departing car: $\qquad f'' = \left(\dfrac{v}{v + v_s}\right)f$

where $\qquad f' = 560\text{ Hz} \qquad$ and $\qquad f'' = 480\text{ Hz}$

Solving the two equations above for f and setting them equal gives:

$$f'(1 - v_s/v) = f''(1 + v_s/v) \qquad \text{or} \qquad f' - f'' = (f' + f'')(v_s/v)$$

so the speed of the source is

$$v_s = \frac{v(f' - f'')}{f' + f''} = \frac{(343\text{ m}/\text{s})(560\text{ Hz} - 480\text{ Hz})}{560\text{ Hz} + 480\text{ Hz}} = 26.4\text{ m}/\text{s} \qquad ◊$$

Finalize: This seems like a reasonable speed (about 50 mi/h) for an ambulance, unless the street is crowded or the car is traveling on an open highway. Of course, this problem is only valid if the siren emits a single tone. A warble could make it difficult or impossible to solve this problem.

41. A tuning fork vibrating at 512 Hz falls from rest and accelerates at 9.80 m/s^2. How far below the point of release is the tuning fork when waves of frequency 485 Hz reach the release point? Take the speed of sound in air to be 340 m/s.

Solution

In order to solve this problem, we must first determine how fast the tuning fork is falling when its frequency is 485 Hz. The tuning fork (source) is moving **away** from a stationary listener.

Therefore, we use the equation $\quad f' = \left(\dfrac{v}{v + v_s} \right) f$

$$485 \text{ Hz} = (512 \text{ Hz})\dfrac{340 \text{ m/s}}{340 \text{ m/s} + v_{fall}}$$

Solving, $\qquad v_{fall} = (340 \text{ m/s})\left(\dfrac{512 \text{ Hz}}{485 \text{ Hz}} - 1 \right) = 18.93 \text{ m/s}$

For the tuning fork we use the particle under constant acceleration model.

From the kinematic equation $\quad v_2{}^2 = v_1{}^2 + 2as,$

we calculate that $\qquad s = \dfrac{v_2{}^2}{2a} = \dfrac{(18.93 \text{ m/s})^2}{2(9.80 \text{ m/s}^2)} = 18.28 \text{ m}$

Since $v = at,$ $\qquad t = \dfrac{v_{fall}}{a} = \dfrac{18.93 \text{ m/s}}{9.80 \text{ m/s}^2} = 1.931 \text{ s}$

At this moment, the fork would appear to ring at 485 Hz to an observer just above the fork. However, it takes some additional time for the waves to reach the point of release. From the traveling wave model,

$$\Delta t = \dfrac{s}{v} = \dfrac{18.28 \text{ m}}{340 \text{ m/s}} = 0.0538 \text{ s}$$

Over the total time $t + \Delta t$, the fork falls a distance

$$s_{total} = \tfrac{1}{2}a(t + \Delta t)^2 = \tfrac{1}{2}(9.80 \text{ m/s}^2)(1.985 \text{ s})^2 = 19.3 \text{ m} \quad \Diamond$$

47. A supersonic jet traveling at Mach 3.00 at an altitude of 20 000 m is directly over a person at time $t = 0$, as in Figure P17.47. (a) How long will it be before the person encounters the shock wave? (b) Where will the plane be when it is finally heard? (Assume that the speed of sound in air is 335 m/s.)

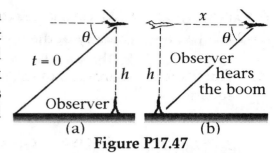

(a)   (b)

Figure P17.47

Solution   Because the shock wave proceeds at a set angle θ from the plane, we solve part (b) first.

(b)   We use $\sin\theta = \dfrac{v}{v_s} = \dfrac{\text{Mach }1}{\text{Mach }3} = \dfrac{1}{3}$,   and solve for $\theta = 19.47°$

$$x = \frac{h}{\tan\theta} = \frac{20\,000 \text{ m}}{\tan 19.47°} = 56\,570 \text{ m} = 56.6 \text{ km} \qquad \Diamond$$

(a)   It takes the plane $t = \dfrac{x}{v_s} = \dfrac{56\,570 \text{ m}}{3(335 \text{ m / s})} = 56.3$ s to travel this distance $\Diamond$

---

57. To permit measurement of her speed, a sky diver carries a buzzer emitting a steady tone at 1 800 Hz. A friend on the ground at the landing site directly below listens to the amplified sound he receives. Assume that the air is calm and that the sound speed is 343 m/s, independent of altitude. While the sky diver is falling at terminal speed, her friend on the ground receives waves of frequency 2 150 Hz. (a) What is the sky diver's speed of descent? (b) **What if?** Suppose the sky diver can hear the sound of the buzzer reflected from the ground. What frequency does she receive?

Solution

Conceptualize: Sky divers typically reach a terminal speed of about 150 mi/h (~ 75 m/s), so this sky diver should also fall near this rate. Since her friend receives a higher frequency as a result of the Doppler shift, the sky diver should detect a frequency with twice the Doppler shift, at approximately

$$f' = 1\,800 \text{ Hz} + 2(2\,150 \text{ Hz} - 1\,800 \text{ Hz}) = 2\,500 \text{ Hz}$$

Categorize: We can use the equation for the Doppler effect to answer both (a) and (b).

Analyze: Call $f_e = 1\,800$ Hz the emitted frequency; v_e, the speed of the sky diver; and $f_g = 2\,150$ Hz, the frequency of the wave crests reaching the ground.

(a) The sky diver source is moving toward the stationary ground, so we rearrange the equation

$$f_g = f_e\left(\frac{v}{v - v_e}\right) \quad \text{to give} \quad v_e = v\left(1 - \frac{f_e}{f_g}\right) = (343 \text{ m / s})\left(1 - \frac{1\,800 \text{ Hz}}{2\,150 \text{ Hz}}\right) = 55.8 \text{ m / s} \;\lozenge$$

(b) The ground now becomes a stationary source, reflecting crests with the 2 150-Hz frequency at which they reach the ground, and sending them to a moving observer:

$$f_{e2} = f_g\left(\frac{v + v_e}{v}\right) = (2\,150 \text{ Hz})\left(\frac{343 \text{ m / s} + 55.8 \text{ m / s}}{343 \text{ m / s}}\right) = 2\,500 \text{ Hz} \qquad \lozenge$$

Finalize: The answers appear to be consistent with our predictions, although the sky diver is falling somewhat slower than expected. The Doppler effect can be used to find the speed of many different types of moving objects, like raindrops (Doppler radar) and cars (police radar).

---

59. Two ships are moving along a line due east. The trailing vessel has a speed relative to a land-based observation point of 64.0 km/h, and the leading ship has a speed of 45.0 km/h relative to that point. The two ships are in a region of the ocean where the current is moving uniformly due west at 10.0 km/h. The trailing ship transmits a sonar signal at a frequency of 1 200.0 Hz. What frequency is monitored by the leading ship? (Use 1 520 m/s as the speed of sound in ocean water.)

Solution When the observer is moving in front of and in the same direction as the source,

$$f' = \left(\frac{v + (-v_o)}{v - v_s}\right)f$$

where v_o and v_s are speeds measured relative to the **medium** in which the sound is propagated. In this case the ocean current is opposite the direction of travel of the ships and

$$v_o = 45.0 \text{ km / h} - (-10.0 \text{ km / h}) = 55.0 \text{ km / h} = 15.3 \text{ m / s}$$

$$v_s = 64.0 \text{ km / h} - (-10.0 \text{ km / h}) = 74.0 \text{ km / h} = 20.6 \text{ m / s}$$

Therefore, $\quad f' = (1\,200 \text{ Hz})\left(\dfrac{1\,520 \text{ m / s} - 15.3 \text{ m / s}}{1\,520 \text{ m / s} - 20.6 \text{ m / s}}\right) = 1\,204 \text{ Hz} \qquad\qquad \lozenge$

---

65. A meteoroid the size of a truck enters the earth's atmosphere at a speed of 20.0 km/s and is not significantly slowed before entering the ocean. (a) What is the Mach angle of the shock wave from the meteoroid in the atmosphere? (Use 331 m/s as the sound speed.) (b) Assuming that the meteoroid survives the impact with the ocean surface, what is the (initial) Mach angle of the shock wave that the meteoroid produces in the water? (Use the wave speed for seawater given in Table 17.1.)

Solution

(a) The Mach angle in the air is

$$\theta_{atm} = \sin^{-1}\left(\frac{v}{v_s}\right) = \sin^{-1}\left(\frac{331 \text{ m/s}}{2.00 \times 10^4 \text{ m/s}}\right) = 0.948° \qquad \lozenge$$

(b) At impact with the ocean,

$$\theta_{ocean} = \sin^{-1}\left(\frac{v}{v_s}\right) = \sin^{-1}\left(\frac{1\,533 \text{ m/s}}{2.00 \times 10^4 \text{ m/s}}\right) = 4.40° \qquad \lozenge$$

---

67. With particular experimental methods, it is possible to produce and observe in a long thin rod both a longitudinal wave and a transverse wave whose speed depends primarily on the tension in the rod. The speed of the longitudinal wave is determined by the Young's modulus and the density of the material as $\sqrt{Y/\rho}$. The transverse wave can be modeled as a wave in a stretched string. A particular metal rod is 150 cm long and has a radius of 0.200 cm and a mass of 50.9 g. Young's modulus for the material is 6.80×10^{10} N/m². What must the tension in the rod be if the ratio of the speed of longitudinal waves to the speed of transverse waves is 8.00?

Solution The longitudinal wave and the transverse wave have respective phase speeds of

$$v_L = \sqrt{\frac{Y}{\rho}} \qquad \text{where} \qquad \rho = \frac{\text{mass}}{\text{volume}} = \frac{m}{\pi r^2 L}$$

and

$$v_T = \sqrt{\frac{T}{\mu}} \qquad \text{where} \qquad \mu = \frac{m}{L}$$

Since $v_L = 8.00 v_T$, $\quad T = \dfrac{\mu Y}{64.0\rho} = \dfrac{\pi r^2 Y}{64.0}$

Evaluating, $\qquad T = \dfrac{\pi (0.002\ 00 \text{ m})^2 (6.80 \times 10^{10} \text{ N/m}^2)}{64.0} = 1.34 \times 10^4 \text{ N} \qquad \lozenge$

---

71. Three metal rods are located relative to each other as shown in Figure P17.71, where $L_1 + L_2 = L_3$. The speed of sound in the rod is given by $v = \sqrt{Y/\rho}$, where ρ is the density and Y is Young's modulus for the rod. Values of density and Young's modulus for the three materials are

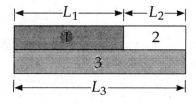

Figure P17.71

$$\rho_1 = 2.70 \times 10^3 \text{ kg/m}^3, \qquad Y_1 = 7.00 \times 10^{10} \text{ N/m}^2,$$
$$\rho_2 = 11.3 \times 10^3 \text{ kg/m}^3, \qquad Y_2 = 1.60 \times 10^{10} \text{ N/m}^2,$$
$$\rho_3 = 8.80 \times 10^3 \text{ kg/m}^3, \quad \text{and} \quad Y_3 = 11.0 \times 10^{10} \text{ N/m}^2.$$

(a) If $L_3 = 1.50$ m, what must the ratio L_1/L_2 be if a sound wave is to travel the length of rods 1 and 2 in the same time as it takes for the wave to travel the length of rod 3? (b) If the frequency of the source is 4.00 kHz, determine the phase difference between the wave traveling along rods 1 and 2 and the one traveling along rod 3.

Solution

(a) The time required for a sound pulse to travel a distance L at a speed v is given by

$$t = \frac{L}{v} = \frac{L}{\sqrt{Y/\rho}}$$

Using this expression, we find

$$t_1 = L_1 \sqrt{\frac{\rho_1}{Y_1}} = L_1 \sqrt{\frac{2.70 \times 10^3 \text{ kg}/\text{m}^3}{7.00 \times 10^{10} \text{ N}/\text{m}^2}} = L_1\left(1.96 \times 10^{-4} \text{ s}/\text{m}\right)$$

$$t_2 = (1.50 - L_1) \sqrt{\frac{11.3 \times 10^3 \text{ kg}/\text{m}^3}{1.60 \times 10^{10} \text{ N}/\text{m}^2}} = 1.26 \times 10^{-3} \text{ s} - \left(8.40 \times 10^{-4} \text{ s}/\text{m}\right)L_1$$

$$t_3 = (1.50 \text{ m}) \sqrt{\frac{8.80 \times 10^3 \text{ kg}/\text{m}^3}{11.0 \times 10^{10} \text{ N}/\text{m}^2}} = 4.24 \times 10^{-4} \text{ s}$$

We require $t_1 + t_2 = t_3$,

or $\qquad \left(1.96 \times 10^{-4} \text{ s}/\text{m}\right)L_1 + \left(1.26 \times 10^{-3} \text{ s}\right) - \left(8.40 \times 10^{-4} \text{ s}/\text{m}\right)L_1 = 4.24 \times 10^{-4} \text{ s}$

This gives $\quad L_1 = 1.30$ m $\quad$ and $\quad L_2 = (1.50 \text{ m}) - (1.30 \text{ m}) = 0.201$ m

The ratio of lengths is $\qquad L_1/L_2 = 6.45$ ◊

(b) The ratio of lengths L_1/L_2 is adjusted in part (a) so that $t_1 + t_2 = t_3$.

Therefore, sound travels the two paths in equal time intervals and the phase difference $\Delta\phi = 0$. ◊

Chapter 18

SUPERPOSITION AND STANDING WAVES

EQUATIONS AND CONCEPTS

The **resultant wave function** of two traveling sinusoidal waves having the same direction, frequency, and amplitude is also harmonic and has the same frequency and wavelength as the individual waves. The two individual waves differ in phase by ϕ.

$$y = 2A\cos\left(\frac{\phi}{2}\right)\sin\left(kx - \omega t + \frac{\phi}{2}\right)$$

The **amplitude of the resultant wave** depends on the phase difference between the two individual waves. Each of the individual waves has an amplitude equal to A.

$$\text{amplitude} = y_{\text{max}} = 2A\cos\left(\frac{\phi}{2}\right)$$

The **path difference** between two interfering waves depends on the phase angle. A path difference occurs when two coherent waves arrive at a common point having traveled along paths of unequal length. Constructive and destructive interference can occur depending on the number of half-wavelengths in the path difference.

$$\Delta r = \frac{\phi}{2\pi}\lambda \qquad (18.1)$$

$$\Delta r = (2n)\frac{\lambda}{2} \qquad \left(\begin{array}{c}\text{constructive} \\ \text{interference}\end{array}\right)$$

$$\Delta r = (2n+1)\frac{\lambda}{2} \qquad \left(\begin{array}{c}\text{destructive} \\ \text{interference}\end{array}\right) \qquad (18.2)$$

The **amplitude of motion** for any element on a string with a standing wave is a function of its position x along the string. **Antinodes** (A) are points of maximum displacement and **nodes** (N) are points of zero displacement. The distance between adjacent antinodes or between adjacent nodes is equal to $\lambda/2$; and the distance between a node and an adjacent antinodes is equal to $\lambda/4$. *The "standing" wave is not a "traveling" wave because the expression for the wave does not contain the function $(kx - \omega t)$.*

$$y = (2A\sin kx)\cos \omega t \qquad (18.3)$$

Normal modes of oscillation (a series of natural patterns of vibration), can be excited in a string of length L fixed at each end. Each mode corresponds to a quantized frequency and wavelength. *The frequencies are integral multiples of a fundamental frequency (when $n=1$) and can be expressed in terms of wave speed and string length or in terms of string tension and linear mass-density.*

$$\lambda_n = \frac{2L}{n} \qquad n = 1, 2, 3, \ldots \qquad (18.6)$$

$$f_n = n\frac{v}{2L} \qquad n = 1, 2, 3, \ldots \qquad (18.7)$$

$$f_n = \frac{n}{2L}\sqrt{\frac{T}{\mu}} \qquad n = 1, 2, 3, \ldots \qquad (18.8)$$

In an **"open" pipe** (open at both ends), the natural frequencies of oscillation form a harmonic series that includes all integral multiples of the fundamental frequency. *All harmonics are possible.*

$$f_n = n\frac{v}{2L} \qquad n = 1, 2, 3, \ldots \qquad (18.11)$$

$$\text{(open pipe)}$$

In a **"closed" pipe** (closed at one end), the natural frequencies of oscillation form a harmonic series that includes only the odd integral multiples of the fundamental frequency. *Only the odd harmonics are possible.*

$$f_n = n\frac{v}{4L} \qquad n = 1, 3, 5, \ldots \qquad (18.12)$$

$$\text{(closed pipe)}$$

Beats are formed by the superposition of two waves of equal amplitude but having slightly different frequencies.

$$y = \left[2A\cos 2\pi\left(\frac{f_1 - f_2}{2}\right)t\right]\cos 2\pi\left(\frac{f_1 + f_2}{2}\right)t$$

$$(18.13)$$

The **amplitude of the resultant wave** described by Equation 18.13 above is time dependent. *There are two maxima in each period of the resultant wave; and each occurrence of maximum amplitude results in a "beat".*

$$A_{resultant} = 2A\cos 2\pi\left(\frac{f_1 - f_2}{2}\right)t \qquad (18.14)$$

The **beat frequency** f_{beat} equals the absolute difference of the frequencies of the individual waves.

$$f_{beat} = |f_1 - f_2| \qquad (18.15)$$

A **Fourier series** is a sum of sine and cosine terms (combination of fundamental and various harmonics) that can be used to represent any non-sinusoidal periodic wave form.

$$y(t) = \sum_n \left(A_n \sin 2\pi f_n t + B_n \cos 2\pi f_n t\right)$$

$$(18.16)$$

REVIEW CHECKLIST

You should be able to:

▷ Write out the wave function which represents the superposition of two sinusoidal waves of equal amplitude and frequency traveling in opposite directions in the same medium at given time and coordinate. Determine the amplitude and frequency of the resultant wave. (Section 18.1)

▷ Calculate the phase difference between two sinusoidal waves for given values of time and position. (Section 18.1)

▷ Determine the angular frequency, maximum amplitude, and determine the values of x which correspond to nodal and antinodal points of a standing wave, given an equation for the wave function. (Section 18.2)

▷ Calculate the normal mode frequencies for a string under tension, and for open and closed air columns. (Sections 18.3 and 18.5)

▷ Describe the time dependent amplitude and calculate the expected beat frequency when two waves of slightly different frequency interfere. (Section 18.7)

ANSWERS TO SELECTED CONCEPTUAL QUESTIONS

1. Does the phenomenon of wave interference apply only to sinusoidal waves?

Answer No. Any waves moving in the same medium can interfere with each other. For example, two pulses moving in opposite directions on a stretched string interfere when they meet each other.

☐   ☐   ☐   ☐

4. When two waves interfere, can the amplitude of the resultant wave be greater than either of the two original waves? Under what conditions?

Answer They can, wherever the two waves are enough in phase that their displacements will add to create a total displacement greater than the amplitude of either of the two original waves.

When two one-dimensional waves of the same amplitude interfere, these conditions are satisfied whenever the absolute value of the phase difference between the two waves is less than 120°.

☐   ☐   ☐   ☐

6. When two waves interfere constructively or destructively, is there any gain or loss in energy? Explain.

Answer No. The energy may be transformed into other forms of energy. For example, when two pulses traveling on a stretched string in opposite directions overlap, and one is inverted, some potential energy is transferred to kinetic energy when they overlap. In fact, if they have the same shape except that one is inverted, they completely cancel each other at one instant in time. In this case, all of the energy is transverse kinetic energy when the resultant amplitude is zero.

☐   ☐   ☐   ☐

17. An airplane mechanic notices that the sound from a twin-engine aircraft rapidly varies in loudness when both engines are running. What could be causing this variation from loud to soft?

Answer Apparently the two engines are emitting sounds having frequencies which differ only by a very small amount from each other. This results in a beat frequency, causing the variation from loud to soft, and back again.

☐   ☐   ☐   ☐

SOLUTIONS TO SELECTED END-OF-CHAPTER PROBLEMS

3. Two pulses traveling on the same string are described by

$$y_1 = \frac{5}{(3x - 4t)^2 + 2} \qquad \text{and} \qquad y_2 = \frac{-5}{(3x + 4t - 6)^2 + 2}$$

(a) In which direction does each pulse travel? (b) At what time do the two cancel everywhere? (c) At what point do the two pulses always cancel?

Solution

(a) At constant phase, $\phi = 3x - 4t$, or $x = \dfrac{\phi + 4t}{3}$

As t increases, x increases, so the first wave moves to the right. ◊

In the same way, in the second case $x = \dfrac{\phi - 4t + 6}{3}$

As t increases, x must decrease, so the second wave moves to the left. ◊

(b) We require that $y_1 + y_2 = 0$: $\dfrac{5}{(3x - 4t)^2 + 2} + \dfrac{-5}{(3x + 4t - 6)^2 + 2} = 0$

This can be written as $(3x - 4t)^2 = (3x + 4t - 6)^2$

Solving for the positive root, $t = 0.750 \text{ s}$ ◊

(c) The negative root yields $(3x - 4t) = -(3x + 4t - 6)$

The time terms cancel, leaving $x = 1.00 \text{ m}$

At this point, the waves **always** cancel.

---

5. Two traveling sinusoidal waves are described by the wave functions

$$y_1 = (5.00 \text{ m})\sin[\pi(4.00x - 1200t)]$$

and $$y_2 = (5.00 \text{ m})\sin[\pi(4.00x - 1200t - 0.250)]$$

where x, y_1, and y_2 are in meters and t is in seconds. (a) What is the amplitude of the resultant wave? (b) What is the frequency of the resultant wave?

Solution   We can represent the waves symbolically as

$$y_1 = A_0 \sin(kx - \omega t) \quad \text{and} \quad y_2 = A_0 \sin(kx - \omega t - \phi)$$

with $A_0 = 5.00$ m, $\omega = 1200\pi\,\text{s}^{-1}$,   and $\phi = 0.250\pi$

According to the principle of superposition, the resultant wave function has the form

$$y = y_1 + y_2 = 2A_0 \cos\left(\frac{\phi}{2}\right) \sin\left(kx - \omega t - \frac{\phi}{2}\right)$$

(a)   with amplitude $A = 2A_0 \cos\left(\dfrac{\phi}{2}\right) = 2(5.00)\cos\left(\dfrac{\pi}{8.00}\right) = 9.24$ m   ◊

(b)   and frequency $f = \dfrac{\omega}{2\pi} = \dfrac{1200\pi}{2\pi} = 600$ Hz   ◊

---

11.   Two sinusoidal waves in a string are defined by the functions

$$y_1 = (2.00\text{ cm})\sin(20.0x - 32.0t) \quad \text{and} \quad y_2 = (2.00\text{ cm})\sin(25.0x - 40.0t)$$

where y_1, y_2, and x are in centimeters and t is in seconds.   (a) What is the phase difference between these two waves at the point $x = 5.00$ cm at $t = 2.00$ s?   (b) What is the positive x value closest to the origin for which the two phases differ by $\pm\pi$ at $t = 2.00$ s? (This is where the two waves add to zero.)

Solution

At any time and place, the phase shift between the waves is found by subtracting the phase of the two waves, $\Delta\phi = \phi_1 - \phi_2$:

$$\Delta\phi = (20.0\text{ rad / cm})x - (32.0\text{ rad / s})t - \left[(25.0\text{ rad / cm})x - (40.0\text{ rad / s})t\right]$$

Collecting terms, $\Delta\phi = -(5.00\text{ rad / cm})x + (8.00\text{ rad / s})t$

(a)   At $x = 5.00$ cm and $t = 2.00$ s,

$$\Delta\phi = (-5.00\text{ rad / cm})(5.00\text{ cm}) + (8.00\text{ rad / s})(2.00\text{ s})$$

$$|\Delta\phi| = 9.00\text{ rad} = 516° = 156°$$ ◊

(b) The phase shift equals $\pm\pi$ whenever $\Delta\phi = \pi + 2n\pi$, for all integer values of n. Substituting this into the phase equation,

$$\pi + 2n\pi = -(5.00 \text{ rad / cm})x + (8.00 \text{ rad / s})t$$

At $t = 2.00$ s, $\qquad \pi + 2n\pi = -(5.00 \text{ rad / cm})x + (8.00 \text{ rad / s})(2.00 \text{ s})$

or $\qquad\qquad (5.00 \text{ rad / cm})x = (16.0 - \pi - 2n\pi) \text{ rad}$

The smallest positive value of x is found when $n = 2$:

$$x = \frac{(16.0 - 5\pi) \text{ rad}}{5.00 \text{ rad / cm}} = 0.058\ 4 \text{ cm} \qquad\qquad \lozenge$$

---

15. Two speakers are driven in phase by a common oscillator at 800 Hz and face each other at a distance of 1.25 m. Locate the points along a line joining the two speakers where relative minima of sound pressure amplitude would be expected. (Use $v = 343$ m/s.)

$\longleftarrow x \text{ m} \longrightarrow\!\!\longleftarrow (1.25 - x) \text{ m} \longrightarrow$

Solution

The wavelength is: $\qquad\qquad \lambda = \dfrac{v}{f} = \dfrac{343 \text{ m / s}}{800 \text{ Hz}} = 0.429 \text{ m}$

The two waves moving in opposite directions along the line between the two speakers will add to produce a standing wave with this distance between nodes:

$$\text{distance N to N} = \lambda/2 = 0.214 \text{ m}$$

Because the speakers vibrate in phase, air compressions from each will simultaneously reach the point halfway between the speakers, to produce an antinode of pressure here. A node of pressure will be located at this distance on either side of the midpoint:

$$\text{distance N to A} = \lambda/4 = 0.107 \text{ m}$$

Therefore nodes of sound pressure will appear at these distances from either speaker:

$$\tfrac{1}{2}(1.25 \text{ m}) + 0.107 \text{ m} = 0.732 \text{ m} \qquad\qquad \tfrac{1}{2}(1.25 \text{ m}) - 0.107 \text{ m} = 0.518 \text{ m}$$

The standing wave contains a chain of equally-spaced nodes at distances from either speaker of

$$0.732 \text{ m} + 0.214 \text{ m} = 0.947 \text{ m},$$

$$0.947 \text{ m} + 0.214 \text{ m} = 1.16 \text{ m},$$

and also at

$$0.518 \text{ m} - 0.214 \text{ m} = 0.303 \text{ m},$$

$$0.303 \text{ m} - 0.214 \text{ m} = 0.0891 \text{ m}$$

The standing wave exists only along the line segment between the speakers. No nodes or antinodes appear at distances greater than 1.25 m or less than 0, because waves add to give a standing wave only if they are traveling in opposite directions and not in the same direction. In order, the distances from either speaker to the nodes of pressure between the speakers are at 0.0891 m, 0.303 m, 0.518 m, 0.732 m, 0.947 m, and 1.16 m. ◊

17. Two sinusoidal waves combining in a medium are described by the wave functions

$$y_1 = (3.0 \text{ cm})\sin \pi(x + 0.60t) \quad \text{and} \quad y_2 = (3.0 \text{ cm})\sin \pi(x - 0.60t)$$

where x is in centimeters and t is in seconds. Determine the **maximum** transverse position of an element of the medium at (a) $x = 0.250$ cm, (b) $x = 0.500$ cm, and (c) $x = 1.50$ cm. (d) Find the three smallest values of x corresponding to antinodes.

Solution According to the waves in interference model, we add y_1 and y_2 using the trigonometry identity

$$\sin(\alpha + \beta) = \sin \alpha \cos \beta + \cos \alpha \sin \beta$$

We get

$$y = y_1 + y_2 = (6.0 \text{ cm})\sin(\pi x)\cos(0.60\pi t)$$

Since $\cos(0) = 1$, we can find the maximum value of y by setting $t = 0$:

$$y_{max}(x) = y_1 + y_2 = (6.0 \text{ cm})\sin(\pi x)$$

(a) At $x = 0.250$ cm, $y_{max} = (6.0 \text{ cm})\sin(0.250\pi) = 4.24$ cm   ◊

(b) At $x = 0.500$ cm, $y_{max} = (6.0 \text{ cm})\sin(0.500\pi) = 6.00$ cm   ◊

(c) At $x = 1.50$ cm, $y_{max} = |(6.0 \text{ cm})\sin(1.50\pi)| = +6.00$ cm   ◊

(d)   The antinodes occur when $x = n\lambda/4$ $(n = 1, 3, 5, \ldots)$

   But $k = 2\pi/\lambda = \pi$, so $\lambda = 2.00$ cm

   and $x_1 = \lambda/4 = (2.00 \text{ cm})/4 = 0.500$ cm   ◊

                          $x_2 = 3\lambda/4 = 3(2.00 \text{ cm})/4 = 1.50$ cm   ◊

                          $x_3 = 5\lambda/4 = 5(2.00 \text{ cm})/4 = 2.50$ cm   ◊

---

19.   Find the fundamental frequency and the next three frequencies that could cause standing-wave patterns on a string that is 30.0 m long, has a mass per length of 9.00×10^{-3} kg/m, and is stretched to a tension of 20.0 N.

Solution

Conceptualize: The string described in the problem is very long, loose, and somewhat heavy, so it should have a very low fundamental frequency, maybe only a few vibrations per second.

Categorize: The tension and linear density of the string can be used to find the wave speed, which can then be used along with the required wavelength to find the fundamental frequency.

Analyze: The wave speed is

$$v = \sqrt{\frac{T}{\mu}} = \sqrt{\frac{20.0 \text{ N}}{9.00 \times 10^{-3} \text{ kg / m}}} = 47.1 \text{ m / s}$$

For a vibrating string of length L fixed at both ends, the wavelength of the fundamental is $\lambda = 2L = 60.0$ m ; and the frequency is

$$f_1 = \frac{v}{\lambda} = \frac{v}{2L} = \frac{47.1 \text{ m / s}}{60.0 \text{ m}} = 0.786 \text{ Hz}$$

The next three harmonics are

$$f_2 = 2f_1 = 1.57 \text{ Hz}, \quad f_3 = 3f_1 = 2.36 \text{ Hz}, \quad \text{and} \quad f_4 = 4f_1 = 3.14 \text{ Hz} \qquad ◊$$

Finalize: The fundamental frequency is even lower than expected, less than 1 Hz. You could watch the string vibrating.   It would weakly broadcast sound into the surrounding air, but all four of the lowest resonant frequencies are below the normal human hearing range (20 to 17 000 Hz), so the sounds are not even audible.

27. A cello A-string vibrates in its first normal mode with a frequency of 220 Hz. The vibrating segment is 70.0 cm long and has a mass of 1.20 g. (a) Find the tension in the string. (b) Determine the frequency of vibration when the string vibrates in three segments.

Solution

Conceptualize: The tension should be less than 500 N (~100 lb) since excessive force on the four cello strings would break the neck of the instrument. If a string vibrates in three segments, there will be three antinodes (instead of one for the fundamental mode), so the frequency should be three times greater than the fundamental.

Categorize: From the string's length, we can find the wavelength. We then can use the wavelength with the fundamental frequency to find the wave speed. Finally, we can find the tension from the wave speed and the linear mass density of the string.

Analyze: When the string vibrates in the lowest frequency mode, the length of string forms a standing wave where $L = \lambda/2$ so the fundamental harmonic wavelength is

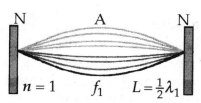

$$\lambda = 2L = 2(0.700 \text{ m}) = 1.40 \text{ m}$$

and the speed is

$$v = f\lambda = (220 \text{ s}^{-1})(1.40 \text{ m}) = 308 \text{ m / s}$$

(a) From the tension equation

$$v = \sqrt{\frac{T}{\mu}} = \sqrt{\frac{T}{m / L}}$$

We get $T = v^2 m / L$, or

$$T = \frac{(308 \text{ m / s})^2 (1.20 \times 10^{-3} \text{ kg})}{0.700 \text{ m}} = 163 \text{ N} \qquad \lozenge$$

(b) For the third harmonic, the tension, linear density, and speed are the same, but the string vibrates in three segments. Thus, that the wavelength is one third as long as in the fundamental.

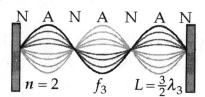

$$\lambda_3 = \lambda / 3$$

From the equation $v = f\lambda$, we find the frequency is three times as high:

$$f_3 = \frac{v}{\lambda_3} = 3\frac{v}{\lambda} = 3f = 660 \text{ Hz} \qquad \lozenge$$

Finalize: The tension seems reasonable, and the third harmonic is three times the fundamental frequency as expected. Related to part (b), some stringed instrument players use a technique to double the frequency of a note by "cutting" a vibrating string in half. When the string is lightly touched at its midpoint to form a node, the second harmonic is formed, and the resulting note is one octave higher (twice the original fundamental frequency).

37. Calculate the length of a pipe that has a fundamental frequency of 240 Hz if the pipe is (a) closed at one end and (b) open at both ends.

Solution

The relationship between the frequency and the wavelength of a sound wave is

$$v = f\lambda \qquad \text{and} \qquad \lambda = v / f$$

Next, we draw the pipes to help us visualize the relationship between λ and L.

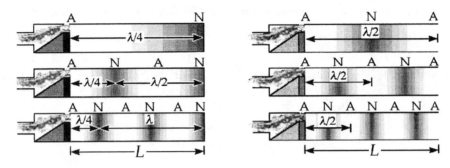

(a) For the fundamental mode in a closed pipe, $\lambda = 4L$

so $\qquad L = \dfrac{\lambda}{4} = \dfrac{v/f}{4} \qquad$ and $\qquad L = \dfrac{(343 \text{ m / s})/(240 \text{ s}^{-1})}{4} = 0.357 \text{ m} \qquad \Diamond$

(b) For the fundamental mode in an open pipe, $\lambda = 2L$

so $\qquad L = \dfrac{\lambda}{2} = \dfrac{v/f}{2} \qquad$ and $\qquad L = \dfrac{(343 \text{ m / s})/(240 \text{ s}^{-1})}{2} = 0.715 \text{ m} \qquad \Diamond$

41. A shower stall measures 86.0 cm $\times$ 86.0 cm $\times$ 210 cm. If you were singing in this shower, which frequencies would sound the richest (because of resonance)? Assume that the stall acts as a pipe closed at both ends, with nodes at opposite sides. Assume that the voices of various singers range from 130 Hz to 2 000 Hz. Let the speed of sound in the hot shower stall be 355 m/s.

Solution

For a closed box, the resonant frequencies will have nodes at both sides, so the permitted wavelengths will be defined by

$$L = \frac{n\lambda}{2} = \frac{nv}{2f}, \quad \text{with} \quad n = 1, 2, 3, \ldots$$

Rearranging, and substituting $L = 0.860$ m, the side-to-side resonant frequencies are

$$f_n = n\frac{v}{2L} = n\frac{355 \text{ m / s}}{2(0.860 \text{ m})} = n(206 \text{ Hz}), \quad \text{for each } n \text{ from 1 to 9} \qquad \Diamond$$

With $\qquad L' = 2.10$ m, the top-to-bottom resonance frequencies are

$$f_n = n\frac{355 \text{ m / s}}{2(2.10 \text{ m})} = n(84.5 \text{ Hz}), \quad \text{for each } n \text{ from 2 to 23} \qquad \Diamond$$

---

43. If two adjacent natural frequencies of an organ pipe are determined to be 550 Hz and 650 Hz, calculate the fundamental frequency and length of this pipe. (Use $v = 340$ m/s.)

Solution

We use the wave under boundary conditions model.

Because harmonic frequencies are given by $f_1 n$ for open pipes, and $f_1(2n-1)$ for closed pipes, the difference between all adjacent harmonics is constant. Therefore, we can find each harmonic below 650 Hz by subtracting

$$\Delta f_{\text{Harmonic}} = (650 \text{ Hz} - 550 \text{ Hz}) = 100 \text{ Hz} \quad \text{from the previous value.}$$

The harmonics are:

$$\{650 \text{ Hz, } 550 \text{ Hz, } 450 \text{ Hz, } 350 \text{ Hz, } 250 \text{ Hz, } 150 \text{ Hz, and } 50 \text{ Hz}\}$$

(a) The fundamental frequency, then, is 50 Hz. $\qquad \Diamond$

(b) The wavelength of the fundamental frequency can be calculated from the velocity:

$$\lambda = \frac{v}{f} = \frac{340 \text{ m / s}}{50.0 \text{ Hz}} = 6.80 \text{ m}$$

Because the step size Δf is twice the fundamental frequency, we know the pipe is closed, with an antinode at the open end, and a node at the closed end. The wavelength in this situation is four times the pipe length, so

$$L = \frac{\lambda}{4} = 1.70 \text{ m} \qquad \Diamond$$

---

45. An air column in a glass tube is open at one end and closed at the other by a movable piston. The air in the tube is warmed above room temperature, and a 384-Hz tuning fork is held at the open end. Resonance is heard when the piston is 22.8 cm from the open end and again when it is 68.3 cm from the open end. (a) What speed of sound is implied by these data? (b) How far from the open end will the piston be when the next resonance is heard?

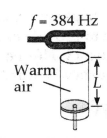

$f = 384$ Hz

Warm air

Solution   For an air column closed at one end, resonances will occur when the length of the column is equal to $\lambda/4, 3\lambda/4, 5\lambda/4$, and so on. Thus, the change in the length of the pipe from one resonance to the next is $\lambda/2$. In this case,

$$\lambda/2 = (0.683 \text{ m} - 0.228 \text{ m}) = 0.455 \text{ m} \quad \text{and} \quad \lambda = 0.910 \text{ m}$$

(a) $v = f\lambda = (384 \text{ Hz})(0.910 \text{ m}) = 349 \text{ m / s}$ ◊

(b) $L = 0.683 \text{ m} + 0.455 \text{ m} = 1.14 \text{ m}$ ◊

---

49. An aluminum rod 1.60 m long is held at its center. It is stroked with a rosin-coated cloth to set up a longitudinal vibration. The speed of sound in a thin rod of aluminum is 5 100 m/s. (a) What is the fundamental frequency of the waves established in the rod? (b) What harmonics are set up in the rod held in this manner? (c) **What if?** What would be the fundamental frequency if the rod were copper, in which the speed of sound is 3 560 m/s?

Solution

(a)   Since the central clamp establishes a node at the center, the fundamental mode of vibration will be ANA. Our first harmonic frequency, then, is

$$f_1 = \frac{v}{\lambda_1} = \frac{v}{2L} = \frac{5\,100 \text{ m / s}}{3.20 \text{ m}} = 1.59 \text{ kHz}$$ ◊

(b)   Since the rod is free at each end, the ends will be antinodes. The next vibration state will be ANANANA, as shown, with a wavelength and frequency of:

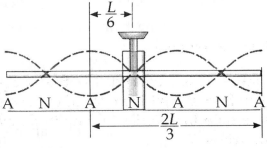

$$\lambda = \frac{2L}{3} \quad \text{and} \quad f = \frac{v}{\lambda} = \frac{3v}{2L} = 3f_1$$

Since $f = 2f_1$ was bypassed, we know that we get odd harmonics.

That is, $$f = \frac{nv}{2L} \quad \text{for} \quad n = 1, 3, 5, \ldots$$ ◊

(c)   For a copper rod, the speed of sound changes: $f_1 = \dfrac{v}{2L} = \dfrac{3\,560 \text{ m / s}}{3.20 \text{ m}} = 1.11 \text{ kHz}$ ◊

---

51. In certain ranges of a piano keyboard, more than one string is tuned to the same note to provide extra loudness. For example, the note at 110 Hz has two strings at this frequency. If one string slips from its normal tension of 600 N to 540 N, what beat frequency is heard when the hammer strikes the two strings simultaneously?

Solution

Conceptualize: Directly noticeable beat frequencies are usually only a few Hertz, so we should not expect a frequency much greater than this.

Categorize: As in previous problems, the two wave speed equations can be used together to find the frequency of vibration that corresponds to a certain tension. The beat frequency is then just the difference in the two resulting frequencies from the two strings with different tensions.

Analyze:

Combining the velocity and the tension equations $v = f\lambda$ and $v = \sqrt{T/\mu}$,

we find that
$$f = \sqrt{\frac{T}{\mu\lambda^2}}$$

Since μ and λ are constant, we can apply that equation to both frequencies, and then divide the two equations to get

$$\frac{f_1}{f_2} = \sqrt{\frac{T_1}{T_2}}$$

With $f_1 = 110$ Hz, $T_1 = 600$ N, and $T_2 = 540$ N:

$$f_2 = (110 \text{ Hz})\sqrt{\frac{540 \text{ N}}{600 \text{ N}}} = 104.4 \text{ Hz}$$

The beat frequency is:
$$f_b = |f_1 - f_2| = 110 \text{ Hz} - 104.4 \text{ Hz} = 5.64 \text{ Hz} \quad \lozenge$$

Finalize: As expected, the beat frequency is only a few cycles per second. This result from the interference of the two sound waves with slightly different frequencies has a tone that varies in amplitude over time, similar to the sound made by saying "wa-wa-wa..."

Note: The beat frequency above is written with three significant figures on the assumption that the original data is known precisely enough to warrant them. This assumption implies that the original frequency is known somewhat more precisely than to the three significant digits quoted in "110 Hz." For example, if the original frequency of the strings were 109.6 Hz, the beat frequency would be 5.62 Hz.

53. A student holds a tuning fork oscillating at 256 Hz. He walks toward a wall at a constant speed of 1.33 m/s. (a) What beat frequency does he observe between the tuning fork and its echo? (b) How fast must he walk away from the wall to observe a beat frequency of 5.00 Hz?

Solution

For an echo, $$f_1 = f_2 \frac{v + v_s}{v - v_s}$$

and the beat frequency is $f_b = |f_1 - f_2|$

Solving for f_b gives $f_b = f_2 \frac{2v_s}{v - v_s}$ when approaching the wall.

(a) $f_b = (256\ \text{Hz}) \dfrac{2(1.33\ \text{m / s})}{343\ \text{m / s} - 1.33\ \text{m / s}} = 1.99\ \text{Hz}$ ◊

(b) When moving away from wall, v_s changes sign. Solving for v_s gives

$$v_s = f_b \frac{v}{2f_2 - f_b} = (5.00\ \text{Hz}) \frac{343\ \text{m / s}}{2(256\ \text{Hz}) - 5.00\ \text{Hz}} = 3.38\ \text{m / s}$$ ◊

---

59. Two train whistles have identical frequencies of 180 Hz. When one train is at rest in the station and the other is moving nearby, a commuter standing on the station platform hears beats with a frequency of 2.00 beats/s when the whistles sound at the same time. What are the two possible speeds and directions that the moving train can have?

Solution

We know from our beat equation $f_b = |f_1 - f_2|$ that the moving train can have apparent frequencies of $f' = 182\ \text{Hz}$ or $f' = 178\ \text{Hz}$. If we assume the train is moving away from the station, the apparent frequency of the moving train is lower (178 Hz):

$$f' = f \frac{v}{v + v_s}$$

and the train is **moving away** at
$$v_s = v\left(\frac{f}{f'} - 1\right) = (343 \text{ m / s})\left(\frac{180 \text{ Hz}}{178 \text{ Hz}} - 1\right)$$

$$v_s = 3.85 \text{ m / s} \qquad \Diamond$$

If the train is pulling into the station, then the apparent frequency is 182 Hz.

Again from the Doppler shift, $\qquad f' = f\dfrac{v}{v - v_s}$

The train is **approaching** at: $\qquad v_s = v\left(1 - \dfrac{f}{f'}\right) = (343 \text{ m / s})\left(1 - \dfrac{180 \text{ Hz}}{182 \text{ Hz}}\right)$

$$v_s = 3.77 \text{ m / s} \qquad \Diamond$$

---

63. Two wires are welded together end to end. The wires are made of the same material, but the diameter of one is twice that of the other. They are subjected to a tension of 4.60 N. The thin wire has a length of 40.0 cm and a linear mass density of 2.00 g/m. The combination is fixed at both ends and vibrated in such a way that two antinodes are present, with the node between them being right at the weld. (a) What is the frequency of vibration? (b) How long is the thick wire?

Solution

(a) Since the first node is at the weld, the wavelength in the thin wire is 2L or 80.0 cm. The frequency and tension are the same in both sections, so

$$f = \frac{1}{2L}\sqrt{\frac{T}{\mu}} = \frac{1}{2(0.400 \text{ m})}\sqrt{\frac{4.60 \text{ N}}{0.002\,00 \text{ kg / m}}} = 59.9 \text{ Hz} \qquad \Diamond$$

(b) Since the thick wire is twice the diameter, it will have 4 times the cross-sectional area, and a linear density μ that is 4 times that of the thin wire.

$$\mu' = 4(2.00 \text{ g / m}) = 0.008\,00 \text{ kg / m}$$

L' varies accordingly: $\quad L' = \dfrac{1}{2f}\sqrt{\dfrac{T}{\mu'}} = \dfrac{1}{2(59.9 \text{ Hz})}\sqrt{\dfrac{4.60 \text{ N}}{0.008\,00 \text{ kg / m}}} = 20.0 \text{ cm} \qquad \Diamond$

Note that the thick wire is half the length of the thin wire.

---

65. A standing wave is set up in a string of variable length and tension by a vibrator of variable frequency. Both ends of the string are fixed. When the vibrator has a frequency f, in a string of length L and under tension T, n antinodes are set up in the string. (a) If the length of the string is doubled, by what factor should the frequency be changed so that the same number of antinodes is produced? (b) If the frequency and length are held constant, what tension will produce $n+1$ antinodes? (c) If the frequency is tripled and the length of the string is halved, by what factor should the tension be changed so that twice as many antinodes are produced?

Solution

Combining the equations $v = f\lambda$ and $v = \sqrt{T/\mu}$ and noting that $\lambda_n = 2L/n$,

(a)   we find that
$$f_n = \frac{n}{2L}\sqrt{\frac{T}{\mu}} \qquad\qquad [1]$$

Keeping n, T, and μ constant, we can create two equations:

$$f_n L = \frac{n}{2}\sqrt{\frac{T}{\mu}} \quad \text{and} \quad f_n' L' = \frac{n}{2}\sqrt{\frac{T}{\mu}}$$

Dividing the equations,
$$\frac{f_n}{f_n'} = \frac{L'}{L}$$

If $L' = 2L$, then $f_n' = \frac{1}{2}f_n$

Therefore, in order to double the length but keep the same number of antinodes, the frequency should be halved.   ◊

(b)   From the same Equation [1], we can hold L and f_n constant to get

$$\frac{n'}{n} = \sqrt{\frac{T}{T'}}$$

From this relation, we see that the tension must be decreased to

$$T' = T\left(\frac{n}{n+1}\right)^2 \quad \text{to produce } n+1 \text{ antinodes.}\qquad ◊$$

(c)   The time, we rearrange Equation [1] to produce

$$\frac{2f_n L}{n} = \sqrt{\frac{T}{\mu}} \quad \text{and} \quad \frac{2f_n' L'}{n'} = \sqrt{\frac{T'}{\mu}}$$

$$\frac{T'}{T} = \left(\frac{f_n'}{f_n}\cdot\frac{n}{n'}\cdot\frac{L'}{L}\right)^2 = \left(\frac{3f_n}{f_n}\right)^2\left(\frac{n}{2n}\right)^2\left(\frac{L/2}{L}\right)^2 = \frac{9}{16}\qquad ◊$$

Chapter 19
TEMPERATURE

EQUATIONS AND CONCEPTS

The **Celsius temperature** T_C is related to the absolute temperature T (in kelvins, K) according to Equation 19.1, for which 0 °C corresponds to 273.15 K. *The size of a degree (unit change in temperature) on the Celsius scale equals the size of a degree on the absolute scale.*

$$T_C = T - 273.15 \tag{19.1}$$

The **Fahrenheit temperature** T_F can be converted to degrees Celsius using Equation 19.2. Note that 0 °C = 32 °F and 100 °C = 212 °F.

$$T_F = \frac{9}{5}T_C + 32 \text{ °F} \tag{19.2}$$

The **change in length** ΔL of a solid, due to a change in temperature, is proportional to the change in temperature and the initial length of the object. *The proportionality constant α is called the average coefficient of linear expansion and equals the factional change in length per degree change in temperature.*

$$\Delta L = \alpha L_i \Delta T \tag{19.4}$$

or

$$\alpha = \frac{1}{L_i}\frac{\Delta L}{\Delta T}$$

The **change in volume** of a solid at constant pressure is proportional to ΔT and to the original volume. *The constant of proportionality β is the average coefficient of volume expansion. For an isotropic solid, $\beta \approx 3\alpha$.*

$$\Delta V = \beta V_i \Delta T \tag{19.6}$$

The **number of moles** in a sample of any substance equals the ratio of the mass of the sample to the molar mass characteristic of that particular substance. *One mole of any substance is that mass which contains Avogadro's number (N_A) of molecules.*

$$n = \frac{m}{M} \tag{19.7}$$

The **equation of state of an ideal gas** can be expressed in terms of number of moles n and the universal gas constant R (as in Equation 19.8) or in terms of the number of molecules N and Boltzmann's constant k_B (as in Equation 19.10).

$$PV = nRT \tag{19.8}$$

or

$$PV = Nk_BT \tag{19.10}$$

$$R = 8.314 \text{ J / mol} \cdot \text{K} \tag{19.9}$$

$$R = 0.082\,14 \text{ L} \cdot \text{atm / mol} \cdot \text{K}$$

$$k_B = \frac{R}{N_A} = 1.38 \times 10^{-23} \text{ J / K} \tag{19.11}$$

Avagadro's number is the number of particles in one mole of a substance.

$$N_A = 6.022 \times 10^{23} \text{ atoms / mole}$$

REVIEW CHECKLIST

You should be able to:

▷ Describe the operation of the constant-volume gas thermometer and the way in which it was used to define the absolute gas temperature scale. (Section 19.3)

▷ Convert among the various temperature scales: from Celsius to Kelvin, Fahrenheit to Kelvin, and Celsius to Fahrenheit. (Section 19.3)

▷ Define the linear expansion coefficient and volume expansion coefficient for an isotropic solid; and make calculations of changes in length and volume. (Section 19.4)

▷ Make calculations using the ideal gas equation, expressed either in terms of n and R or in terms of N and k_B. (Section 19.5)

ANSWERS TO SELECTED CONCEPTUAL QUESTIONS

2. A piece of copper is dropped into a beaker of water. If the water's temperature rises, what happens to the temperature of the copper? Under what conditions are the water and copper in thermal equilibrium?

Answer If the water's temperature increases, that means that energy is being transferred by heat to the water. This can happen either if the copper changes phase from liquid to solid, or if the temperature of the copper is above that of the water, and falling.

In this case, the copper is referred to as a "piece" of copper, so it is already in the solid phase; therefore, its temperature must be falling. When the temperature of the copper reaches that of the water, the copper and water will reach equilibrium, and the subsequent net energy transfer by heat between the two will be zero.

□ □ □ □

18. When the metal ring and metal sphere in Figure Q19.18 are both at room temperature, the sphere can just be passed through the ring. After the sphere is heated, it cannot be passed through the ring. Explain. **What if?** What if the ring is heated and the sphere is left at room temperature? Does the sphere pass through the ring?

Answer The hot sphere has expanded, so it no longer fits through the ring. When the ring is heated, the cool sphere fits through more easily.

Figure Q19.18

Suppose a cool sphere is put through a cool ring and then heated so that it does not come back out. With the sphere still hot, you can separate the sphere and ring by heating the ring as shown in the figure. This more surprising result occurs because the thermal expansion of the ring is not like the inflation of a blood-pressure cuff. Rather, it is like a photographic enlargement; every linear dimension, including the hole diameter, increases by the same factor. The reason for this is that the atoms everywhere, including those around the inner circumference, push away from each other. The only way that the atoms can accommodate the greater distances is for the circumference—and corresponding diameter—to grow. This property was once used to fit metal rims to wooden wagon and horse-buggy wheels.

□ □ □ □

SOLUTIONS TO SELECTED END-OF-CHAPTER PROBLEMS

1. A constant-volume gas thermometer is calibrated in dry ice (that is, carbon dioxide in the solid state, which has a temperature of –80.0 °C) and in boiling ethyl alcohol (78.0 °C). The two pressures are 0.900 atm and 1.635 atm. (a) What Celsius value of absolute zero does the calibration yield? What is the pressure at (b) the freezing point of water and (c) the boiling point of water?

Solution

Since we have a linear graph, the pressure is related to the temperature as $P = A + BT$, where A and B are constants.

To find A and B, we use the given data: $0.900 \text{ atm} = A + (-80.0 \text{ °C})B$

and $\qquad\qquad\qquad\qquad\qquad\qquad\qquad 1.635 \text{ atm} = A + (78.0 \text{ °C})B$

Solving these simultaneously, we find $\quad A = 1.272 \text{ atm}$

and $\qquad\qquad\qquad\qquad\qquad\qquad\qquad B = 4.652 \times 10^{-3} \text{ atm / °C}$

Therefore, $\qquad\qquad\qquad\qquad\qquad P = 1.272 \text{ atm} + \left(4.652 \times 10^{-3} \text{ atm / °C}\right)T$

(a) At absolute zero, $\qquad\qquad\qquad P = 0 = 1.272 \text{ atm} + \left(4.652 \times 10^{-3} \text{ atm / °C}\right)T$

which gives $\qquad\qquad\qquad\qquad T = -273 \text{ °C}$ ◊

(b) At the freezing point of water, $\quad P = 1.272 \text{ atm} + 0 = 1.27 \text{ atm}$ ◊

(c) And at the boiling point,

$\qquad P = 1.272 \text{ atm} + \left(4.652 \times 10^{-3} \text{ atm / °C}\right)(100 \text{ °C}) = 1.74 \text{ atm}$ ◊

---

3. Liquid nitrogen has a boiling point of –195.81 °C at atmospheric pressure. Express this temperature (a) in degrees Fahrenheit and (b) in kelvins.

Solution

(a) By Equation 19.2, $\qquad T_F = \frac{9}{5}T_C + 32 \text{ °F} = \frac{9}{5}(-195.81) + 32 = -320 \text{ °F}$ ◊

(b) Applying Equation 19.1, $\quad 273.15 \text{ K} - 195.81 \text{ K} = 77.3 \text{ K}$ ◊

Related A convenient way to remember Equations 19.1 and 19.2 is to remember the
Comment: freezing and boiling points of water, in each form:

$$T_{freeze} = 32.0 \ ^\circ F = 0 \ ^\circ C = 273.15 \ K$$

$$T_{boil} = 212 \ ^\circ F = 100 \ ^\circ C$$

To convert from Fahrenheit to Celsius, subtract 32 (the freezing point), and then adjust the scale by the liquid range of the water.

$$\text{Scale} = \frac{(100-0) \ ^\circ C}{(212-32) \ ^\circ F} = \frac{5 \ ^\circ C}{9 \ ^\circ F}$$

A kelvin is the same size change as a degree Celsius, but the kelvin scale takes its zero point at **absolute zero**, instead of the freezing point of water. Therefore, to convert from kelvin to Celsius, subtract 273.15 K.

---

9. A copper telephone wire has essentially no sag between poles 35.0 m apart on a winter day when the temperature is –20.0 °C. How much longer is the wire on a summer day when $T_C = 35.0 \ ^\circ C$?

Solution

Conceptualize: Normally, we do not notice a change in the length of the telephone wires. Thus, we might expect the wire to expand by less than a meter.

Categorize: The change in length can be found from the linear expansion of copper wire (we will assume that the insulation around the copper wire can stretch more easily than the wire itself). From Table 19.2, the coefficient of linear expansion for copper is $17.0 \times 10^{-6} \ (^\circ C)^{-1}$.

Analyze: The change in length between cold and hot conditions is

$$\Delta L = \alpha L_i \Delta T = \left[17.0 \times 10^{-6} \ (^\circ C)^{-1}\right](35.0 \ m)(35.0 \ ^\circ C - (-20.0 \ ^\circ C))$$

$$\Delta L = 3.27 \times 10^{-2} \ m = 3.27 \ cm \qquad \qquad \lozenge$$

Finalize: This expansion agrees with our expectation that the change in length is less than a meter. From ΔL we can find that if the wire sags, its midpoint can be displaced downward by 0.757 m on the hot summer day. This also seems reasonable based on everyday observations.

---

13. The active element of a certain laser is made of a glass rod 30.0 cm long by 1.50 cm in diameter. If the temperature of the rod increases by 65.0 °C, what is the increase in (a) its length, (b) its diameter, and (c) its volume? Assume that the average coefficient of linear expansion of the glass is 9.00×10^{-6} °C^{-1}.

Solution

(a) $\Delta L = \alpha L_i \Delta T = (9.00 \times 10^{-6}$ °C$^{-1})(0.300$ m$)(65.0$ °C$) = 1.76 \times 10^{-4}$ m   ◊

(b)   The diameter is a linear dimension, so the same equation applies:

$$\Delta D = \alpha D_i \Delta T = (9.00 \times 10^{-6}$ °C$^{-1})(0.0150$ m$)(65.0$ °C$) = 8.78 \times 10^{-6}$ m \qquad ◊$$

(c)   The original volume is

$$V = \pi r^2 L = \frac{\pi}{4}(0.0150 \text{ m})^2(0.300 \text{ m}) = 5.30 \times 10^{-5} \text{ m}^3$$

Using the volumetric coefficient of expansion, β,

$$\Delta V = \beta V_i \Delta T \approx 3\alpha V_i \Delta T$$

$$\Delta V \approx 3(9.00 \times 10^{-6} \text{ °C}^{-1})(5.30 \times 10^{-5} \text{ m}^3)(65.0 \text{ °C}) = 93.0 \times 10^{-9} \text{ m}^3 \qquad ◊$$

Related Calculation: The above calculation ignores ΔL^2 and ΔL^3 terms. Calculate the change in volume exactly, and compare your answer with the approximate solution above.

The volume will increase by a factor of $\dfrac{\Delta V}{V_i} = \left(1 + \dfrac{\Delta D}{D_i}\right)^2\left(1 + \dfrac{\Delta L}{L_i}\right) - 1$

$$\frac{\Delta V}{V_i} = \left(1 + \frac{8.78 \times 10^{-6} \text{ m}}{0.015 \text{ m}}\right)^2\left(1 + \frac{1.76 \times 10^{-4} \text{ m}}{0.300 \text{ m}}\right) - 1 = 1.76 \times 10^{-3}$$

$$\Delta V = \frac{\Delta V}{V_i} V_i$$

$$\Delta V = (1.76 \times 10^{-3})(5.30 \times 10^{-5} \text{ m}^3) = 93.1 \times 10^{-9} \text{ m}^3 = 93.1 \text{ mm}^3 \qquad ◊$$

The answer is virtually identical; the approximation $\beta \approx 3\alpha$ is a good one.

21. A hollow aluminum cylinder 20.0 cm deep has an internal capacity of 2.000 L at 20.0 °C. It is completely filled with turpentine and then slowly warmed to 80.0 °C. (a) How much turpentine overflows? (b) If the cylinder is then cooled back to 20.0 °C, how far below the cylinder's rim does the turpentine's surface recede?

Solution   When the temperature is increased from 20.0 °C to 80.0 °C, both the cylinder and the turpentine increase in volume by $\Delta V = \beta V_i \Delta T$:

(a)   The overflow, $V_{over} = \Delta V_{turp} - \Delta V_{Al}$

$$V_{over} = \left(\beta V_i \Delta T\right)_{turp} - \left(\beta V_i \Delta T\right)_{Al} = V_i \Delta T \left(\beta_{turp} - 3\alpha_{Al}\right)$$

$$V_{over} = (2.000 \text{ L})(60.0 \text{ °C})\left(9.00 \times 10^{-4} \text{ °C}^{-1} - 0.720 \times 10^{-4} \text{ °C}^{-1}\right)$$

$$V_{over} = 0.0994 \text{ L} \qquad \qquad \lozenge$$

(b)   After warming the whole volume of the turpentine is

$$V' = 2000 \text{ cm}^3 + \left(9.00 \times 10^{-4} \text{ °C}^{-1}\right)\left(2000 \text{ cm}^3\right)(60.0 \text{ °C}) = 2108 \text{ cm}^3$$

The fraction lost is $\dfrac{99.4 \text{ cm}^3}{2108 \text{ cm}^3} = 4.71 \times 10^{-2}$

This also is the fraction of the cylinder that will be empty after cooling:

$$\Delta h = \left(4.71 \times 10^{-2}\right)(20.0 \text{ cm}) = 0.943 \text{ cm} \qquad \qquad \lozenge$$

---

27.   An automobile tire is inflated with air originally at 10.0 °C and normal atmospheric pressure. During the process, the air is compressed to 28.0% of its original volume and the temperature is increased to 40.0 °C. (a) What is the tire pressure? (b) After the car is driven at high speed, the tire air temperature rises to 85.0 °C and the interior volume of the tire increases by 2.00%. What is the new tire pressure (absolute) in pascals?

Solution

(a)   Taking $PV = nRT$ in the initial (i) and final (f) states, and dividing, we have

$$P_i V_i = nRT_i \quad \text{and} \quad P_f V_f = nRT_f \quad \text{yield} \quad \frac{P_f V_f}{P_i V_i} = \frac{T_f}{T_i}$$

So $P_f = P_i \dfrac{V_i T_f}{V_f T_i} = \left(1.013 \times 10^5 \text{ Pa}\right)\left(\dfrac{V_i}{0.280 V_i}\right)\left(\dfrac{273 \text{ K} + 40.0 \text{ K}}{273 \text{ K} + 10.0 \text{ K}}\right) = 4.00 \times 10^5 \text{ Pa} \quad \lozenge$

(b)   Introducing the hot (h) state, $\dfrac{P_h V_h}{P_f V_f} = \dfrac{T_h}{T_f}$

So $P_h = P_f \left(\dfrac{V_f}{V_h}\right)\left(\dfrac{T_h}{T_f}\right) = \left(4.00 \times 10^5 \text{ Pa}\right)\left(\dfrac{V_f (358 \text{ K})}{1.02 V_f (313 \text{ K})}\right) = 4.49 \times 10^5 \text{ Pa} \qquad \lozenge$

29. An auditorium has dimensions 10.0 m × 20.0 m × 30.0 m. How many molecules of air fill the auditorium at 20.0 °C and a pressure of 101 kPa?

Solution

Conceptualize: The given room conditions are close to Standard Temperature and Pressure (STP is 0 °C and 101.3 kPa), so we can use the estimate that one mole of an ideal gas at STP occupies a volume of about 22 L. The volume of the auditorium is 6×10^3 m³ and 1 m³ = 1 000 L, so we can estimate the number of molecules to be:

$$N \approx \left(6\,000\ \text{m}^3\right)\left(\frac{10^3\ \text{L}}{1\ \text{m}^3}\right)\left(\frac{1\ \text{mol}}{22\ \text{L}}\right)\left(\frac{6 \times 10^{23}\ \text{molecules}}{1\ \text{mol}}\right) \approx 1.6 \times 10^{29}\ \text{molecules of air}$$

Categorize: The number of molecules can be found more precisely by applying $PV = nRT$.

Analyze: The equation of state of an ideal gas is $PV = nRT$. We need to solve for the number of moles to find N.

$$n = \frac{PV}{RT} = \frac{(1.01 \times 10^5\ \text{N/m}^2)[(10.0\ \text{m})(20.0\ \text{m})(30.0\ \text{m})]}{(8.314\ \text{J/mol·K})(293\ \text{K})} = 2.49 \times 10^5\ \text{mol}$$

$$N = n(N_A) = (2.49 \times 10^5\ \text{mol})\left(6.022 \times 10^{23}\ \frac{\text{molecules}}{\text{mol}}\right) = 1.50 \times 10^{29}\ \text{molecules} \quad \lozenge$$

Finalize: This result agrees quite well with our initial estimate. The numbers would match even better if the temperature of the auditorium were 0 °C.

---

33. The mass of a hot-air balloon and its cargo (not including the air inside) is 200 kg. The air outside is at 10.0 °C and 101 kPa. The volume of the balloon is 400 m³. To what temperature must the air in the balloon be heated before the balloon will lift off? (Air density at 10.0 °C is 1.25 kg/m³.)

Solution

Conceptualize: The air inside the balloon must be significantly hotter than the outside air in order for the balloon to have a net upward force, but the temperature must also be less than the melting point of the nylon used for the balloon's envelope. (Rip-stop nylon melts around 200 °C.) Otherwise the results could be disastrous!

Categorize: The density of the air inside the balloon must be sufficiently low so that the buoyant force is equal to the weight of the balloon, its cargo, and the air inside. The temperature of the air required to achieve this density can be found from the equation of state of an ideal gas.

Analyze:

The buoyant force equals the weight of the air at 10.0 °C displaced by the balloon:

$$B = m_{air}g = \rho_{air}Vg: \qquad B = \left(1.25 \text{ kg} / \text{m}^3\right)\left(400 \text{ m}^3\right)\left(9.80 \text{ m} / \text{s}^2\right) = 4\,900 \text{ N}$$

The weight of the balloon and its cargo is

$$F_g = m_b g = \left(200 \text{ kg}\right)\left(9.80 \text{ m} / \text{s}^2\right) = 1\,960 \text{ N}$$

Since $B > F_g$, the balloon has a chance of lifting off as long as the weight of the air inside the balloon is less than the difference in these forces:

$$F_{air} < B - F_g = 4\,900 \text{ N} - 1\,960 \text{ N} = 2\,940 \text{ N}$$

The mass of this air is $\qquad m_{air} = \dfrac{F_{air}}{g} = \dfrac{2\,940 \text{ N}}{9.80 \text{ m} / \text{s}^2} = 300 \text{ kg}$

To find the required temperature of this air from $PV = nRT$, we must find the corresponding number of moles of air. Dry air is approximately 20% O_2, and 80% N_2. Using data from a periodic table, we can calculate the molar mass of the air to be approximately

$$M = (0.80)(28.0 \text{ g} / \text{mol}) + (0.20)(32.0 \text{ g} / \text{mol}) = 28.8 \text{ g} / \text{mol}$$

so the number of moles is $n = \dfrac{m}{M} = \left(300 \text{ kg}\right)\left(\dfrac{1\,000 \text{ g} / \text{kg}}{28.8 \text{ g} / \text{mol}}\right) = 1.04 \times 10^4 \text{ mol}$

The pressure of this air is the ambient pressure; from $PV = nRT$, we can now find the minimum temperature required for lift off:

$$T = \frac{PV}{nR} = \frac{\left(1.01 \times 10^5 \text{ N} / \text{m}^2\right)\left(400 \text{ m}^3\right)}{\left(1.04 \times 10^4 \text{ mol}\right)\left(8.314 \text{ J} / \text{mol} \cdot \text{K}\right)} = 466 \text{ K} = 193 \text{ °C} \lozenge$$

Finalize: The average temperature of the air inside the balloon required for lift off appears to be close to the melting point of the nylon fabric, so this seems like a dangerous situation! A larger balloon would be better suited for the given weight of the balloon. (A quick check on the internet reveals that this balloon is only about 1/10 the size of most sport balloons, which have a volume of about $3\,000 \text{ m}^3$).

If the buoyant force were less than the weight of the balloon and its cargo, the balloon would not lift off no matter how hot the air inside might be! If this were the case, then either the weight would have to be reduced or a bigger balloon would be required.

39. The pressure gauge on a tank registers the gauge pressure, which is the difference between the interior and exterior pressure. When the tank is full of oxygen (O_2), it contains 12.0 kg of the gas at a gauge pressure of 40.0 atm. Determine the mass of oxygen that has been withdrawn from the tank when the pressure reading is 25.0 atm. Assume that the temperature of the tank remains constant.

Solution The ideal gas law states $PV = nRT$

At constant volume and temperature, $\dfrac{P}{n} = \text{constant}$ or $\dfrac{P_1}{n_1} = \dfrac{P_2}{n_2}$

and $n_2 = \left(\dfrac{P_2}{P_1}\right)n_1$

However, n is proportional to m, so $m_2 = \left(\dfrac{P_2}{P_1}\right)m_1 = \left(\dfrac{26.0 \text{ atm}}{41.0 \text{ atm}}\right)(12.0 \text{ kg}) = 7.61 \text{ kg}$

The mass removed is $\Delta m = 12.0 \text{ kg} - 7.61 \text{ kg} = 4.39 \text{ kg}$ ◊

---

47. A mercury thermometer is constructed as shown in Figure P19.47. The capillary tube has a diameter of 0.004 00 cm, and the bulb has a diameter of 0.250 cm. Neglecting the expansion of the glass, find the change in height of the mercury column with a temperature change of 30.0 °C.

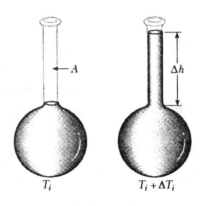

Solution Neglecting the expansion of the glass, the volume of liquid in the capillary will be $\Delta V = A\Delta h$ where A is the cross-sectional area of the capillary. Let V_i represent the volume of the bulb.

Figure P19.47

$$\Delta V = V_i \beta \, \Delta T$$

$$\Delta h = \left(\dfrac{V_i}{A}\right)\beta \, \Delta T = \left[\dfrac{\frac{4}{3}\pi R_{bulb}^3}{\pi R_{cap}^2}\right]\beta \, \Delta T$$

$$\Delta h = \dfrac{4}{3}\dfrac{(0.125 \text{ cm})^3}{(0.002\,00 \text{ cm})^2}\left(1.82 \times 10^{-4} \, (°C)^{-1}\right)(30.0 \text{ °C}) = 3.55 \text{ cm}$$ ◊

---

51. A liquid has a density ρ. (a) Show that the fractional change in density for a change in temperature ΔT is $\Delta\rho/\rho=-\beta\Delta T$. What does the negative sign signify? (b) Fresh water has a maximum density of 1.0000 g/cm^3 at 4.0 °C. At 10.0 °C, its density is 0.9997 g/cm^3. What is β for water over this temperature interval?

Solution   We start with the two equations: $\rho=\dfrac{m}{V}$ and $\dfrac{\Delta V}{V}=\beta\Delta T$

(a) Differentiating the first equation, $d\rho=-\dfrac{m}{V^2}\,dV$

For very small changes in V and ρ, this can be written $\Delta\rho=-\dfrac{m}{V}\dfrac{\Delta V}{V}=-\left(\dfrac{m}{V}\right)\dfrac{\Delta V}{V}$

Substituting both of our initial equations, we find that $\Delta\rho=-\rho\beta\Delta T$

The negative sign means that if β is positive, any increase in temperature causes the density to decrease and vice versa.   ◊

(b) We apply the equation $\beta=-\dfrac{\Delta\rho}{\rho\Delta T}$ for the specific case of water:

$$\beta=-\frac{(1.0000\ \text{g}/\text{cm}^3-0.9997\ \text{g}/\text{cm}^3)}{(1.0000\ \text{g}/\text{cm}^3)(4.00\ °\text{C}-10.0\ °\text{C})}$$

Calculating, we find that $\beta=5.00\times10^{-5}\ °\text{C}^{-1}$ ◊

---

53. A vertical cylinder of cross-sectional area A is fitted with a tight-fitting, frictionless piston of mass m (Fig. P19.53). (a) If n moles of an ideal gas are in the cylinder at a temperature of T, what is the height h at which the piston is in equilibrium under its own weight? (b) What is the value for h if $n=0.200$ mol, $T=400$ K, $A=0.00800$ m^2, and $m=20.0$ kg?

Solution

(a) We suppose that the air above the piston remains at atmospheric pressure, P_0. Model the piston as a particle in equilibrium.

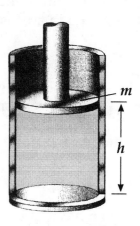

Figure P19.53

$$\sum F_y = ma_y \qquad \text{yields} \qquad -P_0A - mg + PA = 0$$

where P is the pressure exerted by the gas contained.

Noting that $V = Ah$, and that n, T, m, g, A, and P_0 are given,

$$PV = nRT \qquad \text{becomes} \qquad P = \frac{nRT}{Ah}$$

$$\text{so} \qquad -P_0A - mg + \frac{nRT}{Ah}A = 0$$

$$\text{and} \qquad h = \frac{nRT}{P_0A + mg} \qquad\qquad \diamond$$

(b) $$h = \frac{(0.200 \text{ mol})(8.314 \text{ J / mol} \cdot \text{K})(400 \text{ K})}{\left(1.013 \times 10^5 \text{ N / m}^2\right)\left(0.008\,00 \text{ m}^2\right) + (20.0 \text{ kg})(9.80 \text{ m / s}^2)}$$

$$h = \frac{665 \text{ N} \cdot \text{m}}{810 \text{ N} + 196 \text{ N}} = 0.661 \text{ m} \qquad\qquad \diamond$$

---

55. The rectangular plate shown in Figure P19.55 has an area A_i equal to ℓw. If the temperature increases by ΔT, each dimension increases according to $\Delta L = \alpha L_i \Delta T$, where α is the average coefficient of linear expansion. Show that the increase in area is $\Delta A = 2\alpha A_i \Delta T$. What approximation does this expression assume?

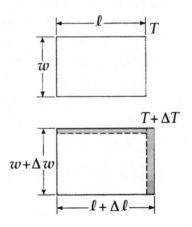

Figure P19.55

Solution   From the diagram in Figure 19.55, we see that the **change** in area is

$$\Delta A = \ell \Delta w + w\Delta \ell + \Delta w \Delta \ell$$

Since $\Delta \ell$ and Δw are each small quantities, the product $\Delta w \Delta \ell$ will be very small.

Therefore, we assume $\Delta w \Delta \ell \approx 0$ $\qquad\qquad \diamond$

Since $\qquad\qquad \Delta w = w\alpha \Delta T$ and $\Delta \ell = \ell \alpha \Delta T$

we then have $\qquad \Delta A = \ell w \alpha \Delta T + w\ell \alpha \Delta T$

Finally, since $A = \ell w$, $\Delta A = 2\alpha A \Delta T$ $\qquad\qquad \diamond$

---

65. Starting with Equation 19.10, show that the total pressure P in a container filled with a mixture of several ideal gases is $P = P_1 + P_2 + P_3 + \dots$, where P_1, P_2, $\dots$, are the pressures that each gas would exert if it alone filled the container (these individual pressures are called the **partial pressures** of the respective gases). This result is known as **Dalton's law of partial pressures**.

Solution For each gas alone, $P_1 = \dfrac{N_1 k_B T}{V_1}$, $P_2 = \dfrac{N_2 k_B T}{V_2}$, $P_3 = \dfrac{N_3 k_B T}{V_3}$, $\dots$

For the gases combined, $N_1 + N_2 + N_3 + \dots = N = \dfrac{PV}{k_B T}$

Therefore, $\dfrac{P_1 V_1}{k_B T} + \dfrac{P_2 V_2}{k_B T} + \dfrac{P_3 V_3}{k_B T} + \dots = \dfrac{PV}{k_B T}$

But $V_1 = V_2 = V_3 = \dots = V$ so $P_1 + P_2 + P_3 + \dots = P$ ◊

71. Review Problem: A steel guitar string with a diameter of 1.00 mm is stretched between supports 80.0 cm apart. The temperature is 0.0 °C. (a) Find the mass per unit length of this string. (Use the value 7.86×10^3 kg/m^3 for the density.) (b) The fundamental frequency of transverse oscillations of the string is 200 Hz. What is the tension in the string? (c) If the temperature is raised to 30.0 °C, find the resulting values of the tension and the fundamental frequency. Assume that both the Young's modulus (Table 12.1) and the average coefficient of expansion (Table 19.1) have constant values between 0.0 °C and 30.0 °C.

Solution

(a) $\mu = \tfrac{1}{4}\rho\left(\pi d^2\right) = \tfrac{1}{4}\pi\left(1.00 \times 10^{-3} \text{ m}\right)^2\left(7.86 \times 10^3 \text{ kg / m}^3\right)$

$\mu = 6.17 \times 10^{-3}$ kg / m ◊

(b) Since $f_1 = \dfrac{v}{2L}$, $v = \sqrt{T/\mu}$ and $f_1 = \dfrac{1}{2L}\sqrt{T/\mu}$

$F = T = \mu\left(2Lf_1\right)^2 = \left(6.17 \times 10^{-3} \text{ kg / m}\right)\left[(2)(0.800 \text{ m})\left(200 \text{ s}^{-1}\right)\right]^2 = 632 \text{ N}$ ◊

(c)   At 0 °C, the length of the guitar string will be

$$L_{0°C} = L_{natural}\left(1 + \frac{F}{AY}\right)$$

We know the string's cross section

$$A = \left(\frac{\pi}{4}\right)\left(1.00 \times 10^{-3} \text{ m}\right)^2 = 7.85 \times 10^{-7} \text{ m}^2$$

and modulus $Y = 20.0 \times 10^{10} \text{ N / m}^2$

Therefore, $\dfrac{F}{AY} = \dfrac{632 \text{ N}}{\left(7.85 \times 10^{-7} \text{ m}^2\right)\left(20.0 \times 10^{10} \text{ N / m}^2\right)} = 4.02 \times 10^{-3}$

$$L_{0°C} = \frac{0.800 \text{ m}}{1 + 4.02 \times 10^{-3}} = 0.796 \text{ 8 m}$$

Then at 30 °C, $L_{30°C} = \left(0.796 \text{ 8 m}\right)\left(1 + (30.0 \text{ °C})\left(11.0 \times 10^{-6} \text{ °C}^{-1}\right)\right) = 0.797 \text{ 1 m}$

Since $0.800 \text{ m} = \left(0.797 \text{ 1 m}\right)\left[1 + \dfrac{F'}{A'Y}\right]$

$$\frac{F'}{A'Y} = \frac{0.800 \text{ 0}}{0.797 \text{ 1}} - 1 = 3.693 \times 10^{-3}$$

and $F' = A'Y\left(3.693 \times 10^{-3}\right)$

$$F' = \left(7.85 \times 10^{-7} \text{ m}^2\right)\left(20.0 \times 10^{10} \text{ N / m}^2\right)\left(3.693 \times 10^{-3}\right)\left(1 + \alpha \, \Delta T\right)^2$$

$$F' = (580 \text{ N})\left(1 + 3.30 \times 10^{-4}\right)^2 = 580 \text{ N} \qquad \Diamond$$

Also, $\dfrac{f_1'}{f_1} = \sqrt{\dfrac{F'}{F}}$

so $f_1' = (200 \text{ Hz})\sqrt{\dfrac{580 \text{ N}}{632 \text{ N}}} = 192 \text{ Hz} \qquad \Diamond$

Chapter 20

HEAT AND THE FIRST LAW OF THERMODYNAMICS

EQUATIONS AND CONCEPTS

The **specific heat** of a substance is the heat capacity per unit mass. Each substance requires a specific quantity of energy to change the temperature of 1 kg of the substance by 1.00 °C. *Heat capacity, C, refers to a sample of material while specific heat, c, refers to a unit mass of a material.*

$$c \equiv \frac{Q}{m\Delta T} \qquad (20.3)$$

The **energy** Q that must be transferred between a system of mass m and its surroundings to produce a temperature change ΔT varies with the substance.

$$Q = mc\Delta T \qquad (20.4)$$

Calorimetry, a technique to measure specific heat, is based on the conservation of energy in an isolated system. Q_{hot} will be a negative quantity because energy leaves that part of the isolated system; and Q_{cold} will be a positive quantity because energy is entering that part of the system. *The negative sign is required so that each side of the equation will be positive.*

$$Q_{cold} = -Q_{hot} \qquad (20.5)$$

Latent heat, L is a thermal property of a material that determines the quantity of energy Q required to change the phase of a unit mass of that substance. *The phase change process occurs at constant temperature and is accompanied by a change in internal energy.* The value of L for a given material depends on the nature of the phase change (e.g. solid to liquid, L_f or liquid to vapor, L_v) and the thermal properties of the substance.

$$Q = \pm mL \qquad (20.6)$$

311

The **work done on a gas** which undergoes an expansion or compression from volume V_i to volume V_i depends on the path between the initial and final states. In order to evaluate the integral, the manner in which pressure varies with volume must be known. In general, the work done on a gas is the negative of the *area under the curve on a PV diagram.*

$$W = -\int_{V_i}^{V_f} P\,dV \qquad (20.8)$$

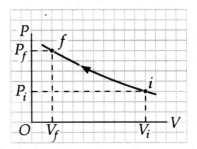

The **first law of thermodynamics** states that the change in internal energy of a system depends on:

Q: the energy transferred into the system by heat, and

W: the work done on the system.

The quantity (Q+W) is independent of the path taken between the initial and final states, i.e. ΔE_{int} is path independent.

$$\Delta E_{int} = Q + W \qquad (20.9)$$

In an **adiabatic process** no energy enters or leaves a system by heat, $Q=0$. *When a gas is compressed adiabatically (W>0), the temperature increases.*

$$\Delta E_{int} = W \qquad (20.10)$$
$$\text{(adiabatic process)}$$

An **isobaric process** is one that occurs at constant pressure.

$$W = -P\left(V_f - V_i\right) \qquad (20.11)$$
$$\text{(isobaric process)}$$

In an **isovolumetric process** (constant volume), zero work is done and all energy added to the system by heat remains in the system and increases the internal energy.

$$\Delta E_{int} = Q \qquad (20.12)$$
$$\text{(isovolumetric process)}$$

An **isothermal process** is one which occurs at constant temperature. *No change in internal energy occurs in an isothermal process: any energy that enters the system by heat is transferred out of the system by work (Q = −W).*

$$W = nRT \ln\left(\frac{V_i}{V_f}\right) \qquad (20.13)$$

$$\text{(isothermal process)}$$

Energy transfer by conduction through a slab or rod of material is proportional to the temperature difference between the hot and cold faces and inversely proportional to the thickness of the slab or length of the rod (see figure). The thermal conductivity, k, is characteristic of a particular material; and *good thermal conductors have large values of k.*

$$\mathcal{P} = kA\left(\frac{T_h - T_c}{L}\right) \qquad (20.15)$$

Conduction through a compound slab of area A is found by using Eq. 20.16. T_c and T_h are the temperatures of the outer surfaces. The thicknesses of the slabs are L_1, L_2, L_3, ... and the respective thermal conductivities are k_1, k_2, k_3, ...

$$\mathcal{P} = \frac{A(T_h - T_c)}{\sum_i (L_i / k_i)} \qquad (20.16)$$

The **R-value** of a material is the ratio of thickness to thermal conductivity. Materials with large R-values are good insulators. In the US, R-values are given in engineering units: $\text{ft}^2 \cdot {}^\circ\text{F} \cdot \text{h}/\text{BTU}$.

$$R = \sum \frac{L_i}{k_i}$$

Stefan's law gives the rate of energy transfer by radiation (radiated power) from a surface of area A. The surface temperature T is measured in kelvins and e (the emissivity of the radiating body) can have a value between 0 and 1 depending on the nature of the surface.

$$\mathcal{P} = \sigma A e T^4 \qquad (20.18)$$

$$\sigma = 5.669\,6 \times 10^{-8} \text{ W} / \text{m}^2 \cdot \text{K}^4$$

313

Net radiated power by a object at temperature T, in an environment at temperature T_0, is the difference between the rate of emission to the environment and rate of absorption.

$$\mathcal{P}_{net} = \sigma Ae\left(T^4 - T_0{}^4\right) \qquad (20.19)$$

SUGGESTIONS, SKILLS, AND STRATEGIES

Many applications of the first law of thermodynamics deal with the work done on a system that undergoes a change in state. Consider the PV diagram showing the path taken by a gas whose initial pressure and volume are P_i, V_i, and whose final pressure and volume are P_f, V_f. The work done on the gas can be calculated directly using Equation 20.8, $W = -\int PdV$, if pressure is some "reasonable" function of the volume. If the path taken by the gas from its initial to its

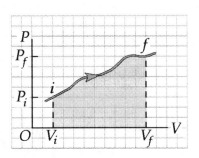

final state can be drawn on a PV diagram, the work done on the gas during the expansion is equal to the negative of the area under the PV curve (the shaded region) shown in the figure. It is important to recognize that the **work depends on the path taken as the gas goes from i to f.** That is, W depends on the specific manner in which the pressure P changes during the process.

PROBLEM SOLVING STRATEGY: CALORIMETRY PROBLEMS

If you are having difficulty with calorimetry problems, consider the following factors:

- Be sure your units are consistent throughout. For instance, if you are using specific heats in cal/g·°C, be sure that masses are in grams and temperatures are Celsius throughout.

- Energy losses and gains are found by using $Q = mc\Delta T$ only for those intervals in which no phase changes are occurring. The equations $Q = mL_f$ and $Q = mL_v$ are to be used only when phase changes **are** taking place.

- Often sign errors occur in calorimetry problems. In Equation 20.5 ($Q_{cold} = -Q_{hot}$), remember to include the negative sign in the equation.

REVIEW CHECKLIST

You should be able to:

▷ Write equations based on Equation 20.5 ($Q_{cold} = -Q_{hot}$)to describe a calorimetry experiment and calculate specific heat or change in temperature. (Section 20.2).

▷ Make calculations of energy exchange involving change of phase. (Section 20.3)

▷ Calculate the work done on (or by) a gas using equation 20.8, (i) when the equation for $P = P(V)$ is given, (ii) for a constant temperature process, (iii) and by evaluating the area under a PV curve. (Sections 20.4, 20.5, and 20.6)

▷ Calculate the change in the internal energy of a system for adiabatic, constant volume, and constant pressure processes. (Section 20.6)

▷ Calculate the work done on a gas and the net energy added by heat during a cyclic process. (Section 20.6)

▷ Make calculations of the rate of energy transfer due to conduction and radiation. (Section 20.7)

ANSWERS TO SELECTED CONCEPTUAL QUESTIONS

6. What is wrong with this statement? "Given any two objects, the one with the higher temperature contains more heat."

Answer The statement shows a misunderstanding of the concept of heat. Heat is a process by which energy is transferred, not a form of energy that is held or contained. If you wish to speak of energy that is "contained", you speak of **internal energy**, not **heat**.

Further, even if the statement used the term "internal energy", it would still be incorrect, since the effects of specific heat and mass are both ignored. A 1-kg mass of water at 20 °C has more internal energy than a 1-kg mass of air at 30 °C. Similarly, the earth has far more internal energy than a drop of molten titanium metal.

Correct statements would be: (1) "given any two bodies in thermal contact, the one with the higher temperature will transfer energy to the other by heat". (2) "given any two bodies of equal mass, the one with the higher product of absolute temperature and specific heat contains more internal energy".

□　　□　　□　　□

8. The air temperature above coastal areas is profoundly influenced by the large specific heat of water. One reason is that the energy released when 1 m³ of water cools by 1 °C will raise the temperature of a much larger volume of air by 1 °C. Find this volume of air. The specific heat of air is approximately 1 kJ/kg·°C. Take the density of air to be 1.3 kg/m³.

Answer   The mass of one cubic meter of water is specified by its density,

$$m = \rho V = \left(1.00 \times 10^3 \text{ kg / m}^3\right)\left(1 \text{ m}^3\right) = 1 \times 10^3 \text{ kg}$$

When one cubic meter of water cools by 1 °C it releases energy

$$Q_c = mc\Delta T = \left(1 \times 10^3 \text{ kg}\right)\left(4\ 186 \text{ J / kg} \cdot {}^\circ\text{C}\right)\left(-1 \ {}^\circ\text{C}\right) = -4 \times 10^6 \text{ J}$$

where the negative sign represents heat output. When $+4 \times 10^6$ J is transferred to the air, raising its temperature 1 °C, the volume of the air is given by $Q_c = mc\Delta T = \rho V c \Delta T$:

$$V = \frac{Q_c}{\rho c \Delta T} = \frac{4 \times 10^6 \text{ J}}{\left(1.3 \text{ kg / m}^3\right)\left(1 \times 10^3 \text{ J / kg} \cdot {}^\circ\text{C}\right)\left(1 \ {}^\circ\text{C}\right)} = 3 \times 10^3 \text{ m}^3$$

The volume of the air is thousands of times larger than the volume of the water.

□   □   □   □

10. Using the first law of thermodynamics, explain why the **total** energy of an isolated system is always constant.

Answer   The first law of thermodynamics says that the net change in internal energy of a system is equal to the energy added by heat, plus the work done on the system.

$$\Delta E_{\text{int}} = Q + W$$

However, an isolated system is defined as an object or set of objects for which there is no exchange of energy with its surroundings. In the case of the first law of thermodynamics, this means that $Q = W = 0$, so the change in internal energy of the system at all times must be zero.

As it stays constant in amount, an isolated system's energy may change from one form to another or move from one object to another within the system. For example, a "bomb calorimeter" is a closed system that consists of a sturdy steel container, a water bath, an item of food, and oxygen. The food is burned in the presence of an excess of oxygen, and chemical energy is converted to internal energy. In the process of oxidation, energy is transferred by heat to the water bath, raising its temperature. The change in temperature of the water bath is used to determine the caloric content of the food. The process works specifically **because** the total energy remains unchanged in a closed system.

□   □   □   □

SOLUTIONS TO SELECTED END-OF-CHAPTER PROBLEMS

1. On his honeymoon James Joule traveled from England to Switzerland. He attempted to verify his idea of the interconvertibility of mechanical energy and internal energy by measuring the increase in temperature of water that fell in a waterfall. If water at the top of an alpine waterfall has a temperature of 10.0 °C and then falls 50.0 m (as at Niagara Falls), what maximum temperature at the bottom of the falls could Joule expect? He did not succeed in measuring the temperature change, partly because evaporation cooled the falling water, and also because his thermometer was not sufficiently sensitive.

Solution

Conceptualize: Water has a high specific heat, so the difference in water temperature between the top and bottom of the falls is probably less than 1 °C. (Besides, if the difference was significantly large, we might have heard about this phenomenon at some point.)

Categorize: The temperature change can be found from the potential energy that is converted to internal energy. The final temperature is this change added to the initial temperature of the water.

Analyze: The change in potential energy is $\Delta E = mgy$. It will produce the same temperature change as the same amount of heat entering the water from a stove, as described by $Q = mc\Delta T$. Thus, $mgy = mc\Delta T$

Isolating ΔT,
$$\Delta T = \frac{gy}{c} = \frac{(9.80 \text{ m/s}^2)(50.0 \text{ m})}{4.186 \times 10^3 \text{ J/kg} \cdot °C} = 0.117 \text{ °C}$$

$$T_g = T_i + \Delta T = 10.0 \text{ °C} + 0.117 \text{ °C} = 10.1 \text{ °C} \qquad \diamond$$

Finalize: The water temperature rose less than 1 °C as expected; however, the final temperature might be less than we calculated since this solution does not account for cooling of the water due to evaporation as it falls. It is interesting to note that the change in temperature is independent of the amount of water.

---

3. The temperature of a silver bar rises by 10.0 °C when it absorbs 1.23 kJ of energy by heat. The mass of the bar is 525 g. Determine the specific heat of silver.

Solution $\qquad \Delta Q = mc_{silver}\Delta T \qquad$ or $\qquad c_{silver} = \dfrac{Q}{m\Delta T}$

$$c_{silver} = \frac{1.23 \times 10^3 \text{ J}}{(0.525 \text{ kg})(10.0 \text{ °C})} = 234 \text{ J/kg} \cdot °C \qquad \diamond$$

---

7. A 1.50-kg iron horseshoe initially at 600 °C is dropped into a bucket containing 20.0 kg of water at 25.0 °C. What is the final temperature? (Ignore the heat capacity of the container, and assume that a negligible amount of water boils away.)

Solution

Conceptualize: Even though the horseshoe is much hotter than the water, the mass of the water is significantly greater, so we might expect the water temperature to rise less than 10 °C.

Categorize: The heat lost by the iron will be gained by the water, and from this heat difference, the change in water temperature can be found.

Analyze:

$$Q_{iron} = -Q_{water} \qquad \text{or} \qquad (mc\,\Delta T)_{iron} = -(mc\,\Delta T)_{water}$$

$$m_{Fe}c_{Fe}(T - 600\ °C) = -m_w c_w(T - 25.0\ °C)$$

$$T = \frac{m_w c_w(25.0\ °C) + m_{Fe}c_{Fe}(600\ °C)}{m_{Fe}c_{Fe} + m_w c_w}$$

$$T = \frac{(20.0\ \text{kg})(4\,186\ \text{J} / \text{kg} \cdot °\text{C})(25.0\ °C) + (1.50\ \text{kg})(448\ \text{J} / \text{kg} \cdot °\text{C})(600\ °C)}{(1.50\ \text{kg})(448\ \text{J} / \text{kg} \cdot °\text{C}) + (20.0\ \text{kg})(4\,186\ \text{J} / \text{kg} \cdot °\text{C})} = 29.6\ °\text{C} \quad \Diamond$$

Finalize: The temperature only rose about 5 °C, so our answer seems reasonable. The specific heat of the water is about 10 times greater than the iron, so this effect also reduces the change in water temperature. In this problem, we assumed that a negligible amount of water boiled away, but in reality, the final temperature of the water would be somewhat less than we calculated, since some of the heat energy would be used to vaporize a bit of water.

---

17. A 3.00-g lead bullet at 30.0 °C is fired at a speed of 240 m/s into a large block of ice at 0 °C, in which it becomes embedded. What quantity of ice melts?

Solution

Conceptualize: The amount of ice that melts is probably small, maybe only a few grams based on the size, speed, and initial temperature of the bullet.

Categorize: We will assume that all of the initial kinetic and excess internal energy of the bullet goes into internal energy to melt the ice, the mass of which can be found from the latent heat of fusion. Because the ice does not all melt, in the final state everything is at 0 °C.

Analyze: At thermal equilibrium, the energy lost by the bullet equals the energy gained by the ice: $\left|\Delta K_b\right|+\left|Q_b\right|=Q_{ice}$ gives

$$\frac{1}{2}m_b v_b^2 + m_b c_{Pb}\left|\Delta T\right| = m_{ice}L_f \qquad \text{or} \qquad m_{ice} = m_b\left(\frac{\frac{1}{2}v_b^2 + c_{Pb}\left|\Delta T\right|}{L_f}\right)$$

$$m_{ice} = \left(3.00\times10^{-3}\text{ kg}\right)\left(\frac{\frac{1}{2}(240\text{ m}/\text{s})^2 + (128\text{ J}/\text{kg}\cdot{}^\circ\text{C})(30.0\text{ }^\circ\text{C})}{3.33\times10^5\text{ J}/\text{kg}}\right)$$

$$m_{ice} = \frac{86.4\text{ J}+11.5\text{ J}}{3.33\times10^5\text{ J}/\text{kg}} = 2.94\times10^{-4}\text{ kg} = 0.294\text{ g} \qquad \Diamond$$

Finalize: The amount of ice that melted is less than a gram, which agrees with our prediction. It appears that most of the energy used to melt the ice comes from the kinetic energy of the bullet (88%), while the excess internal energy of the bullet only contributes 12% to melt the ice. Small chips of ice probably fly off when the bullet makes impact. Some of the energy is transferred to their kinetic energy, so in reality, the amount of ice that would melt should be less than what we calculated. If the block of ice were colder than 0 °C (as is usually the case), then the melted ice would refreeze.

---

21. In an insulated vessel, 250 g of ice at 0 °C is added to 600 g of water at 18.0 °C. (a) What is the final temperature of the system? (b) How much ice remains when the system reaches equilibrium?

Solution

(a) When 250 g of ice is melted,

$$Q_f = mL_f = (0.250\text{ kg})\left(3.33\times10^5\text{ J}/\text{kg}\right) = 83.3\text{ kJ}$$

The energy released when 600 g of water cools from 18.0 °C to 0 °C is

$$\left|Q\right|=\left|mc\Delta T\right|=(0.600\text{ kg})(4\,186\text{ J}/\text{kg}\cdot{}^\circ\text{C})(18.0\text{ }^\circ\text{C}) = 45.2\text{ kJ}$$

Since the energy required to melt 250 g of ice at 0 °C **exceeds** the energy released by cooling 600 g of water from 18.0 °C to 0 °C, the final temperature of the system (water + ice) must be 0 °C. $\qquad \Diamond$

(b) The energy released by the water (45.2 kJ) will melt a mass of ice m, where $Q = mL_f$.

Solving for the mass, $m = \dfrac{Q}{L_f} = \dfrac{45.2 \times 10^3 \text{ J}}{3.33 \times 10^5 \text{ J/kg}} = 0.136 \text{ kg}$

Therefore, the ice remaining is $m' = 0.250 \text{ kg} - 0.136 \text{ kg} = 0.114 \text{ kg}$ ◊

---

23. A sample of ideal gas is expanded to twice its original volume of 1.00 m^3 in a quasi-static process for which $P = \alpha V^2$, with $\alpha = 5.00 \text{ atm/m}^6$, as shown in Figure P20.23. How much work is done on the expanding gas?

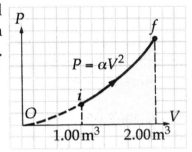

Figure P20.23

Solution

$W = -\int_{V_i}^{V_f} P\,dV$ and $V_f = 2V_i = 2\left(1.00 \text{ m}^3\right) = 2.00 \text{ m}^3$

The work done on the gas is the negative of the area under the curve $P = \alpha V^2$, from V_i to V_f.

$W = -\int_{V_i}^{V_f} \alpha V^2\,dV = -\tfrac{1}{3}\alpha\left(V_f^3 - V_i^3\right)$

$W = -\tfrac{1}{3}\left(5.00 \text{ atm/m}^6\right)\left(1.013 \times 10^5 \text{ Pa/atm}\right)\left[\left(2.00 \text{ m}^3\right)^3 - \left(1.00 \text{ m}^3\right)^3\right] = -1.18 \times 10^6 \text{ J}$ ◊

---

25. An ideal gas is enclosed in a cylinder with a movable piston on top of it. The piston has a mass of 8 000 g and an area of 5.00 cm^2 and is free to slide up and down, keeping the pressure of the gas constant. How much work is done on the gas as the temperature of 0.200 mol of the gas is raised from $20.0 \degree C$ to $300 \degree C$?

Solution

For constant pressure, $W = -P\Delta V = -P\left(\dfrac{nR}{P}\right)(T_h - T_c)$

Therefore, $W = -nR\,\Delta T = -(0.200 \text{ mol})(8.314 \text{ J/mol·K})(280 \text{ K}) = -466 \text{ J}$ ◊

29. A thermodynamic system undergoes a process in which its internal energy decreases by 500 J. At the same time, 220 J of work is done on the system. Find the energy transferred to or from it by heat.

Solution $\Delta E_{int} = Q + W$, where W is positive, because work is done **on** the system:

$$Q = \Delta E_{int} - W = -500 \text{ J} - 220 \text{ J} = -720 \text{ J}$$

+720 J of energy is transferred **from** the system by heat. ◊

35. An ideal gas initially at 300 K undergoes an isobaric expansion at 2.50 kPa. If the volume increases from 1.00 m³ to 3.00 m³ and 12.5 kJ is transferred to the gas by heat, what are (a) the change in its internal energy and (b) its final temperature?

Solution

We use the energy version of the nonisolated system model.

(a) $$\Delta E_{int} = Q + W \quad \text{where} \quad W = -P\Delta V$$

so that $$\Delta E_{int} = Q - P\Delta V$$

$$\Delta E_{int} = 1.25 \times 10^4 \text{ J} - \left(2.50 \times 10^3 \text{ N / m}^2\right)\left(3.00 \text{ m}^3 - 1.00 \text{ m}^3\right) = 7\,500 \text{ J} \quad ◊$$

(b) Since $$\frac{V_1}{T_1} = \frac{V_2}{T_2}, \qquad T_2 = \left(\frac{V_2}{V_1}\right)T_1 = \left(\frac{3.00 \text{ m}^3}{1.00 \text{ m}^3}\right)(300 \text{ K}) = 900 \text{ K} \quad ◊$$

37. How much work is done on the steam when 1.00 mol of water at 100 °C boils and becomes 1.00 mol of steam at 100 °C at 1.00 atm pressure? Assuming the steam to behave as an ideal gas, determine the change in internal energy of the material as it vaporizes.

Solution $$W = -P\Delta V = -P\left(V_s - V_w\right)$$

Substituting, $$PV_s = nRT \quad \text{and} \quad PV_w = nM\left(\frac{P}{\rho}\right)$$

Calculating each work term, $PV_s = (1.00 \text{ mol})\left(8.314 \dfrac{\text{J}}{\text{K}\cdot\text{mol}}\right)(373\text{ K}) = 3\,101\text{ J}$

$$PV_w = (1.00\text{ mol})(18.0\text{ g / mol})\left(\dfrac{1.013\times10^5\text{ N / m}^2}{1.00\times10^6\text{ g / m}^3}\right)$$

$$PV_w = 1.82\text{ J}$$

Thus the work done is $W = -3.10\text{ kJ}$ ◊

$$Q = mL_V = (18.0\text{ g})\left(2.26\times10^6\text{ J / kg}\right) = 40.7\text{ kJ}$$

$$\Delta E_{\text{int}} = Q + W = 37.6\text{ kJ}$$ ◊

---

39. A 2.00-mol sample of helium gas initially at 300 K and 0.400 atm is compressed isothermally to 1.20 atm. Noting that the helium behaves as an ideal gas, find (a) the final volume of the gas, (b) the work done on the gas, and (c) the energy transferred by heat.

Solution

(a) Rearranging $PV = nRT$,

we get $V_i = \dfrac{nRT}{P_i}$

$$V_i = \dfrac{(2.00\text{ mol})(8.314\text{ J / mol}\cdot\text{K})(300\text{ K})}{(0.400\text{ atm})(1.013\times10^5\text{ Pa / atm})}\left(\dfrac{1\text{ Pa}}{\text{N / m}^2}\right) = 0.123\text{ m}^3$$

For isothermal compression, PV is constant, so $P_iV_i = P_fV_f$:

$$V_f = V_i\left(\dfrac{P_i}{P_f}\right) = (0.123\text{ m}^3)\left(\dfrac{0.400\text{ atm}}{1.20\text{ atm}}\right) = 0.041\,0\text{ m}^3$$ ◊

(b) $W = -\int P\,dV$: $W = -\int\dfrac{nRT}{V}\,dV = -nRT\ln\left(\dfrac{V_f}{V_i}\right) = -(4\,988\text{ J})\ln\left(\tfrac{1}{3}\right) = +5.48\text{ kJ}$ ◊

(c) $\Delta E_{\text{int}} = 0 = Q + W$: $Q = -5.48\text{ kJ}$ ◊

43. A bar of gold is in thermal contact with a bar of silver of the same length and area (Fig. P20.43). One end of the compound bar is maintained at 80.0 °C while the opposite end is at 30.0 °C. When the energy transfer reaches steady state, what is the temperature at the junction?

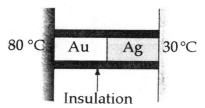

80 °C   Au   Ag   30°C

Insulation

Figure P20.43

Solution Call the gold bar Object 1 and the silver bar Object 2. Each is a nonisolated system in steady state. When energy transfer by heat reaches a steady state, the flow rate through each will be the same:

$$\mathcal{P}_1 = \mathcal{P}_2 \qquad \text{or} \qquad \frac{k_1 A_1 \Delta T}{L_1} = \frac{k_2 A_2 \Delta T}{L_2}$$

In this case, $\qquad L_1 = L_2 \qquad$ and $\qquad A_1 = A_2$

so $\qquad\qquad\qquad\qquad\qquad\qquad\qquad\qquad k_1 \Delta T_1 = k_2 \Delta T_2$

Let T_3 be the temperature at the junction; then $\qquad k_1 (80.0\ °C - T_3) = k_2 (T_3 - 30.0\ °C)$

Rearranging, we find $\qquad\qquad\qquad\qquad T_3 = \dfrac{(80.0\ °C)k_1 + (30.0\ °C)k_2}{k_1 + k_2}$

$$T_3 = \frac{(80.0\ °C)(314\ W\ /\ m \cdot °C) + (30.0\ °C)(427\ W\ /\ m \cdot °C)}{(314\ W\ /\ m \cdot °C) + (427\ W\ /\ m \cdot °C)} = 51.2\ °C \qquad \Diamond$$

55. An aluminum rod 0.500 m in length and with a cross-sectional area of 2.50 cm² is inserted into a thermally insulated vessel containing liquid helium at 4.20 K. The rod is initially at 300 K. (a) If half of the rod is inserted into the helium, how many liters of helium boil off by the time the inserted half cools to 4.20 K? (Assume the upper half does not yet cool.) (b) If the upper end of the rod is maintained at 300 K, what is the approximate boil-off rate of liquid helium after the lower half has reached 4.20 K? (Aluminum has thermal conductivity of 31.0 J/s·cm·K at 4.2 K; ignore its temperature variation. Aluminum has a specific heat of 0.210 cal/g·°C and density of 2.70 g/cm³. The density of liquid helium is 0.125 g/cm³.)

Solution

Conceptualize: Demonstrations with liquid nitrogen give us some indication of the phenomenon described. Since the rod is much hotter than the liquid helium and of significant size (almost 2 cm in diameter), a substantial volume (maybe as much as a liter) of helium will boil off before thermal equilibrium is reached. Likewise, since aluminum conducts rather well, a significant amount of helium will continue to boil off as long as the upper end of the rod is maintained at 300 K.

Categorize: Until thermal equilibrium is reached, the excess internal energy of the rod will be used to vaporize the liquid helium, which is already at its boiling point (so there is no change in the temperature of the helium).

Analyze:

As you solve this problem, be careful not to confuse L (the **conduction length** of the rod) with L_v (the **heat of vaporization** of the helium).

(a) Before heat conduction has time to become important, we suppose the heat energy lost by half the rod equals the heat energy gained by the helium. Therefore,

$$(mL_v)_{He} = (mc\,\Delta T)_{Al} \quad \text{or} \quad (\rho V L_v)_{He} = (\rho V c \,\Delta T)_{Al}$$

So $\quad V_{He} = \dfrac{(\rho V c \,\Delta T)_{Al}}{(\rho L_v)_{He}} = \dfrac{(2.70 \text{ g / cm}^3)(62.5 \text{ cm}^3)(0.210 \text{ cal / g} \cdot {}^\circ\text{C})(295.8 \ {}^\circ\text{C})}{(0.125 \text{ g / cm}^3)(4.99 \text{ cal / g})}$

$$V_{He} = 1.68 \times 10^4 \text{ cm}^3 = 16.8 \text{ liters} \qquad \lozenge$$

(b) Heat will be conducted along the rod at $\qquad \dfrac{dQ}{dt} = \mathcal{P} = \dfrac{k\,A\Delta T}{L}$ [1]

and will boil off helium according to $Q = mL_v$: $\qquad \dfrac{dQ}{dt} = \left(\dfrac{dm}{dt}\right)L_v$ [2]

Combining [1] and [2] gives us the "boil-off" rate: $\qquad \dfrac{dm}{dt} = \dfrac{k\,A\Delta T}{L \cdot L_v}$

Set the conduction length $L = 25$ cm, and use $k = 7.41$ cal/s·cm·K

$$\dfrac{dm}{dt} = \dfrac{(7.41 \text{ cal / s} \cdot \text{cm} \cdot \text{K})(2.50 \text{ cm}^2)(295.8 \text{ K})}{(25.0 \text{ cm})(4.99 \text{ cal / g})} = 43.9 \text{ g / s}$$

or $\quad \dfrac{dV}{dt} = \dfrac{43.9 \text{ g / s}}{0.125 \text{ g / cm}^3} = 351 \text{ cm}^3 \text{ / s} = 0.351 \text{ liters / s} \qquad \lozenge$

Finalize: The volume of helium boiled off initially is much more than expected. If our calculations are correct, that sure is a lot of liquid helium that is wasted! Since liquid helium is much more expensive than liquid nitrogen, most low-temperature equipment is designed to avoid unnecessary loss of liquid helium by surrounding the liquid with a container of liquid nitrogen.

63. A solar cooker consists of a curved reflecting surface that concentrates sunlight onto the object to be warmed (Fig. P20.63). The solar power per unit area reaching the Earth's surface at the location is 600 W/m². The cooker faces the Sun and has a diameter of 0.600 m. Assume that 40.0% of the incident energy is transferred to 0.500 L of water in an open container, initially at 20.0 °C. How long does it take to completely boil away the water? (Ignore the heat capacity of the container.)

Figure P20.63

Solution

If we point the axis of the reflecting surface toward the Sun, the power incident on the solar collector is

$$\mathcal{P}_i = IA = \left(600 \text{ W / m}^2\right)\left[\pi(0.300 \text{ m})^2\right] = 170 \text{ W}$$

For a 40.0%-efficient reflector, the collected power is

$$\mathcal{P}_i = (0.400)(170 \text{ W}) = 67.9 \text{ W} = 67.9 \text{ J / s}$$

The total energy required to increase the temperature of the water to the boiling point and to evaporate it is

$$Q = mc\Delta T + mL_v$$

$$Q = (0.500 \text{ kg})(4\,186 \text{ J / kg} \cdot {}^\circ\text{C})(80.0 \text{ }^\circ\text{C}) + (0.500 \text{ kg})\left(2.26 \times 10^6 \text{ J / kg}\right)$$

$$Q = 1.30 \times 10^6 \text{ J}$$

The time required is

$$\Delta t = \frac{Q}{\mathcal{P}_c} = \frac{1.30 \times 10^6 \text{ J}}{67.9 \text{ J / s}} = 1.91 \times 10^4 \text{ s} = 5.31 \text{ h}$$

◊

71. The passenger section of a jet airliner is in the shape of a cylindrical tube with a length of 35.0 m and an inner radius of 2.50 m. Its walls are lined with an insulating material 6.00 cm in thickness and having a thermal conductivity of 4.00×10^{-5} cal/s·cm·°C. A heater must maintain the interior temperature at 25.0 °C while the outside temperature is −35.0 °C. What power must be supplied to the heater? (Use the result of Problem 70.)

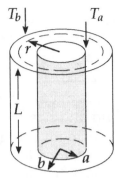

Figure P20.70

Solution

From Problem 70, the rate of heat flow through the wall is:

$$\frac{dQ}{dt} = \frac{2\pi kL(T_a - T_b)}{\ln(b/a)}$$

$$\frac{dQ}{dt} = \frac{2\pi(3\,500\ \text{cm})\left(4.00 \times 10^{-5}\ \text{cal / s} \cdot \text{cm} \cdot {}^\circ\text{C}\right)(60.0\ {}^\circ\text{C})}{\ln(2.56/2.50)}$$

and $\quad \dfrac{dQ}{dt} = 2.23 \times 10^3$ cal / s $= 9.32$ kW $\qquad\qquad \Diamond$

This is the rate of heat loss from the plane, and consequently the rate at which energy must be supplied in order to maintain an equilibrium temperature.

Chapter 21
THE KINETIC THEORY OF GASES

EQUATIONS AND CONCEPTS

The **pressure of an ideal gas** is proportional to the number of molecules per unit volume and the average translational kinetic energy per molecule.

$$P = \frac{2}{3}\left(\frac{N}{V}\right)\left(\frac{1}{2}m\overline{v^2}\right) \tag{21.2}$$

The **absolute temperature of an ideal gas** is a measure of the average molecular kinetic energy. *This result follows from Equation 21.2 and the equation of state for an ideal gas, $PV = Nk_BT$.*

$$T = \frac{2}{3k_B}\left(\frac{1}{2}m\overline{v^2}\right) \tag{21.3}$$

The **internal energy** of a sample of an ideal gas depends only on the absolute temperature. *In Equation 21.10, N is the number of molecules and n is the number of moles.*

$$E_{\text{int}} = \frac{3}{2}Nk_BT = \frac{3}{2}nRT \tag{21.10}$$

The **change in internal energy** of an ideal gas equals the energy transferred by heat at constant volume; and for a given change in temperature, depends on the molar specific heat at constant volume, C_V.

$$\Delta E_{\text{int}} = nC_V\Delta T \tag{21.12}$$

$$C_V = \frac{3}{2}R \qquad \text{(monatomic only)} \tag{21.14}$$

The **molar specific heat of an ideal gas at constant pressure** is greater than the specific heat at constant volume by an amount R. C_P is greater than C_V for the following reason. *An increase in internal energy is a function of temperature only; ΔE_{int} must be the same for a given ΔT whether by constant pressure or by constant volume. When energy is transferred to a gas by heat at constant pressure, the volume increases and* ***a portion of the added energy leaves the system by work.*** *The remaining portion contributes to an increase in internal energy. Therefore the added energy required per mole must be greater at constant pressure to effect the same temperature change.*

$$C_P - C_V = R \quad \text{(any ideal gas)} \qquad (21.16)$$

$$C_P = \frac{5}{2}R \qquad \text{(monatomic)}$$

The **ratio of specific heats**, C_P/C_V is a dimensionless quantity γ.

$$\gamma = \frac{C_P}{C_V} \qquad \text{(any gas)} \qquad (21.17)$$

An **adiabatic process** (expansion or compression) is one in which no energy is transferred by heat between the system and its environment. *Each of the three variables, P, V, and T change during an adiabatic process.*

$$PV^{\gamma} = \text{constant} \qquad (21.18)$$

or

$$P_i V_i^{\gamma} = P_f V_f^{\gamma} \qquad (21.19)$$

A relationship between **temperature and volume** during an adiabatic process can be found by using $PV = nRT$.

$$T_i V_i^{\gamma-1} = T_f V_f^{\gamma-1} \qquad (21.20)$$

The **Boltzmann distribution law** (which is valid for a system of a large number of particles) states that the probability of finding the particles in a particular arrangement (distribution of energy values) varies exponentially as the negative of the energy divided by $k_B T$. *In Equation 21.25, $n_V(E)$ is a function called the number density.*

$$n_V(E) = n_0 e^{-E/k_B T} \qquad (21.25)$$

The **Maxwell - Boltzmann speed distribution function** describes the most probable distribution of speeds of N gas molecules at temperature T (where k_B is the Boltzmann constant). *The number of molecules with speeds between v and $v + dv$ is $dN = N_v dv$. The probability that a molecule has a speed in the range between v and $v + dv$ is equal to dN/N.*

$$N_v = 4\pi N \left(\frac{m}{2\pi k_B T}\right)^{3/2} v^2 e^{-mv^2/2k_B T} \quad (21.26)$$

Root-mean-square speed, average speed, and **most probable speed** can be calculated from the Maxwell-Boltzmann distribution function.

$$v_{rms} = 1.73\sqrt{k_B T / m} \quad (21.27)$$

$$\bar{v} = 1.60\sqrt{k_B T / m} \quad (21.28)$$

$$v_{mp} = 1.41\sqrt{k_B T / m} \quad (21.29)$$

The **mean free path** of a molecule in a gas sample is the average distance between collisions (where d is the molecular "diameter" and n_V is the density of molecules).

$$\ell = \frac{1}{\sqrt{2}\pi d^2 n_V} \quad (21.30)$$

SUGGESTIONS, SKILLS, AND STRATEGIES

CALCULATING SPECIFIC HEATS

It is important to recognize that not all ideal gases are equal. Depending on the type of molecule and the number of degrees of freedom, the ratio of specific heats for a gas (γ) might be 1.67, 1.40, or some other value. However, in any ideal gas, two equations are always valid and can be used to derive the specific heats C_P and C_V from a known value of γ:

$$\gamma = C_P/C_V \qquad \text{and} \qquad C_P - C_V = R$$

Solving the first equation for C_V and substituting into the second equation, we find

$$C_V = \frac{R}{\gamma - 1} \qquad \text{and} \qquad C_P = \frac{\gamma R}{\gamma - 1}$$

These equations may be used for all ideal gases.

REVIEW CHECKLIST

You should be able to:

▷ State and understand the assumptions made in developing the molecular model of an ideal gas.

▷ Calculate pressure, average translational kinetic energy per molecule or rms speed per molecule and quantity (moles) of a confined gas. (Section21.1)

▷ Calculate Q, ΔE_{int}, ΔT, and W for a constant pressure process. (Section 21.2)

▷ Calculate Q, ΔE_{int}, and ΔT for a constant volume process. (Section 21.2)

▷ Make calculations involving an adiabatic process for an ideal gas. (Section 21.3)

▷ Find values for average speed, root-mean-square speed, and most probable speed when given the individual speed for each of a number of particles. (Section 21.6)

▷ Determine the mean free path and density of particles in a gas. (Section 21.7)

Some important points to remember:

▷ The temperature of an ideal gas is proportional to the average molecular kinetic energy. In a gas sample containing molecules with different values of mass, the average speed of the more massive molecules will be smaller.

▷ According to the theorem of equipartition of energy, each degree of freedom of a molecule contributes equally to the total internal energy of a gas.

ANSWERS TO SELECTED CONCEPTUAL QUESTIONS

5. When alcohol is rubbed on your body, it lowers your skin temperature. Explain this effect.

Answer As the alcohol evaporates, high-speed molecules leave the liquid. This reduces the average speed of the remaining molecules. Since the average speed is lowered, the temperature of the alcohol is reduced. This process helps to carry energy away from the skin of the patient, resulting in cooling of the skin. The alcohol plays the same role of evaporative cooling as does perspiration, but alcohol evaporates much more quickly than perspiration.

□   □   □   □

9. If a helium-filled balloon initially at room temperature is placed in a freezer, will its volume increase, decrease, or remain the same?

Answer The helium is far from liquefaction. Therefore, we model it as an ideal gas, described by $PV = nRT$. In this case, the pressure stays nearly constant, being equal to 1 atm. Since the temperature may decrease by 10% (from 293 K to 263 K), the volume should also **decrease** by 10%. This process is called "isobaric cooling", or "isobaric contraction."

□   □   □   □

11. What happens to a helium-filled balloon released into the air? Will it expand or contract? Will it stop rising at some height?

Answer Imagine the balloon rising into the air. The air cannot be uniform in pressure, because the lower layers support the weight of all the air above. Therefore, as the balloon rises, the pressure will decrease. At the same time, the temperature decreases slightly, but not enough to overcome the effects of the pressure.

By the ideal gas law, $PV = nRT$. Since the decrease in temperature has little effect, the decreasing pressure results in an increasing volume; thus the balloon expands quite dramatically. By the time the balloon reaches an altitude of 10 miles, its volume will be 90 times larger.

Long before that happens, one of two events will occur. The first possibility, of course, is that the balloon will break, and the pieces will fall to the earth. The other possibility is that the balloon will come to a spot where the average density of the balloon and its payload is equal to the average density of the air. At that point, buoyancy will cause the balloon to "float" at that altitude, until it loses helium and descends.

People who remember releasing balloons with pen-pal notes will perhaps also remember that replies most often came back when the balloons were low on helium, and were barely able to take off. Such balloons found a fairly low floatation altitude, and were the most likely to be found intact at the end of their journey.

□   □   □   □

SOLUTIONS TO SELECTED END-OF-CHAPTER PROBLEMS

5. A spherical balloon with a volume of $4\,000$ cm³ contains helium at an (inside) pressure of 1.20×10^5 Pa. How many moles of helium are in the balloon if the average kinetic energy of the helium atoms is 3.60×10^{-22} J?

Solution

Conceptualize: The balloon has a volume of 4.00 L and a diameter of about 20 cm, which seems like a reasonable size for a typical toy helium balloon. The pressure of the balloon is only slightly more than 1 atm, and if the temperature is anywhere close to room temperature, we can use the estimate of 22 L/mol for an ideal gas at STP conditions. If this is valid, the balloon should contain about 0.2 moles of helium.

Categorize: The average kinetic energy can be used to find the temperature of the gas, which can be used with $PV = nRT$ to find the number of moles.

Analyze:

The gas temperature must be that implied by $\frac{1}{2}m\overline{v^2} = \frac{3}{2}k_B T$ for a monatomic gas like helium.

$$T = \frac{2}{3}\left(\frac{\frac{1}{2}m\overline{v^2}}{k_B}\right) = \frac{2}{3}\left(\frac{3.60 \times 10^{-22}\ \text{J}}{1.38 \times 10^{-23}\ \text{J / K}}\right) = 17.4\ \text{K}$$

Now $PV = nRT$ gives

$$n = \frac{PV}{RT} = \frac{(1.20 \times 10^5\ \text{N / m}^2)(4.00 \times 10^{-3}\ \text{m}^3)}{(8.314\ \text{J / mol} \cdot \text{K})(17.4\ \text{K})} = 3.32\ \text{mol} \qquad \lozenge$$

Finalize: This result is more than ten times the number of moles we predicted, primarily because the temperature of the helium is much colder than room temperature. In fact, T is only slightly above the temperature at which the helium would liquefy (4.2 K at 1 atm). We should hope this balloon is not being held by a child; not only would the balloon sink in the air, it is cold enough to cause frostbite!

7. (a) How many atoms of helium gas fill a balloon having a diameter of 30.0 cm at 20.0 °C and 1.00 atm? (b) What is the average kinetic energy of the helium atoms? (c) What is the root-mean-square speed of the helium atoms?

Solution

(a) The volume is

$$V = \frac{4}{3}\pi r^3 = \frac{4}{3}\pi(0.150 \text{ m})^3 = 1.41 \times 10^{-2} \text{ m}^3$$

Now $PV = nRT$:

$$n = \frac{PV}{RT} = \frac{(1.013 \times 10^5 \text{ N / m}^2)(1.41 \times 10^{-2} \text{ m}^3)}{(8.314 \text{ N} \cdot \text{m / mol} \cdot \text{K})(293 \text{ K})}$$

$$n = 0.588 \text{ mol}$$

$$N = nN_A = (0.588 \text{ mol})(6.02 \times 10^{23} \text{ molecules / mol})$$

$$N = 3.54 \times 10^{23} \text{ helium atoms} \qquad \diamond$$

(b) The kinetic energy is

$$\overline{K} = \frac{1}{2}m\overline{v^2} = \frac{3}{2}k_B T$$

$$\overline{K} = \frac{3}{2}(1.38 \times 10^{-23} \text{ J / K})(293 \text{ K}) = 6.07 \times 10^{-21} \text{ J} \qquad \diamond$$

(c) An He atom has mass

$$m = \frac{M}{N_A} = \frac{4.002\,6 \text{ g / mol}}{6.02 \times 10^{23} \text{ molecules / mol}}$$

$$m = 6.65 \times 10^{-24} \text{ g} = 6.65 \times 10^{-27} \text{ kg}$$

So the kinetic energy is

$$\frac{1}{2}(6.65 \times 10^{-27} \text{ kg})\overline{v^2} = 6.07 \times 10^{-21} \text{ J}$$

and

$$v_{\text{rms}} = \sqrt{\overline{v^2}} = 1.35 \text{ km / s} \qquad \diamond$$

9. A cylinder contains a mixture of helium and argon gas in equilibrium at 150 °C. (a) What is the average kinetic energy for each type of gas molecule? (b) What is the root-mean-square speed of each type of molecule?

Solution

(a) Both kinds of molecules have the same average kinetic energy.

$$\frac{1}{2}m\overline{v^2} = \frac{3}{2}k_B T = \frac{3}{2}(1.38 \times 10^{-23} \text{ J / K})(273 + 150) \text{ K}$$

$$\overline{K} = 8.76 \times 10^{-21} \text{ J} \qquad \diamond$$

(b)   The root-mean square velocity can be calculated from the kinetic energy:

$$v_{rms} = \sqrt{\overline{v^2}} = \sqrt{\frac{2\overline{K}}{m}}$$

These two gases are noble, and therefore monatomic. The masses of the molecules are:

$$m_{He} = \frac{(4.00 \text{ g} / \text{mol})(10^{-3} \text{ kg} / \text{g})}{6.02 \times 10^{23} \text{ atoms} / \text{mol}} = 6.64 \times 10^{-27} \text{ kg}$$

$$m_{Ar} = \frac{(39.9 \text{ g} / \text{mol})(10^{-3} \text{ kg} / \text{g})}{6.02 \times 10^{23} \text{ atoms} / \text{mol}} = 6.63 \times 10^{-26} \text{ kg}$$

Substituting these values, $v_{rms, \text{ He}} = \sqrt{\dfrac{2(8.76 \times 10^{-21} \text{ J})}{6.64 \times 10^{-27} \text{ kg}}} = 1.62 \times 10^3 \text{ m} / \text{s}$

and $v_{rms, \text{ Ar}} = \sqrt{\dfrac{2(8.76 \times 10^{-21} \text{ J})}{6.63 \times 10^{-26} \text{ kg}}} = 514 \text{ m} / \text{s}$ ◊

---

13.   A 1.00-mol sample of hydrogen gas is heated at constant pressure from 300 K to 420 K. Calculate (a) the energy transferred to the gas by heat, (b) the increase in its internal energy, and (c) the work done on the gas.

Solution   Since this is a constant-pressure process, $Q = nC_P\Delta T$

(a)   The temperature rises by $\Delta T = 420 \text{ K} - 300 \text{ K} = 120 \text{ K}$

$Q = (1.00 \text{ mol})(28.8 \text{ J} / \text{mol} \cdot \text{K})(120 \text{ K}) = 3.46 \text{ kJ}$ ◊

(b)   For any gas $\Delta E_{int} = nC_V\Delta T$

so $\Delta E_{int} = (1.00 \text{ mol})(20.4 \text{ J} / \text{mol} \cdot \text{K})(120 \text{ K}) = 2.45 \text{ kJ}$

(c) $\Delta E_{int} = Q + W$ so $W = \Delta E_{int} - Q = 2.45 \text{ kJ} - 3.46 \text{ kJ} = -1.01 \text{ kJ}$ ◊

---

25. A 2.00-mol sample of a diatomic ideal gas expands slowly and adiabatically from a pressure of 5.00 atm and a volume of 12.0 L to a final volume of 30.0 L. (a) What is the final pressure of the gas? (b) What are the initial and final temperatures? (c) Find Q, W, and ΔE_{int}.

Solution

(a) $P_i V_i^{\gamma} = P_f V_f^{\gamma}$: $P_f = P_i \left(\dfrac{V_i}{V_f}\right)^{\gamma} = (5.00 \text{ atm})\left(\dfrac{12.0 \text{ L}}{30.0 \text{ L}}\right)^{1.40} = 1.39 \text{ atm}$ ◊

(b) $T_i = \dfrac{P_i V_i}{nR} = \dfrac{(5.00 \text{ atm})(1.013 \times 10^5 \text{ Pa / atm})(12.0 \times 10^{-3} \text{ m}^3)}{(2.00 \text{ mol})(8.314 \text{ N} \cdot \text{m / mol} \cdot \text{K})} = 366 \text{ K}$

$T_f = \dfrac{P_f V_f}{nR} = 253 \text{ K}$ ◊

(c) This is an adiabatic process, so by the definition $Q = 0$ ◊

For any process, $\Delta E_{int} = nC_V \Delta T$

and for this gas, $C_V = \dfrac{R}{\gamma - 1} = \dfrac{5}{2} R$

Thus, $\Delta E_{int} = \dfrac{5}{2}(2.00 \text{ mol})(8.314 \text{ J / mol} \cdot \text{K})(253 \text{ K} - 366 \text{ K}) = -4660 \text{ J}$ ◊

Now, $W = \Delta E_{int} - Q = -4660 \text{ J} - 0 = -4660 \text{ J}$ ◊

Note that in this case the work done on the gas is negative, so positive work is done by the gas.

═══════════════════════════════

27. Air in a thundercloud expands as it rises. If its initial temperature was 300 K and if no energy is lost by thermal conduction on expansion, what is its temperature when the initial volume has doubled?

Solution

Conceptualize: The air should cool as it expands, so we should expect $T_f < 300 \text{ K}$.

Categorize: The air expands adiabatically, losing no heat but dropping in temperature as it does work on the air around it, so we assume that $PV^{\gamma} = $ constant (where $\gamma = 1.40$ for an ideal diatomic gas).

Analyze: By Equation 21.20, $\quad T_1 V_1^{\gamma-1} = T_2 V_2^{\gamma-1}$

$$T_2 = T_1 \left(\frac{V_1}{V_2}\right)^{\gamma-1} = 300 \text{ K} \left(\frac{1}{2}\right)^{(1.40-1)} = 227 \text{ K} \qquad \lozenge$$

Finalize: The air does cool, but the temperature is not inversely proportional to the volume. The temperature drops 24% while the volume doubles.

---

33. Consider 2.00 mol of an ideal diatomic gas. (a) Find the total heat capacity at constant volume and at constant pressure assuming the molecules rotate but do not vibrate. (b) **What if?** Repeat, assuming the molecules both rotate and vibrate.

Solution We use the kinetic theory structural model of an ideal gas.

(a) Count degrees of freedom. A diatomic molecule oriented along the y axis can possess energy by moving in x, y and z directions and by rotating around x and z axes. Rotation around the y axis does not represent an energy contribution because the moment of inertia of the molecule about this axis is essentially zero. The molecule will have an average energy $\frac{1}{2} k_B T$ for each of these five degrees of freedom.

The gas will have internal energy $\qquad E_{\text{int}} = N\left(\frac{5}{2}\right) k_B T = n N_A \left(\frac{5}{2}\right)\left(\frac{R}{N_A}\right) T = \frac{5}{2} nRT$

so the constant-volume heat capacity of the whole sample is

$$\frac{\Delta E_{\text{int}}}{\Delta T} = \frac{5}{2} nR = \frac{5}{2}(2.00 \text{ mol})(8.314 \text{ J / mol} \cdot \text{K}) = 41.6 \text{ J / K} \qquad \lozenge$$

For one mole, $\qquad C_P = C_V + R \qquad$ For the sample, $\qquad n C_P = n C_V + nR$

With P constant, $\qquad n C_P = 41.6 \text{ J / K} + (2.00 \text{ mol})(8.314 \text{ J / mol} \cdot \text{K}) = 58.2 \text{ J / K} \qquad \lozenge$

(b) Vibration adds a degree of freedom for kinetic energy and a degree of freedom for elastic energy. Now the molecule's average energy is $\frac{7}{2} k_B T$,

the sample's internal energy is $\qquad N\left(\frac{7}{2}\right) k_B T = \frac{7}{2} nRT$

and the sample's constant-volume heat capacity is

$$\frac{7}{2} nR = \frac{7}{2}(2.00 \text{ mol})(8.314 \text{ J / mol} \cdot \text{K}) = 58.2 \text{ J / K} \qquad \lozenge$$

At constant pressure, its heat capacity is

$$n C_V + nR = 58.2 \text{ J / K} + (2.00 \text{ mol})(8.314 \text{ J / mol} \cdot \text{K}) = 74.8 \text{ J / K} \qquad \lozenge$$

---

37. Fifteen identical particles have various speeds: one has a speed of 2.00 m/s; two have speeds of 3.00 m/s; three have speeds of 5.00 m/s; four have speeds of 7.00 m/s; three have speeds of 9.00 m/s; and two have speeds of 12.0 m/s. Find (a) the average speed, (b) the rms speed, and (c) the most probable speed of these particles.

Solution

(a) $\bar{v} = \dfrac{\sum n_i v_i}{\sum n_i} = \dfrac{1(2.00) + 2(3.00) + 3(5.00) + 4(7.00) + 3(9.00) + 2(12.0)}{1 + 2 + 3 + 4 + 3 + 2}$ m/s

$\bar{v} = 6.80$ m/s ◊

(b) $\overline{v^2} = \dfrac{\sum n_i v_i^2}{\sum n_i}$

$\overline{v^2} = \dfrac{1(2.00^2) + 2(3.00^2) + 3(5.00^2) + 4(7.00^2) + 3(9.00^2) + 2(12.0^2)}{15}$ m²/s²

$\overline{v^2} = 54.9$ m²/s²

$v_{rms} = \sqrt{\overline{v^2}} = \sqrt{54.9 \text{ m}^2/\text{s}^2} = 7.41$ m/s ◊

(c) More particles have $v_{mp} = 7.00$ m/s than any other speed. ◊

---

39. From the Maxwell-Boltzmann speed distribution, show that the most probable speed of a gas molecule is given by Equation 21.29. Note that the most probable speed corresponds to the point at which the slope of the speed distribution curve dN_v/dv is zero.

Solution From the Maxwell speed distribution function,

$$N_v = 4\pi N \left(\frac{m}{2\pi k_B T}\right)^{3/2} v^2 e^{-mv^2/2k_B T}$$

we locate the peak in the graph of N_v versus v by evaluating dN_v/dv and setting it equal to zero, to solve for the most probable speed:

$$\frac{dN_v}{dv} = 4\pi N \left(\frac{m}{2\pi k_B T}\right)^{3/2} \left(\frac{-2mv}{2k_B T}\right) v^2 e^{-mv^2/2k_B T} + 4\pi N \left(\frac{m}{2\pi k_B T}\right)^{3/2} 2v e^{-mv^2/2k_B T} = 0$$

This equation has solutions $v=0$ and $v \to \infty$, but those correspond to minimum-probability speeds, so we divide by v and by the exponential function.

$$v_{mp}\left(-\frac{m(2v_{mp})}{2k_BT}\right)+2=0 \qquad \text{or} \qquad v_{mp}=\sqrt{\frac{2k_BT}{m}}$$

This is Equation 21.29.   ◊

45. In an ultra-high-vacuum system, the pressure is measured to be 1.00×10^{-10} torr (where 1 torr $= 133$ Pa). Assuming the molecular diameter is 3.00×10^{-10} m, the average molecular speed is 500 m/s, and the temperature is 300 K, find (a) the number of molecules in a volume of 1.00 m^3, (b) the mean free path of the molecules, and (c) the collision frequency.

Solution

Conceptualize: Since high vacuum means low pressure as a result of a low molecular density, we should expect a relatively low number of molecules, a long free path, and a low collision frequency compared with the values found in Example 21.6 for normal air. Since the ultrahigh vacuum absolute pressure is 13 orders of magnitude lower than atmospheric pressure, we might expect corresponding values of $N \sim 10^{12}$ molecules/m^3, $\ell \sim 10^6$ m, and $f \sim 0.0001$ /s.

Categorize: The equation of state for an ideal gas can be used with the given information to find the number of molecules in a specific volume. The mean free path can be found directly from Equation 21.30, and this result can be used with the average speed to find the collision frequency.

Analyze:

(a) $PV = \left(\dfrac{N}{N_A}\right)RT$ means $N = \dfrac{PVN_A}{RT}$,   so that

$$N = \frac{(1.00 \times 10^{-10})(133)(1.00)(6.02 \times 10^{23})}{(8.314)(300)} = 3.21 \times 10^{12} \text{ molecules}$$ ◊

(b) $\ell = \dfrac{1}{n_V \pi d^2 \sqrt{2}} = \dfrac{V}{N\pi d^2 \sqrt{2}} = \dfrac{1.00 \text{ m}^3}{(3.21 \times 10^{12} \text{ molecules})\pi(3.00 \times 10^{-10} \text{ m})^2 \sqrt{2}}$

$\ell = 7.79 \times 10^5$ m $= 779$ km   ◊

(c) $f = \dfrac{v}{\ell} = \dfrac{500 \text{ m/s}}{7.78 \times 10^5 \text{ m}} = 6.42 \times 10^{-4} \text{ s}^{-1}$ ◊

Finalize: The calculated results differ from the results in Example 21.6 by about 13 orders of magnitude, as we expected. This ultrahigh vacuum is an environment very different from the ocean floor to which we are accustomed, under the Earth's atmosphere. These conditions can provide a "clean" environment for a variety of experiments and manufacturing processes that would otherwise be impossible.

---

53. A cylinder containing n mol of an ideal gas that undergoes an adiabatic process. (a) Starting with the expression $W = -\int PdV$ and using $PV^\gamma = \text{constant}$, show that the work done on the gas is

$$W = \left(\frac{1}{\gamma - 1}\right)(P_f V_f - P_i V_i)$$

(b) Starting with the first-law equation in differential form, prove that the work done on the gas also is equal to $nC_V(T_f - T_i)$. Show that this result is consistent with the equation in part (a).

Solution

(a) $PV^\gamma = k$ so $W = -\int_i^f PdV = -k\int_{V_i}^{V_f} \frac{dV}{V^\gamma} = \left. \frac{-kV^{1-\gamma}}{1-\gamma} \right|_{V_i}^{V_f}$

$$W = -\frac{P_f V_f^\gamma V_f^{1-\gamma} - P_i V_i^\gamma V_i^{1-\gamma}}{1-\gamma} = \frac{P_f V_f - P_i V_i}{\gamma - 1} \qquad \Diamond$$

(b) For an adiabatic process $dE_{int} = dQ + dW$ and $dQ = 0$

Therefore, $W = \Delta E_{int} = nC_V \Delta T = nC_V(T_f - T_i)$ $\qquad \Diamond$

To show consistency between these two equations,

consider that $\gamma = \dfrac{C_P}{C_V}$ and $C_P - C_V = R$,

so that $\dfrac{1}{\gamma - 1} = \dfrac{C_V}{R}$

Thus part (a) becomes $W = (P_f V_f - P_i V_i)\dfrac{C_V}{R}$

Then, $\dfrac{PV}{R} = nT$ so that $W = nC_V(T_f - T_i)$ $\qquad \Diamond$

---

59. The compressibility κ of a substance is defined as the fractional change in volume of that substance for a given change in pressure:

$$\kappa = -\frac{1}{V}\frac{dV}{dP}$$

(a) Explain why the negative sign in this expression ensures that κ is always positive. (b) Show that if an ideal gas is compressed isothermally, its compressibility is given by $\kappa_1 = 1/P$. (c) Show that if an ideal gas is compressed adiabatically, its compressibility is given by $\kappa_2 = 1/\gamma P$. (d) Determine values for κ_1 and κ_2 for a monatomic ideal gas at a pressure of 2.00 atm.

Solution   The pressure increases as volume decreases (and vice versa), so dV/dP is always negative.

(a)   In equation form, $\dfrac{dV}{dP} < 0$ and $-\left(\dfrac{1}{V}\right)\left(\dfrac{dV}{dP}\right) > 0$ ◊

(b)   For an ideal gas, $V = \dfrac{nRT}{P}$ and $\kappa_1 = -\dfrac{1}{V}\dfrac{d}{dP}\left(\dfrac{nRT}{P}\right)$

For isothermal compression, T is constant: $\kappa_1 = -\dfrac{nRT}{V}\left(\dfrac{-1}{P^2}\right) = \dfrac{1}{P}$ ◊

(c)   For an adiabatic compression, $PV^\gamma = C$ (where C is a constant) and

$$\kappa_2 = -\left(\frac{1}{V}\right)\frac{d}{dP}\left(\frac{C}{P}\right)^{1/\gamma} = \left(\frac{1}{V\gamma}\right)\frac{C^{1/\gamma}}{\left(P^{1/\gamma+1}\right)} = \frac{V}{V\gamma P} = \frac{1}{\gamma P} \qquad ◊$$

(d)   For a monatomic ideal gas, $\gamma = C_P/C_V = 5/3$:

$$\kappa_1 = \frac{1}{P} = \frac{1}{2.00\text{ atm}} = 0.500\text{ atm}^{-1} \qquad \kappa_2 = \frac{1}{\gamma P} = \frac{1}{(5/3)(2.00\text{ atm})} = 0.300\text{ atm}^{-1} \quad ◊$$

63.   For a Maxwellian gas, use a computer or programmable calculator to find the numerical value of the ratio $\{N_v(v)/N_v(v_{mp})\}$ for the following values of v: $v = (v_{mp}/50)$, $(v_{mp}/10)$, $(v_{mp}/2)$, v_{mp}, $2v_{mp}$, $10v_{mp}$, and $50v_{mp}$. Give your results to three significant figures.

Solution   Substitute $v_{mp} = \left(\dfrac{2k_BT}{m}\right)^{1/2}$

into $N_v(v) = 4\pi N\left(\dfrac{m}{2\pi k_BT}\right)^{3/2} v^2\, e^{-mv^2/2k_BT}$

and evaluate

$$N_v(v_{mp}) = 4\pi N\left(\frac{m}{2\pi k_B T}\right)^{3/2} \frac{2k_B T}{m} e^{-1}$$

$$N_v(v_{mp}) = 4\pi^{-1/2} e^{-1} N\left(\frac{m}{2k_B T}\right)^{1/2} = 4\pi^{-1/2} e^{-1} N v_{mp}^{-1}$$

Then,

$$\frac{N_v(v)}{N_v(v_{mp})} = \frac{4\pi^{-1/2} N v_{mp}^{-3} v^2 e^{-(v/v_{mp})^2}}{4\pi^{-1/2} N v_{mp}^{-1} e^{-1}} = \left(\frac{v^2}{v_{mp}^2}\right) e^{1-v^2/v_{mp}^2}$$

For $v = \dfrac{v_{mp}}{50}$,

$$\frac{N_v(v)}{N_v(v_{mp})} = \left(\frac{1}{50}\right)^2 e^{1-(1/50)^2} = 1.09 \times 10^{-3} \qquad \lozenge$$

For $v = \dfrac{v_{mp}}{10}$,

$$\frac{N_v(v)}{N_v(v_{mp})} = \left(\frac{1}{10}\right)^2 e^{1-1/100} = 2.69 \times 10^{-2} \qquad \lozenge$$

For $v = \dfrac{v_{mp}}{2}$,

$$\frac{N_v(v)}{N_v(v_{mp})} = \left(\frac{1}{2}\right)^2 e^{1-1/4} = 0.529 \qquad \lozenge$$

For $v = v_{mp}$,

$$\frac{N_v(v)}{N_v(v_{mp})} = (1)^2 e^{1-1} = 1.00 \qquad \lozenge$$

Similarly,

$\dfrac{v}{v_{mp}}$	$\dfrac{N_v(v)}{N_v(v_{mp})}$
1/50	1.09×10^{-3}
1/10	2.69×10^{-2}
1/2	0.529
1	1.00
2	0.199
10	1.01×10^{-41}
50	$1.25 \times 10^{-1\,082}$

To find the last,

$$\left(50^2\right)e^{1-2\,500} = \left(50^2\right)e^{-2\,499} = 10^{\log 2\,500}\, e^{\ln 10(-2\,499/\ln 10)}$$

$$= 10^{(\log 2\,500 - 2\,499/\ln 10)} = 10^{-1\,081.904}$$

Thus,

$$\left(50^2\right)e^{1-2\,500} = 10^{-1\,081.904} = 10^{0.096\,0}10^{-1\,082} = 1.25 \times 10^{-1\,082} \qquad \lozenge$$

Chapter 22

HEAT ENGINES, ENTROPY, AND THE SECOND LAW OF THERMODYNAMICS

EQUATIONS AND CONCEPTS

The **net work done by a heat engine** during one cycle equals the net heat energy, Q_{net}, flowing into the engine. $|Q_h|$ is the quantity of energy absorbed from a hot reservoir and $|Q_c|$ is the amount of energy expelled to the cold reservoir.

$$W_{eng} = |Q_h| - |Q_c| \qquad (22.1)$$

The **thermal efficiency**, e, of a heat engine is defined as the ratio of the net work done by the engine to the energy input at higher temperature during one cycle of the process. *For any real engine, $e < 1$.*

$$e = \frac{W_{eng}}{|Q_h|} = 1 - \frac{|Q_c|}{|Q_h|} \qquad (22.2)$$

The **effectiveness of a heat pump** is described in terms of the coefficient of performance, COP.

$$\text{COP (heating mode)} \equiv$$
$$\frac{\text{energy transferred at high temp}}{\text{work done on pump}} = \frac{|Q_h|}{W} \qquad (22.3)$$

The **coefficient of performance** of a refrigerator or a heat pump used in the cooling cycle is defined by the ratio of the heat absorbed, Q_c, to the work done. A good refrigerator has a high coefficient of performance.

$$\text{COP (cooling mode)} = \frac{|Q_c|}{W} \qquad (22.4)$$

The **Carnot efficiency** is the maximum theoretical efficiency of a heat engine operating reversibly in a cycle between T_c and T_h. *No real heat engine operating irreversibly between two reservoirs (T_c and T_h) can be more efficient than a Carnot engine operating between the same two reservoirs.*

$$e_C = 1 - \frac{T_c}{T_h} \qquad (22.6)$$

The **efficiency of the Otto cycle** (gasoline engine) depends on the compression ratio (V_1/V_2) and the ratio of molar specific heats. *A diesel engine has a higher compression ratio than a gasoline engine; and therefore operates with greater efficiency.*

$$e = 1 - \frac{1}{(V_1/V_2)^{\gamma-1}} \qquad (22.7)$$

The **change in entropy of a system** during a process between two infinitesimally separated equilibrium states is given by Equation 22.8. *The subscript r is a reminder that dQ_r is energy transfer by heat during a reversible process that connects the two states.* It is assumed that T remains constant since the process is infinitesimal.

$$dS = \frac{dQ_r}{T} \qquad (22.8)$$

The **change in entropy for a finite process** going from one state to another has the same value for all paths connecting the two states; *ΔS depends only on the properties of the initial and final equilibrium states.*

$$\Delta S = \int_i^f \frac{dQ_r}{T} \qquad (22.9)$$

The **change in entropy for any reversible cycle** is identically *zero.*

$$\oint \frac{dQ_r}{T} = 0 \qquad (22.10)$$

The **change in entropy of an ideal gas** which undergoes a quasi-static and reversible process depends only on the initial and final states.

$$\Delta S = nC_V \ln\frac{T_f}{T_i} + nR\ln\frac{V_f}{V_i} \qquad (22.12)$$

In an irreversible process, the change in entropy for a system and its surroundings is always positive. Examples are:

Free expansion of a gas:

$$\Delta S = nR\ln\frac{V_f}{V_i} \qquad (22.13)$$

Calorimetric process:

$$\Delta S = m_1 c_1 \ln\frac{T_f}{T_c} + m_2 c_2 \ln\frac{T_f}{T_h} \qquad (22.15)$$

The **entropy of a given macrostate** depends on the number of microstates, W, of the system corresponding to the macrostate. *The more microstates there are that correspond to a given macrostate, the greater the entropy of the macrostate.*

$$S \equiv k_B \ln W \qquad (22.18)$$

REVIEW CHECKLIST

You should be able to:

▷ State the Kelvin-Planck and Clausius forms of the second law of thermodynamics, and discuss the difference between reversible and irreversible processes. (Sections 22.1, 22.2 and 22.3)

▷ Determine the efficiency of a heat engine operating between two reservoirs of known temperatures and calculate the net work output per cycle. (Section 22.1)

▷ Make calculations involving reversible and irreversible processes: determine Carnot and operating efficiencies, net work per cycle, energy added by heat, and reservoir temperatures. (Sections 22.3 and 22.24)

▷ Determine values of pressure, volume, and temperature at specified points around a closed cycle. (Sections 22.23 and 23.24)

▷ Calculate the COP for a refrigerator and for a heat pump operating in the heating cycle. (Sections 22.3 and 22.4)

▷ Determine the efficiency of an engine operating in the Otto cycle using an ideal gas as the working substance. Find temperature, pressure, volume, and internal energy at specified points in the cycle.

▷ Describe the sequence of processes which comprise the Carnot and Otto cycles. (Sections 22.4 and 22.5)

▷ Calculate energy input, work output, and change in internal energy for an adiabatic compression or expansion. (Section 22.5)

▷ Calculate entropy changes for a reversible process involving an ideal gas and for irreversible processes (e.g. free expansion and calorimetric process). (Sections 22.6 and 22.7)

ANSWERS TO SELECTED CONCEPTUAL QUESTIONS

1. What are some factors that affect the efficiency of automobile engines?

Answer   A gasoline engine does not exactly fit the definition of a heat engine. It takes in energy by mass transfer. Nevertheless, it converts this chemical energy into internal energy, so it can be modeled as taking in energy by heat. Therefore Carnot's limit on the efficiency of a heat engine applies to a gasoline engine. The fundamental limit on its efficiency is $1 - T_c/T_h$, set by the maximum temperature the engine block can stand and the temperature of the surroundings into which exhaust heat must be dumped.

The Carnot efficiency can only be attained by a reversible engine. To run with any speed, a gasoline engine must carry out irreversible processes and have an efficiency below the Carnot limit. Example 22.5 demonstrates this fact explicitly for an Otto cycle. In order for the compression and power strokes to be adiabatic, we would like to minimize irreversibly losing heat to the engine block. We would like to have the processes happen very quickly, so that there is negligible time for energy to be conducted away. But in a standard piston engine the compression and power strokes must take one-half of the total cycle time, and the angular speed of the crankshaft is limited by the time required to open and close the valves to get the fuel into each cylinder and the exhaust out.

Other limits on efficiency are imposed by irreversible processes like friction, both inside and outside the engine block. If the ignition timing is off, then the effective compressed volume will not be as small as it could be. If the burning of the gasoline is not complete, then some of the chemical energy will never enter the process.

The approximation used to obtain Equation 22.7, that the intake and exhaust gases have the same value of γ, hides other possibilities. Combustion breaks up the gasoline and oxygen molecules into a larger number of simpler water and CO_2 molecules, raising γ. By increasing γ and N, combustion slightly improves the efficiency. For this reason a gasoline engine can be more efficient than a methane engine, and also more efficient after the gasoline is refined to remove double carbon bonds (which would raise γ for the fuel).

□   □   □   □

3.   A steam-driven turbine is one major component of an electric power plant. Why is it advantageous to have the temperature of the steam as high as possible?

Answer   The most optimistic limit of efficiency in a steam engine is the ideal (Carnot) efficiency. This can be calculated from the high and low temperatures of the steam, as it expands:

$$e_C = \frac{T_h - T_c}{T_h} = 1 - \frac{T_c}{T_h}$$

The engine will be most efficient when the low temperature is extremely low, and the high temperature is extremely high. However, since the electric power plant is typically placed on the surface of the earth, there is a limit to how much the steam can expand, and how low the lower temperature can be. The only way to further increase the efficiency, then, is to raise the temperature of the hot steam, as high as possible.

□   □   □   □

11. The device shown in Figure Q22.11, called a thermoelectric converter, uses a series of semiconductor cells to convert internal energy to electric potential energy. In the photograph at the left, both legs of the device are at the same temperature, and no electric potential energy is produced. However, when one leg is at a higher temperature than the other, as in the photograph on the right, electric potential energy is produced as the device extracts energy from the hot reservoir and drives a small electric motor. (a) Why does the temperature differential produce electric potential energy in this demonstration? (b) In what sense does this intriguing experiment demonstrate the second law of thermodynamics?

Figure Q22.11

Answer

(a) The semiconductor converter operates essentially like a thermocouple, which is a pair of wires of different metals, with a junction at each end. When the junctions are at different temperatures, a small voltage appears around the loop, so that the device can be used to measure temperature or (here) to drive a small motor.

(b) The second law states that an engine operating in a cycle cannot absorb energy by heat from one reservoir and expel it entirely by work. This exactly describes the first situation, where both legs are in contact with a single reservoir, and the thermocouple fails to produce electrical work. To expel energy by work, the device must transfer energy by heat from a hot reservoir to a cold reservoir, as in the second situation.

□   □   □   □

13. Discuss the change in entropy of a gas that expands (a) at constant temperature and (b) adiabatically.

Answer

(a) The expanding gas is doing work. If it is ideal, its constant temperature implies constant internal energy, and it must be taking in energy by heat equal in amount to the work it is doing. As energy enters the gas by heat its entropy increases. The derivation of Equation 22.12 shows that the change in entropy is $\Delta S = nR\ln(V_f/V_i)$

(b) In a reversible adiabatic expansion there is no entropy change. We can say this is because the heat input is zero, or we can say it is because the temperature drops to compensate for the volume increase. In an irreversible adiabatic expansion the entropy increases. Equation 22.13 shows the entropy change in a free expansion.

□   □   □   □

SOLUTIONS TO SELECTED END-OF-CHAPTER PROBLEMS

3. A particular heat engine has a useful power output of 5.00 kW and an efficiency of 25.0%. The engine expels 8 000 J of exhaust energy in each cycle. Find (a) the energy taken in during each cycle and (b) the time interval for each cycle.

Solution   We are given that $|Q_c| = 8000$ J

(a)   We have $e = \dfrac{W_{eng}}{|Q_h|} = \dfrac{|Q_h| - |Q_c|}{|Q_h|} = 1 - \dfrac{|Q_c|}{|Q_h|} = 0.250$

   Isolating $|Q_h|$, we have $|Q_h| = \dfrac{|Q_c|}{1-e} = \dfrac{8\,000\text{ J}}{1-0.250} = 10.7$ kJ   ◊

(b)   The work per cycle is $W_{eng} = |Q_h| - |Q_c| = 2\,667$ J

   From $\mathcal{P} = \dfrac{W_{eng}}{\Delta t}$, $\Delta t = \dfrac{W_{eng}}{\mathcal{P}} = \dfrac{2\,667\text{ J}}{5\,000\text{ J / s}} = 0.533$ s   ◊

11. One of the most efficient heat engines ever built is a steam turbine in the Ohio valley, operating between 430 °C and 1 870 °C on energy from West Virginia coal to produce electricity for the Midwest. (a) What is its maximum theoretical efficiency? (b) The actual efficiency of the engine is 42.0%. How much useful power does the engine deliver if it takes in 1.40×10^5 J of energy each second from its hot reservoir?

Solution

The engine is a steam turbine in an electric generating station:

$T_c = 430$ °C $= 703$ K   and $T_h = 1870$ °C $= 2143$ K

(a) $e_C = \dfrac{\Delta T}{T_h} = \dfrac{1440\text{ K}}{2143\text{ K}} = 0.672$ or   67.2%   ◊

(b) $|Q_h| = 1.40 \times 10^5$ J $W_{eng} = 0.420|Q_h| = 5.88 \times 10^4$ J

   $\mathcal{P} = \dfrac{W_{eng}}{\Delta t} = \dfrac{5.88 \times 10^4\text{ J}}{1\text{ s}} = 58.8$ kW   ◊

13. An ideal gas is taken through a Carnot cycle. The isothermal expansion occurs at 250 °C, and the isothermal compression takes place at 50.0 °C. The gas takes in 1 200 J of energy from the hot reservoir during the isothermal expansion. Find (a) the energy expelled to the cold reservoir in each cycle and (b) the net work done by the gas in each cycle.

Solution

(a) For a Carnot cycle,
$$e_C = 1 - \frac{T_c}{T_h}$$

For any engine,
$$e = \frac{W_{eng}}{|Q_h|} = 1 - \frac{|Q_c|}{|Q_h|}$$

Therefore, for a Carnot engine,
$$1 - \frac{T_c}{T_h} = 1 - \frac{|Q_c|}{|Q_h|}$$

Then, since $|Q_c| = |Q_h|\left(\dfrac{T_c}{T_h}\right)$, $|Q_c| = (1\,200\text{ J})\left(\dfrac{323\text{ K}}{523\text{ K}}\right) = 741\text{ J}$ ◊

(b) The work is calculated as $W_{eng} = |Q_h| - |Q_c| = 1\,200\text{ J} - 741\text{ J} = 459\text{ J}$ ◊

---

25. An ideal refrigerator or ideal heat pump is equivalent to a Carnot engine running in reverse. That is, energy Q_c is taken in from a cold reservoir and energy Q_h is rejected to a hot reservoir. (a) Show that the work that must be supplied to run the refrigerator or heat pump is $W = Q_c(T_h - T_c)/T_c$. (b) Show that the coefficient of performance of the ideal refrigerator is $COP = T_c/(T_h - T_c)$.

Solution

(a) For a complete cycle $\Delta E_{int} = 0$, and $W = |Q_h| - |Q_c| = |Q_c|\left(\dfrac{|Q_h|}{|Q_c|} - 1\right)$

For a Carnot cycle (and only for a Carnot cycle), $|Q_h|/|Q_c| = T_h/T_c$

Then, $W = |Q_c|(T_h - T_c)/T_c$ ◊

(b) From Equation 22.4, $COP = \dfrac{|Q_c|}{W}$, so $COP = \dfrac{T_c}{T_h - T_c}$ ◊

---

27. How much work does an ideal Carnot refrigerator require to remove 1.00 J of energy from helium at 4.00 K and reject this energy to a room-temperature (293-K) environment?

Solution $(\text{COP})_{\text{Carnot, refrig}} = \dfrac{T_c}{\Delta T} = \dfrac{4.00\ \text{K}}{293\ \text{K} - 4.00\ \text{K}} = 0.013\,8 = \dfrac{|Q_c|}{W}$

$$W = \dfrac{|Q_c|}{\text{COP}} = \dfrac{1.00\ \text{J}}{0.013\,8} = 72.2\ \text{J}$$ ◊

31. In a cylinder of an automobile engine, just after combustion, the gas is confined to a volume of 50.0 cm³ and has an initial pressure of 3.00×10^6 Pa. The piston moves outward to a final volume of 300 cm³ and the gas expands without energy loss by heat. (a) If $\gamma = 1.40$ for the gas, what is the final pressure? (b) How much work is done by the gas in expanding?

Solution

Conceptualize: The pressure will decrease as the volume increases. If the gas were to expand at constant temperature, the final pressure would be

$$\tfrac{50}{300}\left(3 \times 10^6\ \text{Pa}\right) = 5 \times 10^5\ \text{Pa}$$

But the temperature must drop a great deal as the gas spends internal energy in doing work. Thus the final pressure must be lower than 500 kPa. The order of magnitude of the work can be estimated from the average pressure and change in volume:

$$W \sim \left(10^6\ \text{N} / \text{m}^2\right)\left(100\ \text{cm}^3\right)\left(10^{-6}\ \text{m}^3 / \text{cm}^3\right) = 100\ \text{J}$$

Categorize: The gas expands adiabatically (there is not enough time for significant heat transfer), so equation 21.18 can be applied to find the final pressure. With $Q = 0$, the amount of work can be found from the change in internal energy.

Analyze:

(a) For adiabatic expansion, $\qquad P_i V_i^{\gamma} = P_f V_f^{\gamma}$

Therefore, $\qquad P_f = P_i\left(\dfrac{V_i}{V_f}\right)^{\gamma} = \left(3.00 \times 10^6\ \text{Pa}\right)\left(\dfrac{50.0\ \text{cm}^3}{300\ \text{cm}^3}\right)^{1.40} = 2.44 \times 10^5\ \text{Pa}$ ◊

(b) Since $Q = 0$,

we have $W_{eng} = Q - \Delta E = -\Delta E = -nC_V \Delta T = -nC_V(T_f - T_i)$

From $\gamma = \dfrac{C_P}{C_V} = \dfrac{C_V + R}{C_V}$ we get $(\gamma - 1)C_V = R$

So that $C_V = \dfrac{R}{1.40 - 1} = 2.50R$

$W_{eng} = n(2.50R)(T_i - T_f) = 2.50P_iV_i - 2.50P_fV_f$

$W_{eng} = 2.50\left[(3.00 \times 10^6 \text{ Pa})(50.0 \times 10^{-6} \text{ m}^2) - (2.44 \times 10^5 \text{ Pa})(300 \times 10^{-6} \text{ m}^2)\right]$

$W_{eng} = 192 \text{ J}$ ◊

Finalize: The final pressure is about half of the 500 kPa, in agreement with our qualitative prediction. The order of magnitude of the work is as we predicted.

From the work done by the gas in part (b), the average power of the engine could be calculated if the time for one cycle was known. Adiabatic expansion is the power stroke of our industrial civilization.

---

37. Calculate the change in entropy of 250 g of water heated slowly from 20.0 °C to 80.0 °C. (**Suggestion:** Note that $dQ = mc\,dT$.)

Solution We use the energy version of the nonisolated system model. To do the heating reversibly, put the water pot successively into contact with reservoirs at temperatures 20.0 °C + δ, 20.0 °C + 2δ, ... 80.0 °C, where δ is some small increment.

Then, $\Delta S = \displaystyle\int_i^f \frac{dQ}{T} = \int_{T_i}^{T_f} mc\,\frac{dT}{T}$

Here T means the absolute temperature. We would ordinarily think of dT as the change in the Celsius temperature, but one Celsius degree of temperature change is the same size as one kelvin of change, so dT is also the change in absolute T.

Then $\Delta S = mc\ln T\Big|_{T_i}^{T_f} = mc\ln\left(\dfrac{T_f}{T_i}\right)$

$\Delta S = (0.250 \text{ kg})(4\,186 \text{ J / kg} \cdot \text{K})\ln\left(\dfrac{353 \text{ K}}{293 \text{ K}}\right) = 195 \text{ J / K}$ ◊

---

41. A 1 500-kg car is moving at 20.0 m/s. The driver brakes to a stop. The brakes cool off to the temperature of the surrounding air, which is nearly constant at 20.0 °C. What is the total entropy change?

Solution The original kinetic energy of the car,

$$K = \frac{1}{2}mv^2 = \frac{1}{2}(1\,500 \text{ kg})(20.0 \text{ m/s})^2 = 300 \text{ kJ}$$

becomes irreversibly 300 kJ of extra internal energy in the brakes, the car, and its surroundings. Since their total heat capacity is so large, their equilibrium temperature will be approximately 20.0 °C. To carry them reversibly to this same final state, imagine putting 300 kJ into the car and its environment from a heater at 20.001 °C.

Then $$\Delta S = \int_i^f \frac{dQ}{T} = \frac{1}{T}\int dQ_r = \frac{Q}{T} = \frac{300 \text{ kJ}}{293 \text{ K}}$$

$$\Delta S = 1.02 \text{ kJ/K} \qquad \Diamond$$

---

45. A 1.00-mol sample of H_2 gas is contained in the left-hand side of the container shown in Figure P22.45, which has equal volumes left and right. The right-hand side is evacuated. When the valve is opened, the gas streams into the right-hand side. What is the final entropy change of the gas? Does the temperature of the gas change?

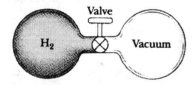

Figure P22.45

Solution

(a) This is an example of free expansion; from Equation 22.13 we have

$$\Delta S = nR \ln\left(\frac{V_f}{V_i}\right) \qquad\qquad \Delta S = (1.00 \text{ mole})(8.314 \text{ J/mole}\cdot\text{K}) \ln\left(\frac{2}{1}\right)$$

$$\Delta S = 5.76 \text{ J/K} \qquad \text{or} \qquad \Delta S = 1.38 \text{ cal/K} \qquad \Diamond$$

(b) The gas is expanding into an evacuated region. Therefore, $W=0$; and it expands so fast that heat has no time to flow; $Q=0$. But $\Delta E_{int}=Q-W$, so in this case $\Delta E_{int}=0$. For an ideal gas, the internal energy is a function of the temperature, so if $\Delta E_{int}=0$, the temperature remains constant.

---

51. Repeat the procedure used to construct Table 22.1 (a) for the case in which you draw three marbles from your bag rather than four and (b) for the case in which you draw five rather than four.

Solution

(a)

Result	Possible combinations	Total
All Red	RRR	1
2R, 1G	RRG, RGR, GRR	3
1R, 2G	RGG, GRG, GGR	3
All Green	GGG	1 ◊

(b)

Result	Possible combinations	Total
All Red	RRRRR	1
4R, 1G	RRRRG, RRRGR, RRGRR, RGRRR, GRRRR	5
3R, 2G	RRRGG, RRGRG, RGRRG, GRRRG, RRGGR, RGRGR, GRRGR, RGGRR, GRGRR, GGRRR	10
2R, 3G	GGGRR, GGRGR, GRGGR, RGGGR, GGRRG, GRGRG, RGGRG, GRRGG, RGRGG, RRGGG	10
1R, 4G	GGGGR, GGGRG, GGRGG, GRGGG, RGGGG	5
All Green	GGGGG	1 ◊

53. A house loses energy through the exterior walls and roof at a rate of $5\,000$ J/s $= 5.00$ kW when the interior temperature is 22.0 °C and the outside temperature is −5.00 °C. Calculate the electric power required to maintain the interior temperature at 22.0 °C for the following two cases. (a) The electric power is used in electric resistance heaters (which convert all of the energy transferred in by electrical transmission into internal energy). (b) **What if?** The electric power is used to drive an electric motor that operates the compressor of a heat pump, which has a coefficient of performance equal to 60.0% of the Carnot-cycle value.

Solution

Conceptualize: The electric heater should be 100% efficient, so $\mathcal{P} = 5$ kW in part (a). It sounds as if the heat pump is only 60% efficient, so we might expect $\mathcal{P} = 9$ kW in (b).

Categorize: Power is the amount of energy transferred per unit of time, so we can find the power in each case by examining the energy input as heat required for the house as a nonisolated system in steady state.

Analyze:

(a) We know that $\mathcal{P}_{electric} = H_{ET}/\Delta t$, so if all of the energy transferred into the heater by electrical transmission is stored in the heater as internal energy, then

$$\mathcal{P}_{electric} = \frac{H_{ET}}{\Delta t} = 5.00 \text{ kW} \qquad \Diamond$$

(b) For a heat pump, $\quad (COP)_{Carnot} = \dfrac{T_h}{\Delta T} = \dfrac{295 \text{ K}}{27.0 \text{ K}} = 10.93$

$$\text{Actual COP} = (0.600)(10.93) = \frac{|Q_h|}{W} = \frac{|Q_h|/\Delta t}{W/\Delta t}$$

Therefore, to bring 5 000 W of heat into the house only requires input power

$$\mathcal{P}_{heat\ pump} = \frac{W}{\Delta t} = \frac{|Q_h|/\Delta t}{COP} = \frac{5\,000 \text{ W}}{6.56} = 763 \text{ W} \qquad \Diamond$$

Finalize: The result for the electric heater's power is consistent with our prediction, but the heat pump actually requires **less** power than we expected. Since both types of heaters use electricity to operate, we can now see why it is more cost effective to use a heat pump even though it is less than 100% efficient!

---

57. In 1816 Robert Stirling, a Scottish clergyman, patented the **Stirling engine**, which has found a wide variety of applications ever since. Fuel is burned externally to warm one of the engine's two cylinders. A fixed quantity of inert gas moves cyclically between the cylinders, expanding in the hot one and contracting in the cold one. Figure P22.57 represents a model for its thermodynamic cycle. Consider n mol of an ideal monatomic gas being taken once through the cycle, consisting of two isothermal processes at temperatures $3T_i$ and T_i and two constant-volume processes. Determine, in terms of n, R, and T_i, (a) the net energy transferred by heat to the gas and (b) the efficiency of the engine. A Stirling engine is easier to manufacture than an internal combustion engine or a turbine. It can run on burning garbage. It can run on the energy of sunlight and produce no material exhaust.

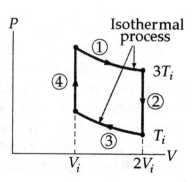

Figure P22.57

Solution   The Sterling engine need have no material exhaust, but it has energy exhaust.

(a)   For an isothermal process,

$$Q = nRT \ln\left(\frac{V_f}{V_i}\right)$$

Therefore,

$$Q_1 = nR(3T_i)\ln 2 \quad \text{and} \quad Q_3 = nRT_i \ln\left(\tfrac{1}{2}\right)$$

The internal energy of a monatomic ideal gas is $E_{int} = \frac{3}{2}nRT$.

In the constant-volume processes, $Q_2 = \Delta E_{int,2} = \frac{3}{2}nR(T_i - 3T_i)$

and

$$Q_4 = \Delta E_{int,4} = \frac{3}{2}nR(3T_i - T_i)$$

The net energy transferred by heat is then $Q = Q_1 + Q_2 + Q_3 + Q_4$,

or

$$Q = 2nRT_i \ln 2 \qquad\qquad\qquad\qquad \Diamond$$

(b) $|Q_h|$ is the sum of the positive contributions to Q.

$$|Q_h| = Q_1 + Q_4 = 3nRT_i(1 + \ln 2)$$

Since the change in temperature for the complete cycle is zero,

$$\Delta E_{int} = 0 \quad \text{and} \quad W_{eng} = Q$$

Therefore, the efficiency is

$$e = \frac{W_{eng}}{|Q_h|} = \frac{Q}{|Q_h|} = \frac{2\ln 2}{3(1+\ln 2)} = 0.273 = 27.3\% \quad \Diamond$$

---

65. A 1.00-mol sample of a monatomic ideal gas is taken through the cycle shown in Figure P22.65. At point A, the pressure, volume, and temperature are P_i, V_i, and T_i, respectively. In terms of R and T_i, find (a) the total energy entering the system by heat per cycle, (b) the total energy leaving the system by heat per cycle, (c) the efficiency of an engine operating in this cycle, and (d) the efficiency of an engine operating in a Carnot cycle between the same temperature extremes.

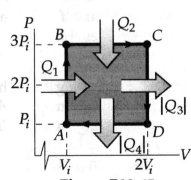

Figure P22.65

Solution   At point A, $\quad P_i V_i = nRT_i$ $\qquad$ with $\qquad n = 1.00$ mol

$\qquad\qquad$ At point B, $\quad 3P_i V_i = nRT_B$ $\qquad$ so $\qquad T_B = 3T_i$

$\qquad\qquad$ At point C, $\quad (3P_i)(2V_i) = nRT_C$ and $\qquad T_C = 6T_i$

$\qquad\qquad$ At point D, $\quad P_i(2V_i) = nRT_D$ and $\qquad T_D = 2T_i$

We find the energy transfer by heat for each step in the cycle using

$$C_V = \frac{3}{2}R \qquad\qquad \text{and} \qquad\qquad C_P = \frac{5}{2}R$$

$$Q_1 = Q_{AB} = C_V(3T_i - T_i) = 3RT_i \qquad\qquad Q_2 = Q_{BC} = C_P(6T_i - 3T_i) = 7.5RT_i$$

$$Q_3 = Q_{CD} = C_V(2T_i - 6T_i) = -6RT_i \qquad\qquad Q_4 = Q_{DA} = C_P(T_i - 2T_i) = -2.5RT_i$$

(a) $\qquad$ Therefore, $\qquad\qquad\qquad\qquad Q_{in} = |Q_h| = Q_{AB} + Q_{BC} = 10.5RT_i$ ◊

(b) $\qquad\qquad\qquad\qquad\qquad\qquad\qquad Q_{out} = |Q_c| = |Q_{CD} + Q_{DA}| = 8.5RT_i$ ◊

(c) $\qquad\qquad\qquad\qquad\qquad\qquad\qquad e = \dfrac{|Q_h| - |Q_c|}{|Q_h|} = 0.190 = 19.0\%$ ◊

(d) $\qquad$ Carnot efficiency, $\qquad\qquad e_C = 1 - \dfrac{T_c}{T_h} = 1 - \dfrac{T_i}{6T_i} = 0.833 = 83.3\%$ ◊

---

67.   A system consisting of n mol of an ideal gas undergoes two reversible processes. It starts with pressure P_i and volume V_i, expands isothermally, and then contracts adiabatically to reach a final state with pressure P_i and volume $3V_i$. (a) Find its change in entropy in the isothermal process. The entropy does not change in the adiabatic process. (b) **What if?** Explain why the answer to part (a) must be the same as the answer to Problem 66.

Solution   The diagram shows the isobaric process considered in Problem 66 as AB. The processes considered in this problem are AC and CB.

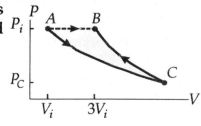

(a)   For isotherm (AC), $\qquad P_A V_A = P_C V_C$

$\qquad$ For adiabat (CB) $\qquad P_C V_C{}^{\gamma} = P_B V_B{}^{\gamma}$

Combining these gives $\qquad V_C = \left(\dfrac{P_B V_B^{\gamma}}{P_A V_A}\right)^{1/(\gamma-1)} = \left[\left(\dfrac{P_i}{P_i}\right)\dfrac{(3V_i)^{\gamma}}{V_i}\right]^{1/(\gamma-1)} = \left(3^{\gamma/(\gamma-1)}\right)V_i$

Therefore, $\qquad \Delta S_{AC} = nR\ln\left(\dfrac{V_C}{V_A}\right) = nR\ln\left[3^{\gamma/(\gamma-1)}\right] = \dfrac{nR\gamma\ln 3}{\gamma-1}$ ◊

(b) Since the change in entropy is path independent, $\qquad \Delta S_{AB} = \Delta S_{AC} + \Delta S_{CB}$

But because (CB) is adiabatic, $\qquad\qquad\qquad\qquad \Delta S_{CB} = 0$

Then $\qquad\qquad\qquad\qquad\qquad\qquad\qquad\qquad \Delta S_{AB} = \Delta S_{AC}$

The answer to Problem 66 was stated as $\qquad \Delta S_{AB} = nC_P\ln 3$

Because $\gamma = C_P/C_V$, $\ C_V = C_P/\gamma$, and $\qquad C_P - C_V = R$

gives $\qquad\qquad\qquad\qquad\qquad\qquad\qquad\qquad C_P - C_P/\gamma = R$

so $\qquad\qquad\qquad \gamma C_P - C_P = \gamma R \qquad$ and $\qquad C_P = \gamma R/(\gamma-1)$

Thus, the answers to problem 66 and 67 are in fact equal.

---

69. An idealized diesel engine operates in a cycle known as the **air-standard diesel cycle**, shown in Figure 22.14. Fuel is sprayed into the cylinder at the point of maximum compression, B. Combustion occurs during the expansion $B \to C$, which is modeled as an isobaric process. Show that the efficiency of an engine operating in this idealized diesel cycle is

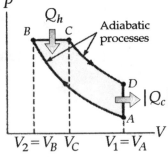

Figure 22.14

$$e = 1 - \dfrac{1}{\gamma}\left(\dfrac{T_D - T_A}{T_C - T_B}\right)$$

Solution   The heat transfer over the paths CD and BA are zero since they are adiabats.

Over path BC: $\qquad\qquad Q_{BC} = nC_P(T_C - T_B) > 0$

Over path DA: $\qquad\qquad Q_{DA} = nC_V(T_A - T_D) < 0$

Therefore, $\qquad\qquad |Q_c| = |Q_{DA}| \qquad$ and $\qquad Q_h = Q_{BC}$

Hence, the efficiency is $\qquad e = 1 - \dfrac{|Q_c|}{Q_h} = 1 - \left(\dfrac{T_D - T_A}{T_C - T_B}\right)\dfrac{C_V}{C_P}$

$$e = 1 - \dfrac{1}{\gamma}\left(\dfrac{T_D - T_A}{T_C - T_B}\right)$$ ◊
